Electrical Systems for Computer Installations

Electrical Systems for Computer Installations

Edited by **Robert J. Lawrie**
and the staff of Electrical Construction and Maintenance Magazine

McGraw-Hill Book Company
New York St. Louis San Francisco Auckland
Bogotá Hamburg London Madrid Mexico
Milan Montreal New Delhi Panama
Paris São Paulo Singapore
Sydney Tokyo Toronto

Library of Congress Cataloging in Publication Data

Lawrie, Robert J.
Electrical systems for computer installations.

Includes index.
1. Electronic digital computers–Power supply. I. Title.
TK7895.P68L38 1988 621.39 88-3014
ISBN 0-07-036729-9

1234567890 EDW/EDW 8954321098

ISBN 0-07-036729-9

Printed and bound by Edwards Brothers.

Contents

Introduction

This book presents, in detail, all essential aspects of design, installation, operation and maintenance of electrical systems that serve computers and data-processing equipment. All material is derived from the best and most timely articles published in EC&M, and presents the most complete, state-of-the-art technical data on the subject available.

The intent of this book is to provide in-depth guidance for consulting engineers, facility electrical engineers, computer facility managers, computer installation engineers and technicians, electrical contractors, and others who are concerned with computers and data-processing equipment.

The greatest value of this book to most readers will be derived from the unique treatment and presentation of the material contained. The material contained combines generic discussions of computer power technology with actual, on-the-spot case histories of problems, solutions and experiences of electrical people as they faced and met challenges posed by electrical systems serving computers. This broad range of information is then made easily and quickly available to the reader via the book's Table of Contents and comprehensive Index.

Thus, an extremely useful feature is the detailed Table of Contents, and the "*highlighted headings*" that correlate with the Table of Contents. This unique treatment enables the reader to locate any subject of interest quickly. Also of great value is the thorough index that leads the reader quickly to precise points referenced.

Hundreds of ideas and practical techniques are presented throughout the book, which is illustrated with plenty of photos, diagrams, and sketches to aid fast understanding of the complexities involved. It is also important to note the pervasive interrelationship of the many facets of electrical systems for computer installations, and that all aspects of design, construction or maintenance are interdependent, and as a result, these subjects are discussed throughout the book.

Thanks are due to the many contributors to this book, whose names are listed with each pertinent article, as well as to the editors of EC&M, which include J. F. McPartland, A. Berutti, A. Freund and J. Knisely. A special thanks is due John Behmke, publisher of EC&M, whose encouragement and guidance made this book possible.

January, 1988
New York, NY

ROBERT J. LAWRIE, *Senior Editor*
Electrical Construction & Maintenance

Chapter 1

Fundamentals of UPS Systems

Uninterruptible Power Supplies

Typical UPS systems

An uninterruptible power system is an electrical/electronic device specifically designed to maintain total continuity of AC power to a critical load. To accomplish this, a typical UPS functions in three basic modes: normal, AC power failure, and bypass (Fig. 1).

Normal mode: In normal operation, the UPS battery charger/rectifier section is powered from the commercial AC line. It, in turn, provides regulated DC output to power the UPS inverter, while simultaneously float-charging the standby battery plant. The inverter converts the regulated DC input into a clean, continuous supply of AC. The inverter output then flows through the output transfer switch and to the critical load.

Functioning in the normal mode, the UPS filters out both intermediate- and short-term aberrations on the incoming AC line. And, as a result, the critical load is effectively isolated from intermediate and transient voltage aberrations, frequency aberrations, and both common-mode and transverse noise.

AC power-failure mode: In the event of a commercial AC line failure, power is maintained to the critical load through the UPS inverter. Operating in this mode, the inverter, *without switching*, obtains DC input from the standby battery plant. The battery plant continues to supply DC input to the inverter until orderly shutdown of the load can be effected and the auxiliary power generation source comes on-line, or commercial AC power is restored.

Bypass mode: Virtually every UPS is designed with an output transfer switch, through which inverter output flows to the critical load. Such switches, which can be either manually or automatically actuated, are designed to protect the load from interruption by permitting a transfer of the load to the bypass source in the event of an overload, inrush surge, load short circuit, or a battery or inverter failure.

The bypass source, which can be the commercial AC line, another inverter, or an auxiliary power generation source, then maintains power to the critical load. Operation in the bypass mode is continued until the problem that necessitated the transfer is corrected.

Fig. 1. Typical UPS system

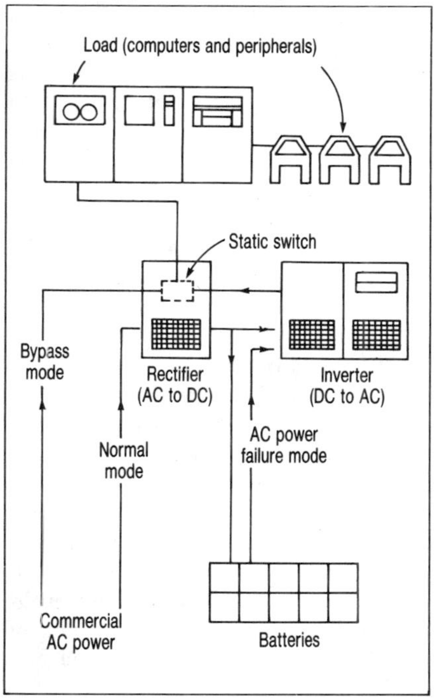

Fig. 2. Foreward and reverse bias with a rectifier diode

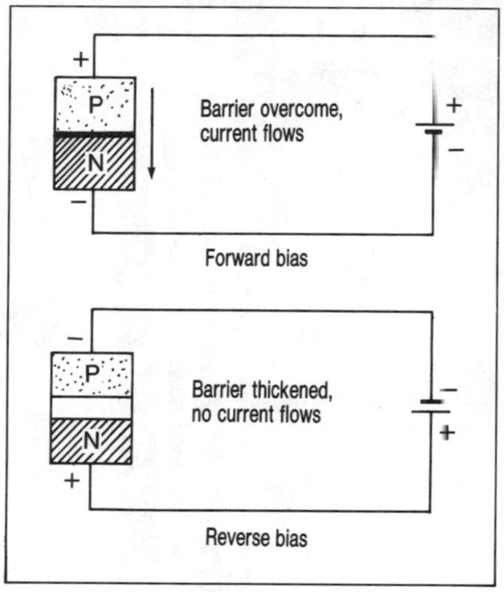

Functioning in one or another of these three modes, a UPS maintains total continuity of AC power to a critical load. Obviously, the effectiveness of this power protection method will be contingent upon the design specifications of the individual components comprising the system. A UPS system normally involves a battery charger/rectifier, an inverter, an output transfer switch, a standy battery plant, and auxiliary power generation sources. An examination of each of these components will lead to an understanding of a complete system.

With the selection of a suitable turns ratio, a transformer can be used to convert an AC supply voltage into a lower or higher AC output voltage. By including some additional components in such a device, the transformer can be used to convert the AC supply voltage into a lower or higher DC output voltage.

Rectification

The circuit of a very simple rectifier includes a transformer, a single diode, and a capacitor. Such a circuit can provide half-wave rectification of the AC input voltage. Half-wave rectification simply means that the circuit passes one-half of each AC cycle as DC while suppressing the other half-cycle. Since this is accomplished by means of a diode, a review of the function of a diode is in order.

Diodes and rectifiers

A diode is an electronic device constucted so that it conducts current in only one direction. If a large-enough voltage (about 0.7 V) is applied, making the anode positive with respect to the cathode, current will flow through the semiconductor element. Voltage applied in this manner is known as *forward bias.*

Similarly, if an external voltage with the opposite polarity is applied, no current will flow through the device, and the diode is said to be "blocking" the current flow. Voltage applied in this manner is known as *reverse bias.* Devices can be rated to withstand a reverse bias of up to several thousand volts. (See Fig. 2.) When a semiconductor device exhibits the characteristics of forward *and* reverse bias, it is known as a diode.

In a *half-wave rectification circuit,* the input voltage is applied to a diode and a series output load resistor. For the positive half-cycles of AC input, the diode is forward-biased. This enables the diode to conduct current and provide a fluctuating DC voltage across the load resistor. On negative half-cycles of the AC input, the polarity produces a reverse bias across the diode. As a result, the negative half-cycle AC voltages are suppressed, preventing current flow. (See Fig. 3.)

The output waveform of such a circuit, while not a steady DC value, is a fluctuating DC voltage with only one polarity. The fluctuations in the output waveform are the AC ripple. Assuming a 60-Hz, AC voltage from the power line, the ripple frequency will be 60 Hz. The amplitude of the AC ripple can be significantly reduced through filtering.

If a capacitor is added to the circuit, the AC ripple in the output can be reduced, since the capacitor will discharge on each suppressed or negative half-cycle. The resulting DC output voltages are then maintained for a longer period.

Fig. 4. Full-wave rectifier circuit

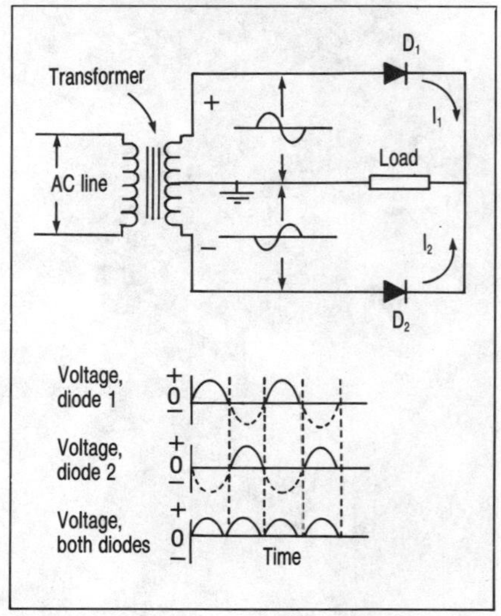

Full-wave circuits

By reversing the connections of the diode, the now-inverted diode will conduct on the negative half-cycle of AC input. The DC output voltage will then have a negative polarity. By coupling a standard half-wave rectifier circuit with such an inverted half-wave rectifier circuit, a full-wave rectifier circuit can be created.

In a *full-wave rectifier* for single-phase applications using two diodes, both alternations of the AC input produce DC output. On the first alternation, diode 1 will conduct, while diode 2

Fig. 3. Half-wave rectifier circuits for positive (top) and negative (bottom) output

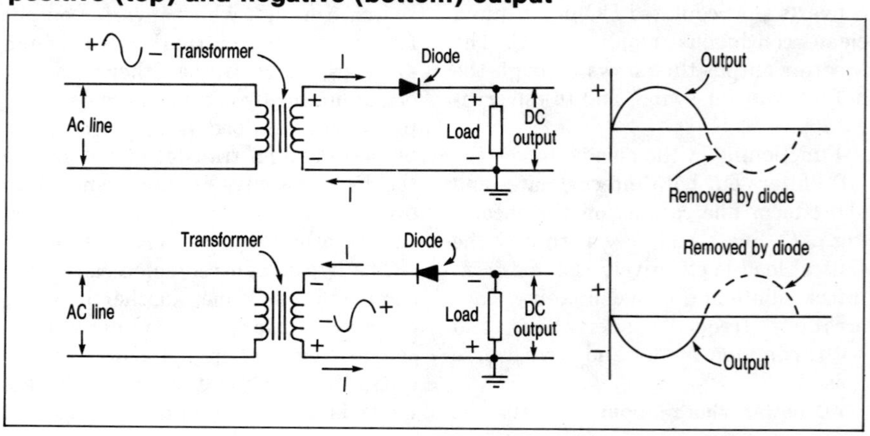

Fig. 5. Full-wave rectifier circuit and diode conduction for a given half-cycle of AC input (uncontrolled rectifier)

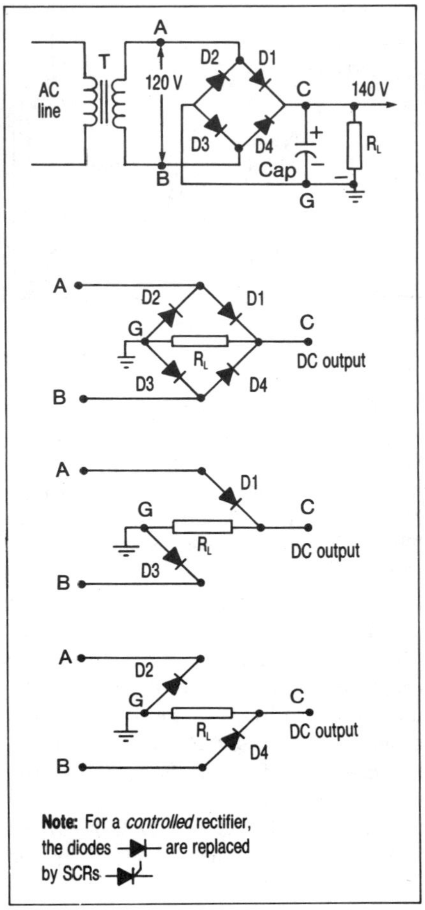

blocks. On the next alternation, the AC input voltage reverses polarity and diode 2 conducts, while diode 1 blocks. The AC input supplies equal and opposite voltages, usually with a center-tapped secondary winding in the power transformer. (See Fig. 4.) *In any full-wave rectifier, while the diodes conduct on opposite half-cycles, the polarity of the DC output is the same.*

The ripple frequency of the DC output for a full-wave rectifier is double the frequency of the AC input (since each half-cycle produces a fluctuation in the DC output voltage). With a 60-Hz power line providing the AC input, the ripple frequency in the DC output will be 120 Hz. This is an even closer approximation of a steady DC value than that found with a half-wave rectifier. And, by adding a capacitor, the DC output can be further smoothed, resulting in a very close approximation of a steady DC value.

A *full-wave bridge rectifier* can be likened to the full-wave rectifier discussed previously. The difference is that here, the center tap is omitted and the circuit incorporates four rather than two diodes.

In Fig. 5, four leads are labeled A, B, C and G. Two leads are for AC input and two are for DC output. The bridge design allows two diodes to conduct in series with the load resistance between them on each half-cycle. The output of such devices has a ripple frequency of 120 Hz with full-wave rectification.

The type of single-phase rectifier circuit (full-wave center-tapped or full-wave bridge) selected for a given application depends on a number of factors. In high-voltage applications, the full-wave bridge circuit is typically necessary. In lower-voltage applications, such as those involving single-phase uninterruptible power systems, the full-wave, center-tapped circuit is usually employed.

The critical component in a rectifier circuit is the solid-state switching device providing the rectification. Such devices are designed to carry current flow in one direction while blocking current flow in the reverse direction. These devices can be either the uncontrolled rectifier diodes that have been discussed or silicon controlled rectifiers. The type of switching device employed determines whether a rectifier circuit is controlled or uncontrolled.

In an uncontrolled rectifier, all of the switching devices are simple rectifier diodes. The average DC output of such devices is fixed by the AC input and the circuit configuration. In a controlled rectifier, some or all of the diodes are replaced with silicon controlled rectifiers (SCRs). Before we examine the action of an SCR in a rectifier circuit, let's briefly review their design and function.

Silicon controlled rectifiers

A silicon controlled rectifier (SCR), or thyristor, is a diode with an additional electrode called a gate. If a bias voltage is applied to the gate to keep it at the same potential as the cathode of the diode, the SCR acts as if it's working with reverse voltage for both directions of applied voltage. As a result, the SCR only allows a very small leakage current to flow.

If the gate is biased to be more positive than the cathode, the SCR acts like a normal diode. Thus, the gate on the SCR can be used to turn the rectifier on (positive bias on the gate), permitting control of the forward current. Once gated on, an SCR will remain on, without further gate drive, until the SCR's anode-to-cathode current is reduced to near zero.

In devices employing SCRs it is important to determine whether the SCRs are naturally commutated or force-commutated. Basically, commutation refers to the process by which forward current is interrupted and the SCR is switched off. In most circuits where power is supplied from an AC source, turn-off occurs naturally, since current flow ceases when the polarity across a given device reverses. Such circuits are said to have "natural commutation."

Other circuits, particularly those supplied from a DC source, must be able to interrupt a nonzero current. It is often necessary to accomplish this by turning off the SCRs. This requires auxiliary components in the circuit to reverse bias momentarily and divert the current from an SCR, allowing it to re-establish its blocking state. This turning off of SCRs is known as "forced commutation."

A controlled rectifier may be created by substituting four SCRs for the diodes in the bridge circuit of Fig. 5. The DC output voltage or current can then be controlled by delaying the application of a gate drive to the SCRs.

3-phase rectifier designs

In applications involving 3-phase rectifier circuits, there is again the

Fig. 6. 3-phase half-wave rectifier

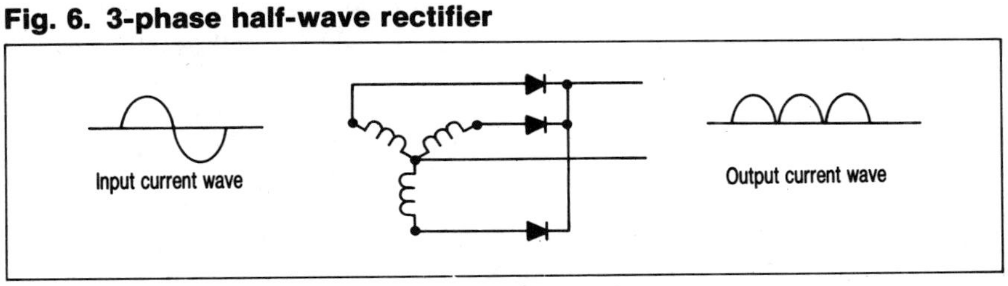

Fig. 7. Six-pulse rectifier

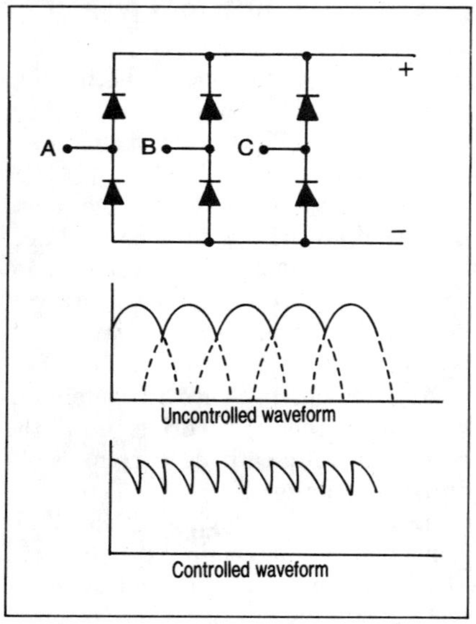

Fig. 8. Boost circuit

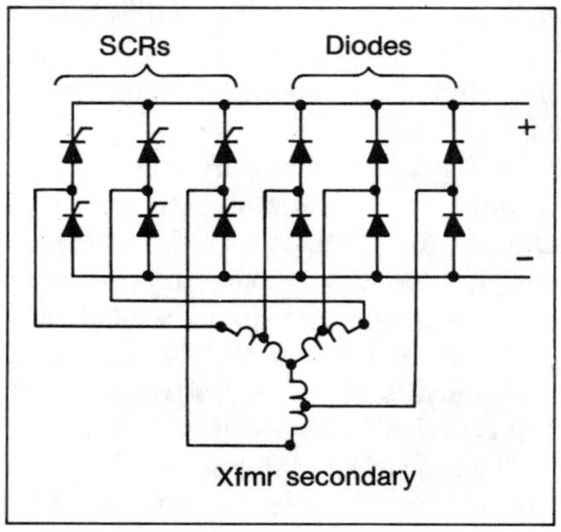

Fig. 9. Ferroresonant transformer

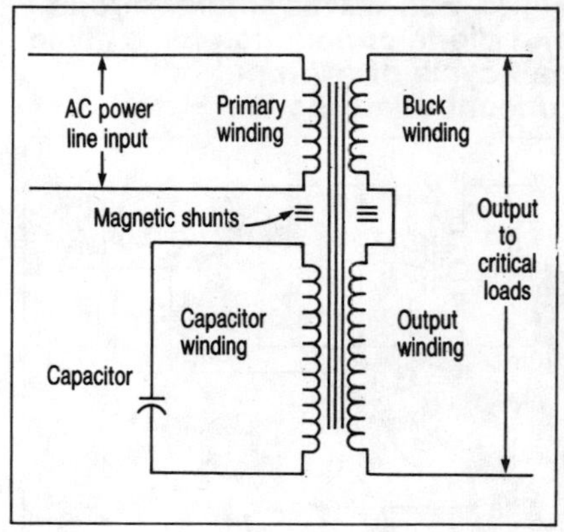

distinction between full-wave and half-wave rectification. But in addition, the number of pulses of output power for a given cycle of the input frequency must be considered, since the number of pulses in the output waveform impacts both the output and input waveforms.

Regarding the quality of the output waveform, if a greater number of pulses are used to produce the output waveform, then the waveform will more closely approximate a steady DC value. The more the output waveform resembles a steady DC value, the less filtering is required of the output, and the lower the losses will be due to such filtering.

Similarly, in terms of the input waveform, the greater the number of pulses present in the output, the higher the input power factor. This is a concern because the lower the input power factor, the greater the capacity of the source required. In addition, a greater number of pulses will result in a smoother (more nearly sinusoidal) input current waveform, which will minimize the distortion of the AC input waveform. Such distortions could adversely affect other loads operating from the same line.

In the single-phase full-wave rectifier examined earlier, the two diodes produced two pulses on the output, for a given cycle of the input frequency. In single-phase applications, this is the maximum number of pulses possible. But in 3-phase applications, where there are three input waveforms displaced from each other by 120°, then 3, 6, 12 or even 24 pulses can be created on the output.

3-phase, half-wave rectifier. The simplest 3-phase rectifier circuit is a 3-phase half-wave rectifier. Essentially, such a circuit consists of three single-phase half-wave rectifier circuits that are shifted 120° and linked together. The three diodes in such a circuit produce three pulses for each cycle of the input frequency. Fig. 6 shows the output of such a simple 3-phase, half-wave rectifier circuit. Such circuits have very limited applications and are rarely used in UPS rectifiers.

3-phase, full-wave rectifiers. More commonly, rectifier circuits for 3-phase power supply and UPS applications will employ a 3-phase, full-wave design. Such devices, which can be either controlled or uncontrolled, provide 6, 12 or 24 pulses in the output waveform. In its simplest form, a 3-phase, full-wave rectifier consists of a 3-phase bridge circuit providing two diodes for each phase of AC input. Each pair of diodes conducts the positive and negative half-cycles of their respective phases to provide a total of six pulses for each cycle of the input frequency. Even in its semicontrolled form, consisting of three SCRs and three diodes, such circuits are rarely used in UPS applications. They are usually limited to less-demanding applications, such as small power supplies.

In UPS applications to about 200 kW, the rectifier will usually employ a controlled 6-pulse rectifier circuit. In such devices, the six diodes, arranged in a 3-phase bridge circuit, are replaced with six SCRs. The resulting output waveform, shown in Fig. 7, is a very close approximation of a steady DC value. Output voltage or current can be controlled by delaying the gate drive to the SCRs.

In applications over about 200 kW, 12- or 24-pulse rectifiers are common. The additional phases are created in a transformer by adding various amounts of existing phases to create new phases. In Fig. 8, equal amounts of phase A and phase B will result in the creation of a new phase, D, which is the opposite of phase C. Similarly, equal amounts of phases B and C result in the creation of phase E, which is the opposite of phase A. Finally, equal amounts of phases A and C result in the creation of phase F, which is the opposite of phase B.

With the new windings added to the transformer, the six phases, conducted through two separate 3-phase bridge circuits, provide 12 pulses on the output waveform for each cycle of the input frequency. By continuing the above process, combining the now six phases of input to create additional phases, 24 or more pulses can be provided in the output waveform.

While 12-pulse rectifiers are not cost-effective for applications below 200 kW, the improvement of the input power factor and current waveform are sometimes desirable. In such cases, a "boost circuit" is often used. Basically, the boost circuit consists of a full-wave, 3-phase SCR bridge connected to a wye-configured transformer secondary. A second bridge, consisting of six diodes, is then connected to low-voltage taps on the same transformer (Fig. 8). With such a circuit in use and the SCRs fully on, their output is at a higher voltage than the diode bridge. As a result, the diodes are reverse-biased and turned off. When the SCRs are partially on, the diodes are free to conduct. The diodes then improve the quality of the output waveform, since the minimum level of the output waveform is maintained at the voltage of the transformer taps to which the diodes are connected.

The boost circuit is particularly valuable in UPS applications below 200 kW. It provides many of the benefits of the 12-pulse rectifier circuit without the additional cost and complexity.

Finally, alternative rectifier designs employ *ferroresonant transformers* (Fig. 9) instead of the standard rectifier transformer. Consisting of a linear reactor (or high-reactance transformer), a capacitor, a saturable transformer, and capacitor input filters, ferroresonant transformers correct line-voltage variations and transform the input voltage to the levels needed by the rectifier circuit. In addition to providing voltage regulation, ferroresonant transformers also produce a higher harmonic content output resembling a square wave. When rectified, minimal filtering is required. A ferroresonant rectifier, which can be either controlled or uncontrolled, is used primarily in single-phase UPS applications where the power requirements are above 10 kW. They prove only marginally advantageous in 3-phase applications and single-phase applications below 10 kW.

UPS inverters

THE OUTPUT of an uninterruptible power system is normally supplied by an inverter, the design of which governs quality (distortion, regulation, freedom from transients, etc.). Inverter reliability is critical. Failure will transfer the critical load to a bypass source, which could be noisy and unreliable.

Fig. 1. Pulse-width modulation

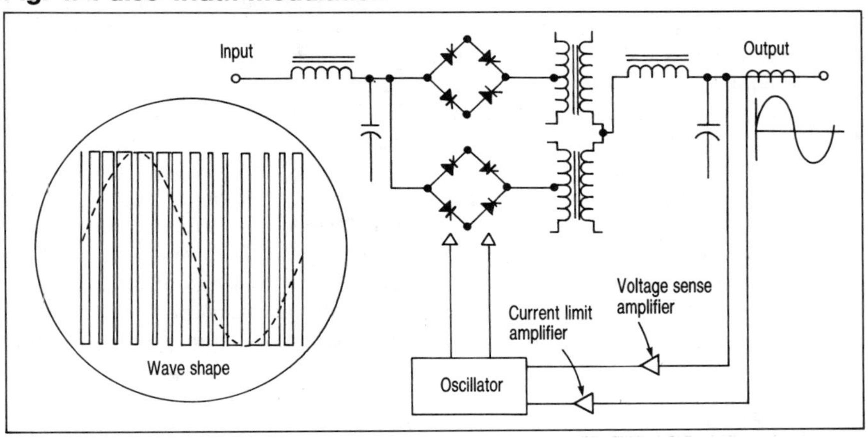

Fig. 2. Step-wave techniques

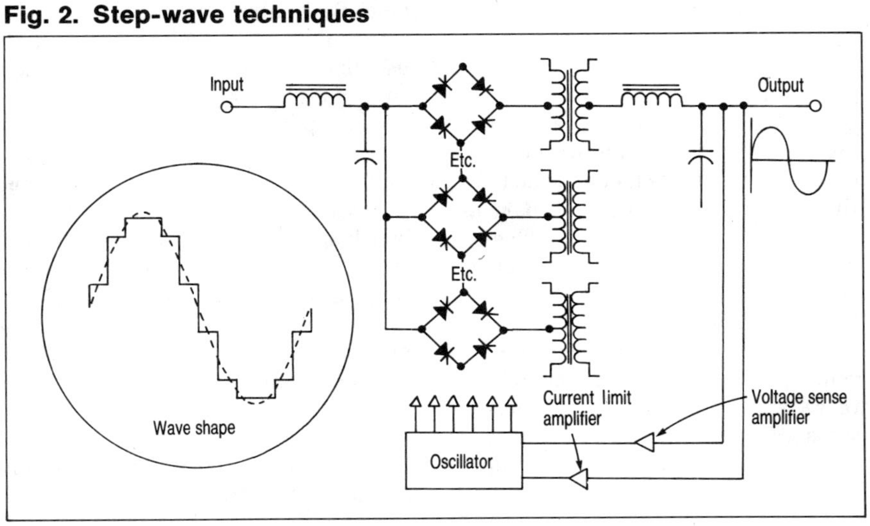

Any inverter intended for UPS applications must reliably produce a clean, stable, regulated, low-distortion sine-wave output. Actually, there are three such sine waves 120° apart for a 3-phase UPS. This is despite variations in load magnitude or power factor, and despite the fact that load current may be irregular, pulsating, and/or high in harmonics.

Several power-conversion techniques are used to produce 3-phase inverters. Each meets the above requirements, but with varying degrees of success. AC inverters use solid-state electronic components to convert DC power to the inverter's output. These techniques can be broadly classified as being "electronic" or "magnetic."

Electronic control

In electronic inverters, feedback and logic circuitry is used to control synthesis of the AC output waveform. This is done using pulse-width modulation (PWM) or step-wave techniques.

Pulse-width modulation. PWM technologies produce waveforms as indicated in Fig. 1. The magnitude of these waveforms is uncontrolled, being a function of the inverter's DC input voltage. By varying waveform duty cycles, the effective output voltage can be varied and controlled. PWM waveform duty cycle must be controlled, not only to regulate inverter output voltage, but also to minimize the filtering required to produce an acceptably low-distortion sinusoidal output.

Step-wave technologies. Step-wave inverter techniques change transformer taps or add multiple square waves to

Fig. 3. Ferroresonant regulation

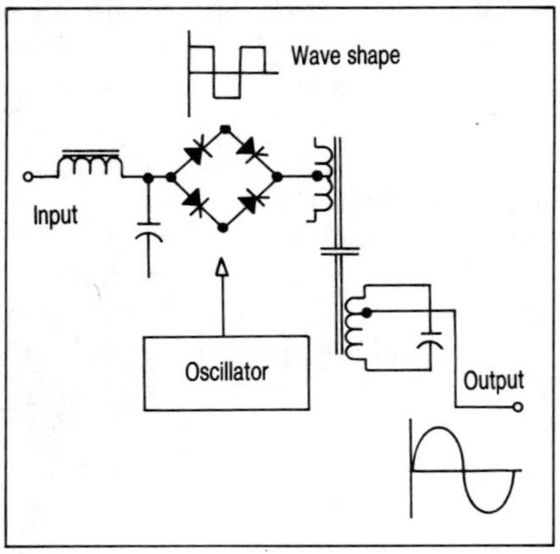

Fig. 4. Delta magnetic system

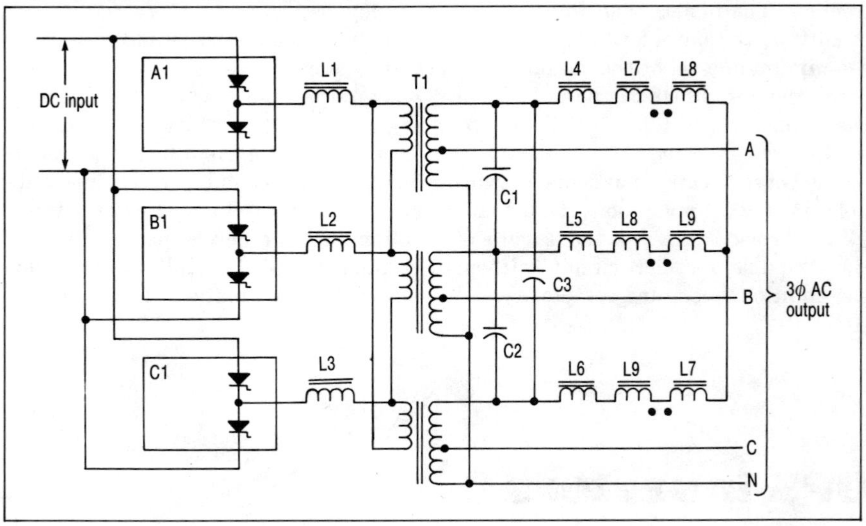

generate a step wave, or staircase-like approximation of a sine wave (Fig. 2). This waveform requires relatively little filtering to produce an acceptably low-distortion output. Being a function of the DC input voltage, the actual magnitude of each step is unregulated. Several techniques can be used to control step-wave-inverter output voltages. One method is to produce two separate inverter outputs that can be added together. By varying phase displacement between two inverters, output voltage can be varied and controlled.

Another method is to use a DC-to-DC converter at the inverter input. By varying DC input voltage supplied to the inverter, output voltage can be controlled. Either way, two complete conversion circuits are required.

Practically speaking, both PWM and step-wave inverters share several common problems. Both are complex. They are not inherently protected against short circuits on their output and must rely on their control circuitry to cut back output power quickly to a level that will not damage the inverter. The control loop must remain stable despite changes in the inverter's DC input voltage and changes in the load's magnitude, power factor, duty cycle, harmonic content, and even direction of power flow.

Magnetic control

There are two basic types of magnetically controlled inverters—those employing ferroresonant regulation techniques and those using delta magnetic regulation techniques.

Ferroresonant inverters. The most common magnetic regulation and filtering method involves the use of a ferroresonant transformer (ferro). Commonly used for single-phase inverters and line-voltage regulators, this device has been adapted for 3-phase UPS use.

In a single-phase ferro, a simple square wave (generated within the inverter) is applied to the primary of a ferroresonant transformer. This transformer primary is inductively coupled, via the shunt structure of the unit, to the transformer secondary. The transformer core is designed so that the secondary section is magnetically saturated at the desired output voltage. As a result, the output voltage remains relatively constant over a wide range of input voltages and loads. Capacitors, connected across the secondary, help drive this core section into saturation and, in conjunction with the inductive coupling, provide harmonic filtering.

In a 3-phase ferroresonant inverter, three single-phase inverters (Fig. 3) are tied together, with their square waves 120° apart. The common point of their secondaries becomes neutral, while one inverter output becomes phase A, another phase B, and a third phase C. The output of a ferro is inherently current-limited, due to the inductive coupling between the primary and secondary of the transformer. The output voltage of a ferro is not dependent on a closed control loop but on transformer design.

While ferroresonant inverters are generally rugged, simple and reliable, there are disadvantages to such designs. First of all, they tend to be larger and heavier than electronically controlled inverters. Secondly, there is a phase shift between the inverter's square wave and the output sine wave. This phase shift varies with load magnitude and power factor. When an unbalanced load is applied to a 3-phase ferroresonant inverter, the output phases can be considerably shifted from their normal 120° phase relationships. This brings a resultant change in line-to-line voltages, even if individual line-to-neutral voltages are perfectly regulated. This voltage imbalance cannot be tolerated by many loads. Finally, the square waves applied to the transformer primaries contain all of the odd-order harmonics (third, fifth, seventh, etc.), necessitating further filtering on the inverter output to remove the harmonics and produce a low-distortion output.

Delta magnetic regulation. Like ferroresonant inverters, delta magnetic inverters are simple, rugged, and reliable. But, unlike the ferro, which is a single-phase device adapted to 3-phase use, the delta magnetic circuit is inherently a 3-phase device.

A simplified circuit schematic of a typical delta magnetic system is shown in Fig. 4. Inverter modules A1, B1 and C1 produce square-wave outputs, which are phase-shifted from each other by 120°. These square waves are coupled to the primaries of transformer T1 through linear inductors L1, L2 and L3. Note that T1 is a conventional 3-phase isolation transformer, not the three single-phase saturating transformers employed in a ferroresonant inverter. The primaries of T1 are connected in a delta configuration, eliminating the third harmonic and all other harmonics that are odd-order multiples of the third. As a result, the delta magnetic design requires significantly less filtering than that demanded by a ferroresonant system.

Secondaries of T1 are connected in a wye configuration to provide a 4-wire, 3-phase output. Inductors L4-L9 form a network connected in a delta configuration to high-voltage taps on the secondaries of T1. Inductors L4-L6 are single-winding saturating reactors, while L7-L9 are double-winding saturating reactors.

Current drawn by this saturating-reactor network is nearly sinusoidal and varies in magnitude in an extremely nonlinear manner, with voltage variations. For example, if an increase in load tended to pull inverter output voltages down, the reduced voltage applied to the reactor network would result in a relatively large decrease in the current drawn by the network. This, in turn, would decrease the voltage drop across inductors L1-L3 to keep the inverter output voltage at the proper level. Thus, the delta magnetic regulation technique is essentially a 3-phase shunt regulator. Capacitors C1-C3 help drive the reactor network into saturation, as well as providing harmonic filtering in conjunction with inductors L1-L3.

There are a number of advantages afforded by the delta magnetic regulation technique over other inverter designs. (1) Output voltage is determined by the design and construction of the saturating reactors. This means that the output voltage remains fixed for the life of the system. (2) There is no control circuit to become unstable or fail. (3) The inverter is inherently short-circuit-protected because inductors L1-L3 will limit inverter current to a safe value, even with a short circuit applied to the inverter output. (4) The delta-magnetic regulation technique is able to maintain regulation with a totally unbalanced load. (5) Since the inductors and capacitors in the delta regulation technique can store considerable energy, use of this stored energy adds significantly to the inverter's ability to handle high inrush current loads and loads with peaky or discontinuous current waveforms. This stored energy gives the delta-regulated inverter a further advantage in that it allows the UPS to provide an actual negative transfer time.

Obviously, the basic design or regulation employed has tremendous impact on the reliability of the system.

Output transfer switches

THE OUTPUT transfer switch, as part of a UPS system, is a device that functions to control power output to the critical load. In doing so, such devices are designed to protect the load from interruption by continuously monitoring both inverter and load parameters. In the event of an inverter failure, the output transfer switch permits a nonbreak transfer to the bypass line source.

There are two different categories of transfers possible with a UPS: break-before-make, and make-before-break.

Transfer of power to bypass source

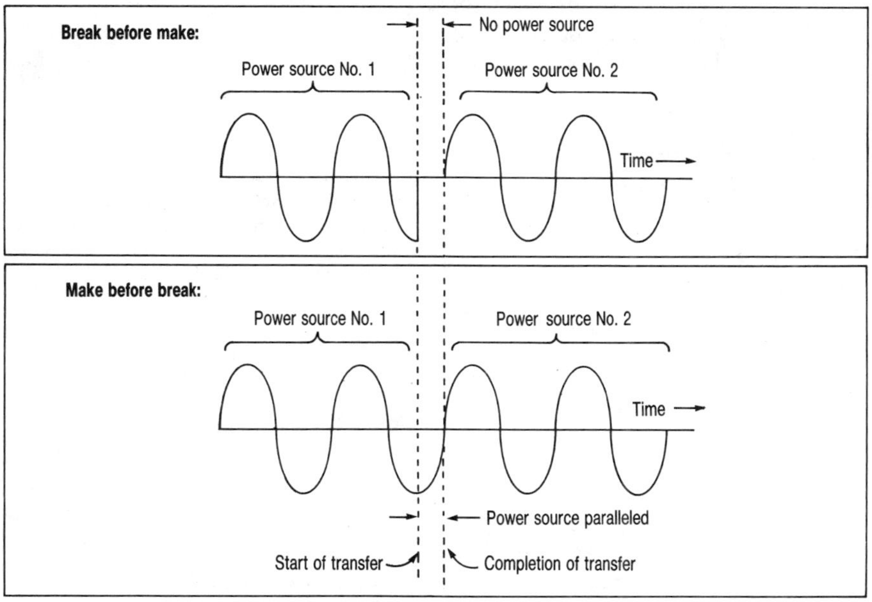

Break-before-make

In a break-before-make transfer, the power output is broken, or interrupted, before the transfer to the new source is made. There is a definite break in power as the load is taken off one source and connected to another.

While this type of transfer is fairly simple, the resulting time delay between break and make creates a definite interruption in the operation of the critical load. If the load is a computer system, for example, the possible damage the system is seeking to avoid by initiating the transfer to begin with is brought about through the transfer itself. Since break-before-make transfers do not result in truly uninterrupted power, this type of transfer is most often employed in noncritical applications involving standard emergency lighting.

Make-before-break

For critical-load applications and true UPS performance, a make-before-break transfer is necessary. With this arrangement, the load is transferred to an alternate power source *without* interruption.

A UPS will normally employ what is known as *reverse transfer*. The load is normally powered by the inverter, with transfer to the bypass source initiated in the event of a battery or inverter

Electromechanical transfer

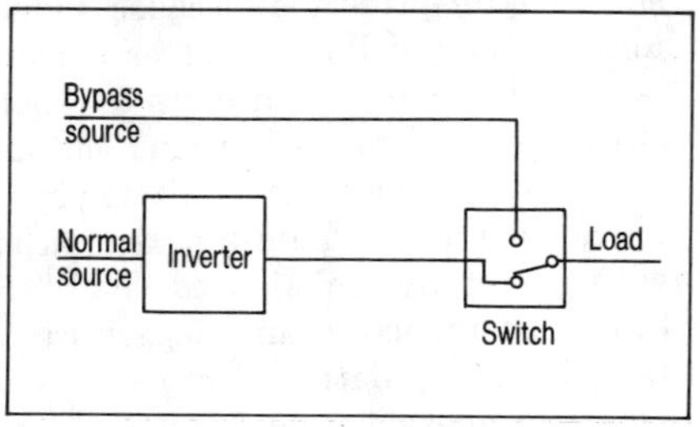

failure, if the system is overloaded, or for maintenance on the inverter.

When using "make-before-break," both sources will be connected to the load before break occurs, and therefore the two sources must be in synchronism. The UPS thus must incorporate a sync circuit to synchronize the inverter to the bypass source to put them "in phase." When the transfer is initiated, there is only a very minimal change in the output to the critical load, usually because of a small difference in the line and inverter outputs.

The alternative to *reverse transfer* is *forward transfer*, where the critical load normally operates off the incoming AC line rather than off the UPS inverter. This method is not preferred because (1) the load is not protected against intermediate and short-term voltage aberrations, and (2) there is a definite break on transfer that could last 1 to 10 cycles on one hookup or as much as one second on another.

Types of UPS output transfer switches

Three types of output transfer switches are commonly employed: electromechanical, static, and a hybrid of the two. Each has its strengths and weaknesses.

Electromechanical type: These switches are motor-actuated circuit breakers on both the inverter side and the line side. In initiating a transfer, the breaker on the line side is closed while the breaker on the inverter side is opened. The disadvantage of this type of transfer is that, while motors are generally very reliable, they are called into service only infrequently in a typical UPS application. They may sit unused for months at a time, during which periods they tend to collect dust, their lubricants dry out, etc. When called upon to initiate a transfer, they often fail.

Static type: These switches consist of multiple SCRs on both the line and inverter side of the UPS. The static switch functions in essentially the same manner as a motor-operated breaker. But, since there are no moving parts, reliability is greatly enhanced;

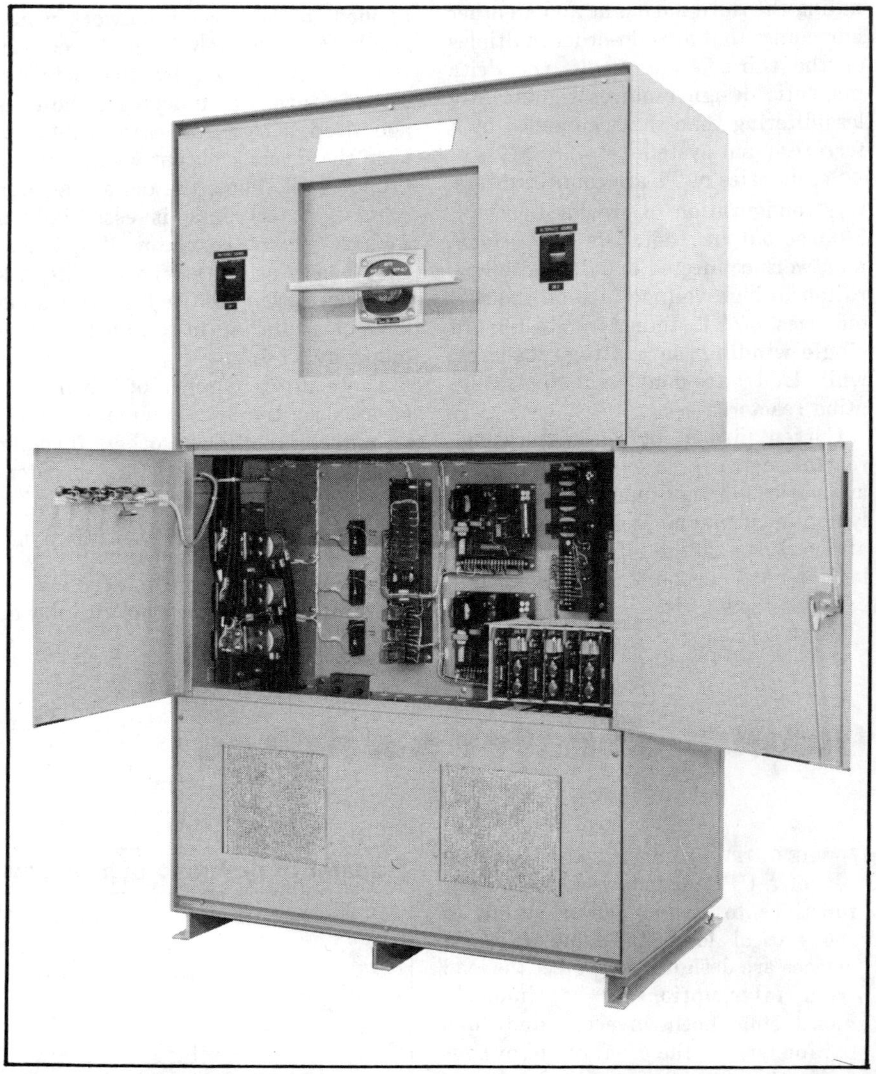

STATIC TRANSFER SWITCH includes manual bypass switch (top, center) flanked by isolation switches for preferred and alternate sources; panel (doors open) containing controls, indicators, fuses, alarm relays, and regulated power supply; and filtered pressurized forced-air cooling (bottom).

Static output transfer and hybrid configurations

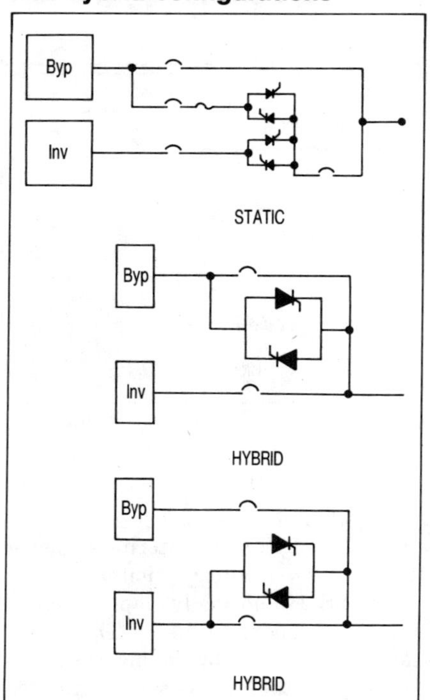

and an SCR cannot fail because of contamination or because of lubricants drying out. Finally, the SCRs are much quicker reacting to inverter and load conditions, resulting in true no-break transfer capability.

Hybrid type: a hybrid switch is a cross between the electromechanical and the static switch, consisting of both motor-operated CBs and SCRs. The SCRs maintain power to the load until the inverter breaker opens and the bypass breaker closes.

While such designs greatly enhance the responsiveness of the switching mechanism (because of the SCRs), the reliability problems inherent in the moving parts of the motor-operated CBs remain. In addition, low-power SCRs are generally used, which fail if the transfer requires more than a minimum amount of time for completion, or if the transfer is due to an overload condition.

Thus, the optimum output transfer switch is of the static variety using a make-before-break reverse transfer.

Batteries and auxiliary power

THE STANDBY battery plant of a UPS provides DC input to the inverter when the rectifier output drops below an acceptable range. The battery size controls the amount of reserve time available during a power outage (blackout or severe brownout). UPS batteries are usually sized to provide adequate reserve time to permit orderly shutdown of the critical load, auxiliary power generation sources to come online, or commercial power to be restored. Since the reserve time available is a function of the battery construction, an examination of batteries commonly employed in UPS applications is in order. Basically, all batteries can be identified as being either primary or secondary. Primary batteries are characterized by the inability to recharge electrically and as such are usually discharged once and discarded. Secondary batteries are capable of being recharged electrically after discharging. They can be recharged to their original condition by passing current through them in the opposite direction to that of the discharge current.

Such being the case, secondary batteries are most often used in UPS applications. Of the various secondary batteries available, two basic types commonly employed in standby battery plants are lead-acid and nickel-cadmium. Lead-acid batteries are by far the most common in UPS applications.

Lead-acid batteries

Lead-acid batteries are widely used because of their low cost, reliability, and excellent performance characteristics in float applications. They use highly reactive sponge lead for the negative electrode, lead dioxide as the positive material, and a sulfuric-acid solution for the electrolyte. As the lead-acid cell discharges, the reactive material of both electrodes is converted into lead sulfate. The electrolyte is also involved in the reaction and produces water. When charged, the opposite action occurs. The state of charge of a lead-acid battery can be determined by measuring the specific gravity, or electrolyte concentration, which increases on charge and decreases on discharge.

The most common constructions for lead-acid batteries are the pasted plate, or Faure designs, and the Plante design. In the Faure design, the active material for each electrode is prepared as a paste by mixing finely divided lead oxides and suitable expander materials with sulfuric acid. The paste is spread onto a lead-alloy grid, which provides the electrical conductivity and structure necessary to hold the active materials.

The plates are then soldered to connecting straps to form positive and negative interwoven groups. Separaters are then placed between the electrodes, and the completed element is placed in a container. The container is designed with a sediment space under the element to collect any of the active material that dislodges. The container also provides sufficient room above the plates to hold excess electrolyte. Faure-type lead-acid batteries are usually classified by the alloy used in the grid. The most common alloys used are calcium and antimony. Hence, the most common lead-acid batteries are the lead-calcium and lead-antimony types. Generally, the lead-calcium batteries prove better suited to UPS applications because of their low gassing characteristics, requiring significantly less ventilation, and their low maintenance requirements.

Similarly, the Plante type of lead-acid batteries use a design that is similar to the Faure for the negative plates. The positive plates, however, employ active material in the form of buttons of soft lead ribbon. Like the Faure, the electrolyte used in the Plante-type lead-acid battery is sulfuric acid. The Plante design is sometimes used in applications where extremely high temperatures are a problem.

Nickel-cadmium batteries

Nickel-cadmium batteries are alkaline rather than acid. This type of battery is noted for its high-rate/short-term discharge capability and good low-temperature performance. The active materials in a charged nickel-cadmium cell are trivalent nickel oxide for the positive electrode and cadmium for the negative electrode. The alkaline electrolyte is a solution of potassium hydroxide.

During discharge, the nickel oxide is reduced to the divalent state and the cadmium is oxidized. During charging, hydrogen and oxygen are evolved by the positive and negative electrodes. Unlike the lead-acid designs, specific gravity measurements provide no indication of the state of charge in a nickel-cadmium battery.

While there are various constructions available for nickel-cadmium batteries, the design commonly applied to UPS applications is known as the vented-pocket plate type. Here the active materials are in powdered form and are contained in perforated, rectangular pockets formed from thin nickel-plate steel ribbon. The pockets are then welded into steel frames to form electrodes of various sizes and capacities. The perforations provide access for the electrolyte but are small enough to retain active material particles.

The positive and negative plates are

then respectively bolted together, insulated from each other, and assembled in a plastic or nickel-plated steel container. A 20% potassium-hydroxide electrolyte is then added, with higher concentrations used to provide better low-temperature performance. NiCd batteries generally are only applicable in UPS applications where their improved low-temperature or extremely short-duration/high-rate characteristics offset the cost disadvantages of this design. Careful analysis of such characteristics as battery life, cost, maintenance, cycle duty, and temperature is necessary to determine the optimum battery construction for a given installation.

Battery selection

The primary considerations in selecting the optimum battery for a given installation include battery life, cost, cycle duty, performance at various temperatures, number of cells, cell capacity, and the resulting configuration and battery rack that will be required to mount the standby battery plant.

Battery life: The effective life of a particular lead-acid or nickel-cadmium cell will depend on the actual conditions, both environmental and functional, to which it will be subjected. However, battery manufacturers have developed fairly acceptable projections as to what one should expect in terms of an acceptable effective life. In the case of lead-acid batteries, lead-calcium constructions can provide up to a 20-year life, lead-antimony constructions can maintain effective operation for 15 to 17 years, and Plante constructions provide an effective life span of 30 years. The projected effective life of nickel-cadmium batteries is somewhere around 25 years.

Cycle duty: There is also variation in the cycle-duty characteristics of the various battery designs. Lead-acid cells have a nominal voltage of 2.0 V per cell, with the voltage on open circuit being a direct function of the specific gravity, ranging from 2.12 V for a cell with a specific gravity of 1.28 to 2.05 V for a cell with a specific gravity of 1.21. The typical specific gravity of a lead-acid cell, determined by its active materials, is 5.2 W-hr/lb. Generally, lead-acid batteries can be discharged safely to a voltage level of 1.75 V per cell. Discharge below this level can result in damage to the battery.

In the case of the nickel-cadmium battery, the nominal voltage of a nickel-cadmium is 1.35 V per cell. NiCds have a typical capacity of approximately 6.5 W-hr/lb per cell. Unfortunately, the discharge levels possible with a NiCd are not functional in UPS applications.

The capacity-versus-discharge rates of typical lead-acid and nickel-cadmium cells at various environmental temperatures are shown in accompanying illustrations.

Battery charging: There are two factors impacting battery charging in UPS applications—the amount of current the charger provides, and how fast the battery can accept this current. The charger should be capable of recharging the battery in no more than 5 times the discharge period. The bat-

Capacity vs discharge rate for lead-acid storage batteries

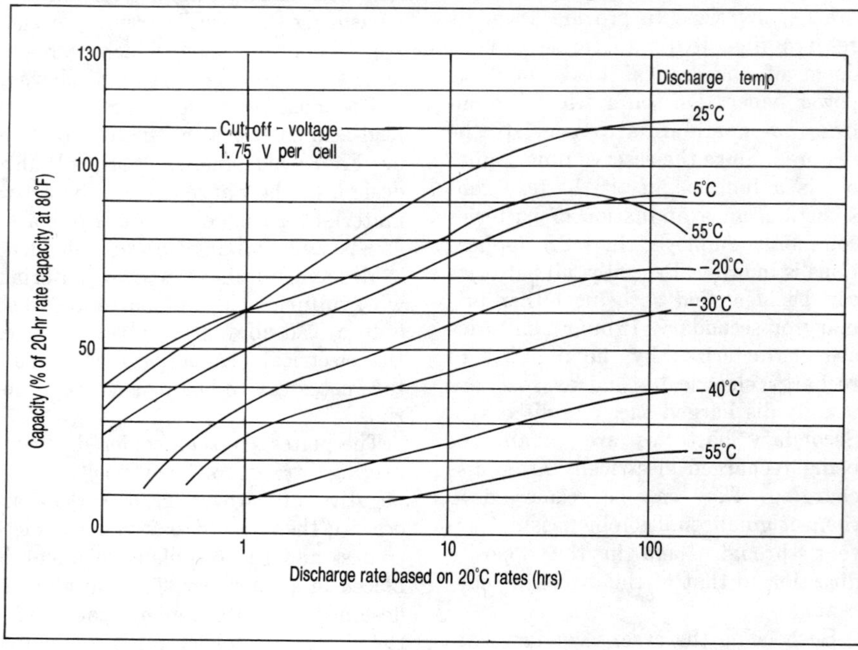

Discharge of lead-acid batteries at different hour rates

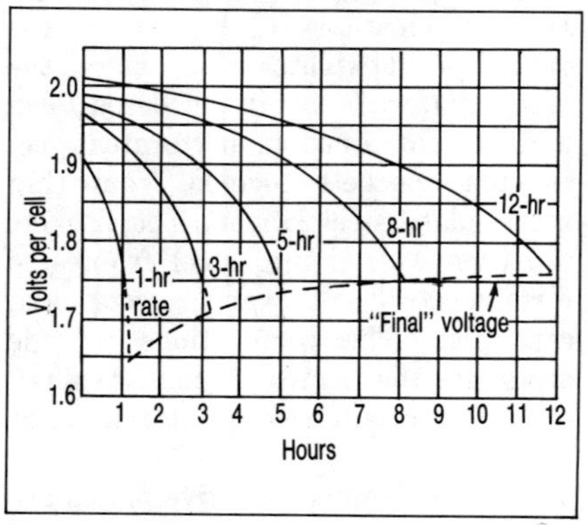

Discharge of pocket-plate NiCd cell at 25°C

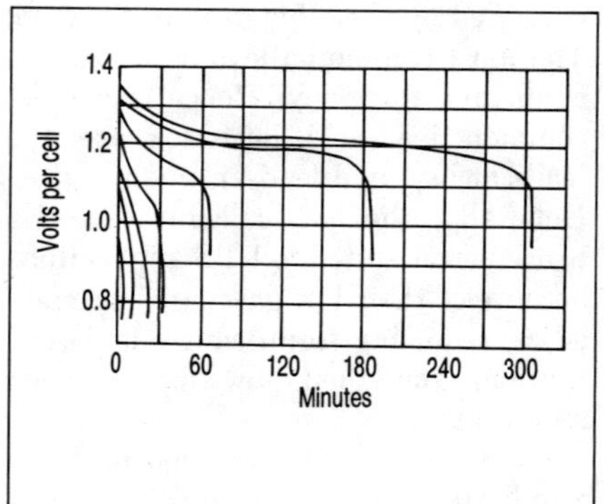

Effect of temperature on capacity, pocket-plate NiCd cell

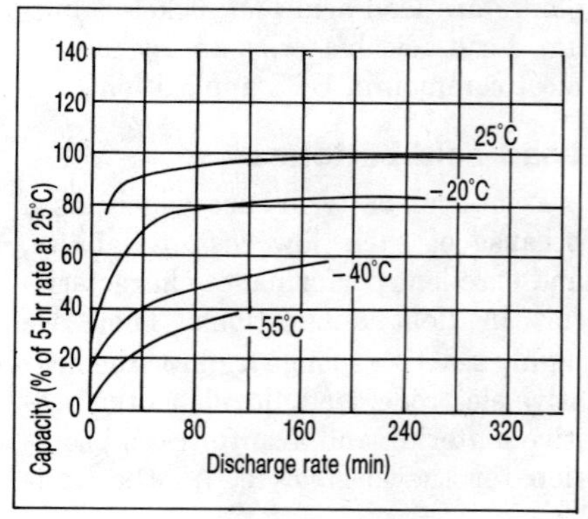

Typical UPS battery racks

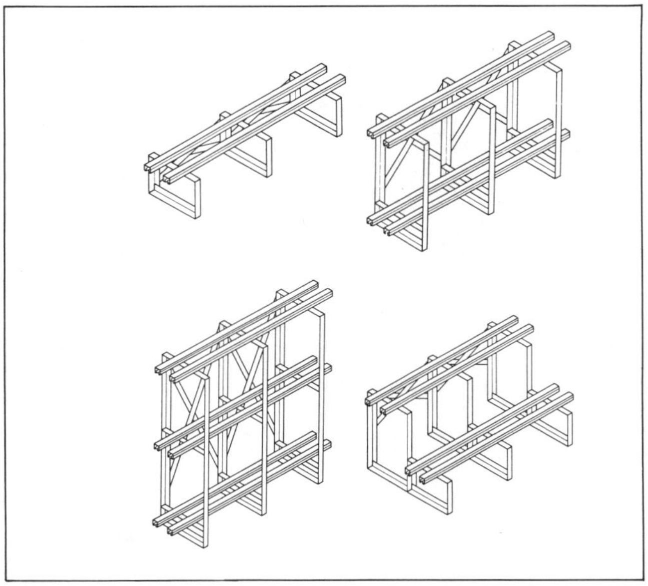

SELECTION of standby batteries is an important step in ensuring a reliable UPS system.

tery, however, may require a longer time because of its inability to accept a rapid recharge.

It is important that the UPS battery charger be designed to equalize the battery voltages, since the tolerances applied to battery manufacturing allow the capacity of a given cell to vary. If the UPS battery charger is not capable of equalizing the voltages, making certain that all cells are charged to a nominal level, premature failure of one or more cells in the plant can occur.

Number of cells: The number of cells required for a given UPS installation is a function of the voltage at which the UPS is designed to operate. As such, the number of cells is dependent upon the designed DC ouput of the battery charger/rectifier and the DC input of the inverter.

Capacity of cells: The capacity of the cells required for a given battery plant is a function of the reserve time desired and the estimated frequency at which the battery plant will have to operate. The latter is a function of the number and proximity of outages occurring during the course of a year in a particular locality. The battery must be sized to handle multiple outages, occurring one right after the other, without the benefit of recharging. The former depends on whether the battery plant will be required to maintain power for the duration of the outage until orderly shutdown of the load can be affected or until an auxiliary power generation source comes on line. Typically, a UPS battery plant will be sized to provide anywhere from 10 minutes to one hour of reserve time.

Battery configuration: With the number and size of the cells required for a given installation determined, the battery configuration can be specified. Most battery manufacturers offer standard configurations involving multiple tiered or stepped racks.

The battery plant ultimately specified for a UPS installation is one of the most critical components in the UPS installation. It is the battery plant that provides long-term ride-through in the event of a blackout or brownout. Inadequate sizing, care, or maintenace of battery plant can render it totally ineffective and reduce the UPS to little more than a very expensive line conditioner.

Auxiliary power sources

The final component of a typical UPS is the only part of the system that is optional. Depending on the long-term outage protection required by a particular installation, it may or may not be necessary to include an auxiliary power generation source. When long-term auxiliary power is critical, as in the case of an area where long-term outages are frequent or 100% reliability is necessary, an auxiliary power generation source can often prove more cost-effective than attempting to provide sufficient battery backup to ride through a long-term outage.

Generally, auxiliary power generation sources consist of either a diesel engine or a gas turbine driving an AC generator. Such devices provide an alternative power generation source when utility power fails. Usually, such systems are designed to cut over after a preset time period following the advent of an outage. This time period can be anywhere from 10 seconds to an hour or more. The transfer to the auxiliary power generation source is accomplished ahead of the rectifier so that the UPS can provide adequate filtering of such sources' typically noisy output.

The addition of an auxiliary power generation source to a particular installation can significantly reduce battery cost, since batteries will be required only to bring the auxiliary power generation source on-line and get its output stabilized.

Chapter 2

Specifying UPS Systems

Specification Factors

Here are factors that must be considered in selecting uninterruptible power supplies, with equal attention to load hardware and load applications.

DEVELOPING specifications for an uninterruptible power supply requires careful analysis of variable factors before a system is chosen. Key factors are related to the load hardware and load applications. Electrical power required by a sensitive load may vary widely, depending on the configuration of the load. The power demand of a computer system, for example, depends on the number of central processing units and the number and types of peripheral devices, such as disk and tape drives.

The job of developing specifications would be simpler if a specific piece of load hardware always required the same amount of electrical power and is always found in a specific application. But such is not the case. A certain popular minicomputer, for example, is often used in a telephone company's toll-ticketing applications. It draws from 3 to 15 kVA, depending on the configuration of the peripherals. This same computer system is used for many applications other than toll-ticketing. Further, it is not the only minicomputer that is used for toll-ticketing. Thus, *nearly any piece of load equipment can be found in nearly any application.* Specification writing, then, becomes more involved than it first appears.

Load power supply factors

There are characteristics of the connected load that dictate the design, performance limits, and capabilities of the UPS to be specified. The principal factors determined by hardware, which vary from load to load, are:

- Power requirements
- Power factor
- Load configuration
- Number of phases
- Voltage regulation
- Phase displacement
- Frequency stability
- Slew rate
- Unbalanced loading
- Filtering and distortion
- Inrush handling capabilities
- Overload capacity
- Bypass capability
- Transfer time
- Reserve time

Each will be discussed in turn.

Power requirements of the project must be defined clearly and "critical" requirements identified. Potential future expansion must also be considered, because the UPS can be sized either to accommodate the total anticipated critical load growth at the outset or to accommodate the initial critical load, with provisions for expansion. Secondary electrical systems such as lighting, air conditioning, and alarm systems should be examined carefully. Although such loads may not be classified as critical—that is, they can effectively operate with relatively untreated power—their operation may be imperative to the continuous operation of the critical load.

The specific power requirement, the specified kW and kVA, can be determined in one of two ways. The preferred method uses actual load measurements to determine the exact power requirements of each device that will be powered by the UPS. When actual load measurements are not available, it becomes necessary to use the nameplate ratings found on various devices, in instruction manuals, or in the computer installation planning manual. Nameplate ratings typically include significant safety margins. Moreover, the load normally will include a diversity factor, which means that all individual loads will not be operating, drawing UPS power, at the same time. As a result, the nameplate ratings, taken in conjunction with a diversity factor of 100%—with all equipment running simultaneously—can result in specification of a substantially oversized UPS. Overall system efficiency thus will be decreased and operating costs increased, compared with a UPS sized for the actual measured load.

But there is another side to this coin. After conscientiously sizing the UPS for the actual measured load, it is common for other critical loads to be "discovered," especially during the first few months after the UPS begins operating. To account for oversight, a modest "pad" is often applied to the final load power requirement calculations.

Power factor is important in specifying a UPS application for two reasons. First, kVA, in addition to kW, determines sizing of such equipment as the engine-generator, the transformer, distribution wire, and circuit breakers. Second, the regulation of the UPS can be affected adversely by the power factor. Many UPS units are capable of proper operation only with power fac-

By THOMAS S. O'NEILL

TOM O'NEILL is President and Chief Executive Officer of LorTec Power Systems, Inc., North Ridgeville, OH. He has a bachelor of electrical engineering degree from Fenn College and a J.D. from Cleveland Marshall College of Law.

tors from unity to lagging and cannot function with leading power factors. As a result, the load's power factor can have a devastating impact on UPS operation.

Increasingly, computer manufacturers are going to high-frequency-switching power supplies with leading power factors. This change is occurring because such power supplies are more efficient, smaller, and less expensive. Therefore, the ability to handle leading power factors will become an increasingly important consideration in specifying uninterruptible power systems.

Load configuration. In 3-phase applications, it is important to determine whether the load uses a delta or a wye connection, or perhaps a mixture of the two. This determination can be a critical concern because some UPS designs cannot handle delta-connected loads for various reasons, including the power factor presented to the UPS. In this case, the manufacturer may need to provide a delta-to-wye isolation transformer. The addition of this element will detract from efficiency and voltage regulation, as well as serving to add another potential failure point in the system.

Number of phases. A critical load usually is a combination of several individual loads, and these loads may be a mixture of single-phase and 3-phase loads or all single-phase or 3-phase loads.

A 3-phase load *cannot* be powered by a single-phase UPS, but a single-phase load *can* be powered by a 3-phase UPS. Many large (greater than 10 kVA) loads are actually multiple single-phase loads or a combination of 3-phase and single-phase loads and thus can be powered by a 3-phase UPS. Extremely important in this context is the unbalanced load capability of the UPS. If the critical load—a large computer system or electric motor, for example—requires 3-phase input, a 3-phase UPS must be specified, regardless of the number of 120-V, single-phase peripherals. For computer systems requiring 3-phase input, with some of the peripherals wired for 208-V single-phase input, a 3-phase UPS is ideal because it provides 208-V 3-phase, 208-V single-phase (line-to-line), and 120-V single-phase (line-to-neutral) inputs. Additionally, due to simpler filtering and other factors, a 3-phase system is generally more efficient than a single-phase system. For this reason, as well as for the added

UPS CHARACTERISTICS are determined by the components designed into the system, as evidenced by the rectifier/inverter section views shown here. Specifications must match load UPS characteristics and characteristics of the available power service.

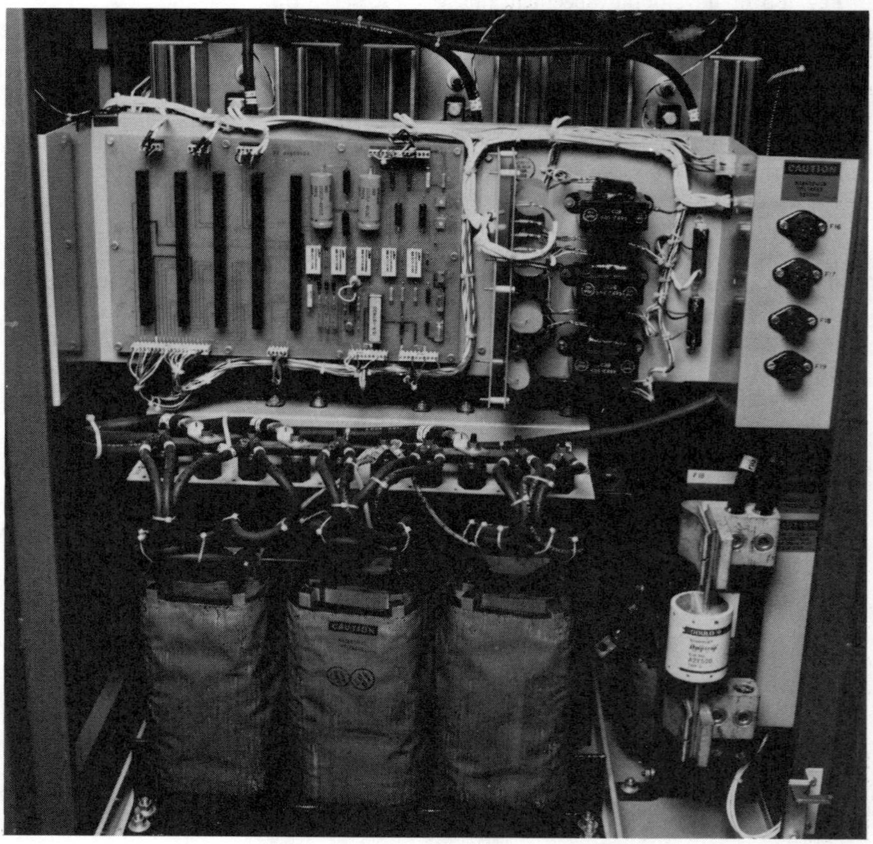

load versatility provided, a 3-phase system is preferred in sizes above 10 kVA.

Voltage regulation. Regulation is the deviation from a nominal voltage, expressed as a percentage of the nominal voltage. For a UPS, voltage regulation is measured over all ranges of AC input and DC input voltage, load power and power factor, and ambient temperature. Most loads are designed to be compatible with the voltage limits of American National Standards Institute (ANSI) Standard X4.11-1973. This standard sets voltage regulation to within +5.833% to −13.333%. In some cases, exceptionally critical loads may require ±5% regulation. Regulation beyond this, say in the 1% range, is extremely rare.

Most regulation problems encountered in applications are the result of factors external to the UPS itself. The actual usable voltage at the load distribution panel should be determined. Such factors as long runs, bad connections, undersized cables, and severe overloading of one phase contribute to poor overall regulation at the load.

Phase displacement. The angle between phases of a 3-phase voltage supply should be exactly 120 electrical degrees. The voltage regulating scheme used by the UPS inverter dictates how well each phase remains fixed in relation to the other phases. The deviation from nominal is most often stated at ±1 or 2 degrees from the nominal 120° phase displacement. One method regulates the line-to-line voltages of the inverter output, rather than the line-to-neutral voltages. This results in the angle between phases being inherently regulated to 120°.

Maintaining the proper phase displacement is important to a 3-phase load, over all ranges of AC and DC input, load power and power factor, and unbalanced load conditions. Failure to do so will result in high circulating currents and increased core losses in 3-phase transformers and motors.

Frequency stability. Frequency variations can have an adverse effect on a critical load. Frequency stability is the deviation from a nominal frequency, expressed as a percentage of the nominal frequency. Therefore, 1% frequency stability based on 60 Hz represents a ±0.6-Hz deviation. Most 60-Hz loads are designed to tolerate a frequency of 59.5 to 60.5 Hz, or 60 ±0.5 Hz.

Most 3-phase UPS systems use a crystal-controlled oscillator to control the output frequency very precisely, and control to ±.001% is not uncommon. However, the UPS normally has the capability to synchronize its output frequency and phase to that of the AC bypass source. This capability is important for two reasons. First, when the load must be transferred to the bypass source, minimum transfer disturbances to the load will result if the inverter voltage and bypass voltage are in phase with each other. Second, computer systems often contain clock circuits, which may be used for billing purposes. Because the AC utility line frequency will average precisely 60 Hz over a 24-hr period, a slight deviation from the nominal frequency can accumulate a large total error over the course of several months with a higher frequency indicating a longer time period used.

INVERTER MODULES, shown here under test, must be capable of handling sudden inrush requirements without relying on the bypass source to provide sufficient overload capacity, one of the load factors that must be considered in specifying a UPS.

Slew rate, defined as the rate of change of frequency, in Hertz per second, is especially important for installations employing motors, such as the disk drives in a computer system. Because motor speed is proportional to the frequency, a sudden change in frequency will result in a high inrush to the motor until the speed of the motor "catches up" to the frequency. Most critical loads can tolerate a maximum slew rate of 0.5 Hz/sec.

Unbalanced loading. In most installations requiring 3-phase power sources, at least 75% of the loads powered by the system can actually be single-phase devices. Although the 20% unbalanced load capability available with most units may appear adequate, this capability can be exceeded easily. For example, in a 15-kW UPS system, phase A may be loaded with 38 A, phase B with 40 A, and phase C with 42 A, which is a reasonably well-balanced condition. However, if a small load is removed from service on phase A, reducing its load to 33 A, the UPS has a load that is unbalanced by more than the typical limit of 20%.

In a 15-kW unit, a mere 1-kW unbalance in one phase represents a 20% total unbalance. Where loading exceeds the specified maximum unbalance, UPS reliability suffers as components are stressed, regulation begins to deteriorate rapidly, and distortion increases. Problems with unbalanced loading are common with ferroresonant and electronic inverter designs.

Filtering

Filtering capabilities and distortion requirements. Every load has a limited tolerance for noise and harmonic distortion. Total harmonic distortion (THD), which is the measure of the quality of the waveform applied to the load, is calculated by geometrically summing the harmonic voltages present in the waveform and by relating this sum as a percentage of the fundamental voltage of the waveform. Typically, critical loads can withstand 5% THD, where no single harmonic exceeds 3%.

The UPS must provide this high-quality output waveform to the load, regardless of the level of noise and distortion produced by the AC input service. The AC utility voltage is relatively noise-free and well within distortion limits, but lightning or downed power lines can create input waveform problems. Less obvious is localized noise from arc-welding units in the plant, a large motor in a nearby plant, or nearby electrically noisy power supplies. The UPS must filter all input disturbances before they reach the critical load.

One type of noise that is difficult for the UPS to attenuate is load-produced noise. If a critical load is very noisy or has an extremely nonsinusoidal waveform, the result will be reflected in the UPS output. Special filters may be required to prevent a false trigger or misfiring circuit element or to prevent an adverse effect by reflected noise on other loads connected to the UPS output. UPS designs that already have inherently large amounts of filtering on their output are less likely to need the addition of an external filter.

Inrush handling capabilities. Many types of loads, such as incandescent lights and some types of motors, require as much as 12 to 15 times the normal current at startup. With incandescent lights, the cold filament offers decreased resistance. With motors, increased power is required to start rotation.

A surge, lasting from ½ cycle to a few cycles, can produce an effect similar to that of a short circuit. In specifying a UPS for such applications, the short-circuit capabilities of the unit must be sufficient to accommodate such an inrush surge, particularly when the bypass source is unavailable. When the inrush is greater than 125% of the normal current, the UPS will attempt to transfer to the bypass source until the current demands return to acceptable limits. Without a bypass source, the UPS will be forced to shut down, unless it has been designed to tolerate a short circuit or severe overloads.

For a single, large 3-phase motor, for example, the startup current demand can be several times the rating of the UPS. Where the inverter is designed to work into short circuits, the output will provide current well in excess of its rating, at a drastically reduced voltage. If this current is sufficient to overcome frictional losses in the motor, the motor will begin rotating, slowly at first. As the counter-EMF of the rotating motor increases, the current demand will decrease. The inverter's output voltage will increase, and so on, until the motor is rotating at full speed, drawing nominal voltage and current.

However, any other device connected to the inverter's output during that time would also see low (or zero) voltage, until the motor comes up to speed. Normally, the bypass source is available and is used to start the heavy loads automatically, via the UPS static transfer switch. Occasionally, an installation is equipped to start heavy loads in sequence to prevent a severe, momentary overload. Where a UPS design does not tolerate short circuits but relies on an output fuse or breaker for inverter protection, load sequencing becomes necessary.

Overload capacity. Depending on the installation, its projected growth, and ongoing peak demands, a UPS sometimes may be required to handle peak demands exceeding 100% of its rated capacity. Overload capability can have a significant effect on the size and cost of the unit ultimately specified for the application. The UPS must be capable of handling occasional moderate overload conditions. Unless extreme care is exercised in the selection of the UPS for a particular installation, however, the user may be swayed to purchase an oversized UPS capable of withstanding the projected overloads. The user will be forced to pay a substantial premium in purchasing a larger unit than required by the application to handle marginal overload conditions on an infrequent basis.

Bypass capability. In analyzing a particular application, the necessary bypass capability should be determined at the outset. Depending on the load and configuration, it may be necessary in the event of a severe overload or UPS failure to bypass the UPS and obtain power from a commercial AC line, another UPS, or an engine generator.

The bypass source must be sized to power the load, and its output must be maintained within parameters acceptable to the load. When a commercial AC line is used, a line-voltage regulator may be required. The bypass source should be independent of the AC input source used by the battery charger/rectifier in the UPS. Using the same AC source for both functions, assuming the AC rectifier input service is adequately fused, would result in underprotecting the AC bypass line. The loss of the AC rectifier input service would mean automatic loss of bypass service also.

Transfer requirements

Transfer time. The UPS configuration and transfer mechanism must be adequate in terms of transfer time. It is important to determine several factors: type of transfer necessitated by the load (break-before-make or make-before-break); the configuration that will best fulfill this requirement (forward or reverse transfer); and the speed with which the transfer can be accomplished.

A decision regarding the type of output transfer switch employed, electromechanical, solid-state, or hybrid, depends on the load device's ride-through capability. Some devices can withstand 3 to 6 cycles of interruption; others cannot tolerate as little as ½ cycle. Specifying a UPS requires stating whether voltage must be present at the load during the transfer. Some output transfer switches allow the voltage to remain at zero during transfer. Other types present usable voltage throughout the transfer period. Most computer loads need usable voltage continuously during a transfer.

Reserve time. Every critical load has a minimum period, after a blackout or severe brownout, during which it requires power to prevent damage. This time may vary from microseconds to several minutes. For a computer system, this requirement is the time required to accomplish an orderly shutdown, usually a couple of minutes. Such reserve time is necessary to prevent the loss of memory and data.

This load-hardware-related reserve time must not be confused with application-related reserve time, where a specific application may require much more reserve time to finish the current production, to start and bring on-line an engine-generator, to complete a series of instructions involving satellite control, or to satisfy desire to continue operating in any event.

Most often, the standby battery plant employed by the UPS is sized to provide 15 to 30 minutes of reserve time (at full UPS load) with a standby engine-generator being considered for longer outages. Occasionally, a second utility feed can minimize the size of the battery plant.

Application factors

An application-related factor is a requirement dictated by the particular project or application, unrelated to the load hardware. Key application factors that can influence UPS specifications are

- Reliability requirements
- Efficiency requirements
- Geographical considerations
- Compatibility requirements
- Environmental requirements
- Physical characteristics
- Code requirements
- Meters, alarms, and controls
- Installation requirements
- Service and maintenance requirements
- Cost considerations.

These factors must be considered independently on each project, even for an identical application.

Reliability requirements. If a UPS application fails to operate with high reliability, the result often can mean more than mere inconvenience. Such elements as health, safety, lost revenue or inventory, or massive hardware damage are usually threatened.

Reliability must be considered from three different viewpoints. First, the reliability of the commercial power in the area in which the facility is located must be documented. Second, the impact of line aberrations on the critical loads located within the facility must be determined—that is, major faults, minor faults, lightning strikes, transients, brownouts, or surges. Finally, the ability of the UPS to maintain reliable operation when subjected to the above conditions must be considered.

In some locations, major faults may occur 10 or more times a year, minor faults 500 to 1000 times a year, and transients, of one cycle or less, as often as 10,000 times a year. Among questions to be answered are: What personnel and equipment are affected by such events? How much time typically transpires before normal operation is restored? What are the costs associated with bringing the system back to normal? What is the total cost to the facility in terms of lost revenues, downtime/recovery time, increased service and maintenance costs, and manpower expenses? The cost for system checkout, restart and program reruns can run from $200 to $1000 per hr for a computer system.

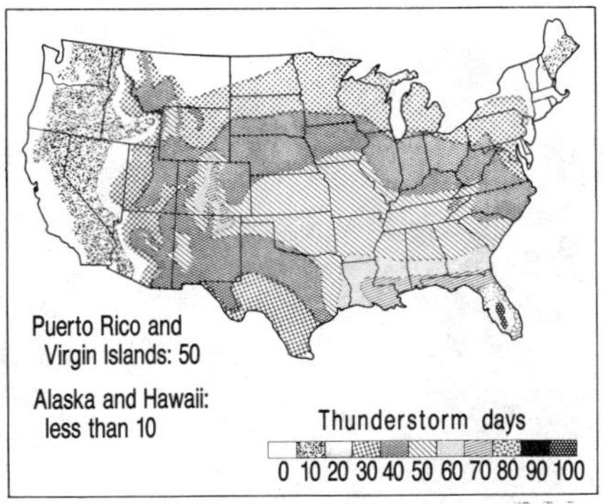

ISOKERAUNIC map prepared by the National Weather Service indicates frequency of thunderstorm days per year across the country. This data should be taken into account in assessing the reliability of commercial power in a given area.

Efficiency requirements. Efficiency of the UPS, defined as the output kW divided by the input kW, is expressed as a percentage. Efficiency is an application-related factor determined both in terms of the efficiency of the inverter section of the UPS and of the system as a whole. The inverter efficiency dictates the battery size required to provide a given reserve time, which is most often application-driven, rather than a function of the length of time the load hardware requires. Total system efficiency determines how much power is lost as heat and how much the user will pay for input power.

Certain trade-offs with efficiency sometimes are required to ensure reliability of the UPS, however. Reliability can be sacrificed by inadequate filter design, hardware packed in small cabinets, ambient temperature limits that are unrealistically low, components that are not derated, and so on. Power required for the air-handling equipment that is used to keep the UPS cool also should be considered as a power loss.

Geographic considerations. A national thunderstorm map will show that some areas of the U.S.A. are as much as 15 times more prone than others to disruptions from thunderstorm activity. Building surge suppression is often inadequate to dissipate completely the energy produced in a transient surge caused by lightning. Temperature can vary widely also, especially if the UPS is to be located in an outdoor enclosure.

Seismic risk to the installation must be determined. Seismic zone maps of the United States show five different zones of seismic intensity, ranging from Zone 0, where no damage is anticipated, to Zone 4, where major damage is possible because of the proximity to certain major fault systems. The UPS should be designed as a rugged device or should be braced accordingly. The battery plant should contain the appropriate earthquake bracing for the zone involved.

Compatibility requirements. A UPS system must be compatible with other systems operating within the installation. The UPS should be capable of interfacing with existing or planned batteries, chargers, power sources, and alarm systems. Moreover, the UPS must not disrupt sensitive electronic devices used in processing or monitoring non-UPS-related functions. Such electromagnetic interference can effectively disrupt the operation of a critical load or, as in the case of load-generated EMI, the operation of the UPS. Thus, the UPS specification must ensure compliance with Federal Communication Commission regulations governing EMI.

Environmental requirements. Environmental conditions to which the UPS will be subjected are highly variable. The UPS being considered must be specified to accommodate extremes of temperature, altitude and humidity. Extreme conditions can precipitate premature component failure and unnecessary system shut-down. UPS output can deteriorate as the thermal limits of the semiconductors or capacitors are exceeded. Most UPS units are rated at 40°C (104°F), but during a commercial power failure the temperature in the UPS room can exceed such a minimal rating quickly. Additional air-handling equipment in the UPS room may be required, thus increasing the cost of the installation and reducing the overall efficiency of the UPS. Some well-designed UPS systems are rated at 50°C (122°F).

Although the UPS itself may need only a method to move the expelled heat out of the UPS area, the battery plant may warrant special consideration. It is designed for nameplate operation at 25°C (77°F). At higher temperatures, chemical activity increases and available reserve time appears higher, but life is decreased. In a lead-acid battery, for example, each 18°F rise above 77°F will result in a loss of half of its useful life. The battery typically is placed in an area where the ambient temperature is maintained at 70 to 85°F. Although not required by most codes, the battery plant is sometimes placed in a separate room.

The ventilation requirements for the battery also include provisions for the exhaust of gas produced. Ordinarily, air movement provided by the equipment maintaining the temperature around the battery plant is sufficient to remove any dangerous gas concentration. The level of gas produced depends on the type of battery cell and on whether it is being floated or recharged.

A second environmental consideration is atmospheric pressure. The altitude of the site at which the UPS is located can have a significant impact on the reliability of the unit. Most UPS units are designed to operate without derating at elevations up to 400 ft above sea level. Above that altitude, extreme care must be exercised in the selection of a UPS. A good design calls for full ratings up to 7000 ft.

The relative humidity present within the UPS room is an important environmental consideration, because electrical devices cannot tolerate excessive moisture. A UPS should be designed to tolerate 100% (noncondensing) humidity levels.

Physical characteristics. Floor space must be considered well in advance of the actual installation. Clearance above the unit must be maintained for proper cooling. Access space, either front only or front and rear, must be allowed. Floor loading capacity must be adequate. The UPS and battery plant can weigh up to 250 and 4090 lb/sq ft, respectively.

Some UPS equipment with multiple cabinets can be installed in a physical configuration other than side-by-side. Versatile designs include back-to-back or separation, with space between units. The battery plant can be installed on more than one rack, and racks are available in 2- and 3-tier styles as well as 2- and 3-step designs.

The access route the equipment will follow from the off-loading point to the installation area must also be considered. Narrow hallways, low doors, tight corners, and lack of freight elevators can present difficult problems.

Code requirements. Various building and safety codes may affect an installation. Basically, a facility may be restricted in the integration of a UPS by statutes and regulations, national standards (such as the National Electrical Code), consumer and equipment standards (such as UL and CSA certification), or the requirements of a certifying authority (such as municipal code approvals). These requirements can have a significant impact on the ultimate UPS specification. While most UPS manufacturers design their units to meet such regulations, there may be cases where a particular unit fails to meet such standards.

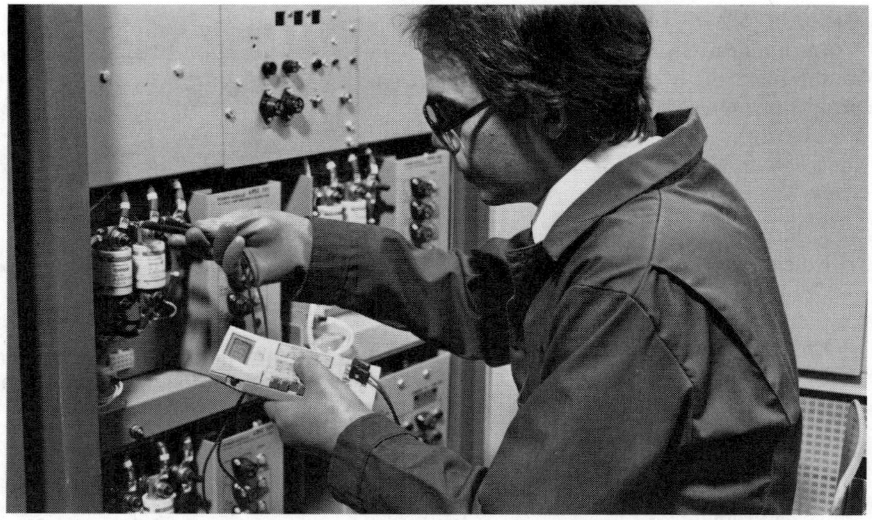

SERVICE AND MAINTENANCE requirements must be considered in choosing UPS equipment. Such application factors as personnel training, spare-part availability, and manufacturer service capabilities should be investigated.

Meters, alarms and controls provided with the UPS may vary widely from project to project. The type of monitoring depends chiefly on the level of maintenance skill of the customer's personnel and the method of alarm reporting and acknowledgement/response. Normally, the UPS manufacturer will offer a standard meter package mounted in the UPS, in addition to alarms, with the capability of providing extended alarms by means of remote panels or relay extenders. Remote alarms may be needed in such locations as the office of the director of management information systems, the security desk, and the maintenance room.

Some UPS users may prefer to differentiate between major alarm and minor alarm indications. Thus, the alarms must be integrated with the total building alarm system.

Although a UPS is usually designed for unattended operation, manual control is required for certain functions, such as starting and stopping, bypassing for maintenance, and programming the battery recharge circuit.

Installation requirements. Such installation requirements as cabling, fusing, grounding, AC service requirements, and cooling and ventilation capabilities must be considered. Building power must be adequate to support the UPS. The AC input service must be sufficient to provide for battery recharge current in addition to the fully loaded UPS.

Proper cabling must be specified to

RELIABILITY of equipment under such adverse conditions as transients, brownouts or other commercial power faults is vital. Thorough testing by the manufacturer before shipment is a must.

prevent excessive voltage drop between the UPS and the critical load and between the UPS and the battery plant. The cabling and protection should be checked for adherence to safety codes. In applications where there is an engine-generator on-site, additional considerations include verifying that the capacity of the engine generator is adequate to support the UPS and that the UPS is electrically positioned between the engine generator and the critical load. The review should also determine whether battery recharge should be inhibited (delayed) while the engine-generator is operating.

While a UPS is thoroughly factory-tested before shipment, the installed UPS must be rechecked for possible damage during shipping, handling and installation. Some manufacturers offer on-site tests, using dummy load boxes, to identify damaged components.

Service and maintenance requirements. Critical service and maintenance considerations include the degree of training of an operator, availability of operator and maintenance training from the UPS supplier, clear instructions on degree and nature of maintenance required by the UPS and battery plant, availability of maintenance training and maintenance services by the UPS manufacturer, provision of spare parts for the UPS and their availability from the manufacturer, and most important, the ability of the UPS manufacturer to respond to service and maintenance problems at the facility.

Cost considerations. The installed cost of a typical UPS will include the purchase price of the UPS and battery plant and the expenses incurred in installation. The installed cost of a UPS may vary from $650 to $1500 per kVA. The cost of operating a UPS usually varies between $20 and $40 per kVA per year, and maintenance expense varies between 3 to 8% of the installed cost of the installation per year.

Additional expenses include the cost of an engine-generator and delivery expenses of the UPS and battery plant. Most engine-generators are available at a cost of from $125 to $175 per kVA. Depending on the distance, speed and special handling requirements involved in the application, delivery expense is ordinarily less than 5% of the installed cost of the unit.

In developing UPS specification and performance parameters, the specification writer will have developed a precise profile of the specific load-related and application-related factors to be considered for a given installation. From that point, UPS selection becomes little more than a logical process of elimination as the various UPS designs are compared to the specific performance requirements of the application.

BATTERY PLANT associated with the UPS can be affected by the environment. Battery life decreases at high temperatures, and some air movement is required to dissipate gases developed.

UPS systems checklist

SOPHISTICATED electronic equipment depends on constant, uniform power to function properly. It does not take much—just a millisecond interruption in power or a barely visible variation in voltage or frequency—almost any power inconsistency at all—can cause critical loads to malfunction. The cost of such malfunctions in computers, telecommunications equipment, process controllers, and all types of data processors can be enormous in terms of lost production, equipment downtime and damage, idle personnel, and lost data that must be regenerated. The ultimate protection against brownouts, blackouts, and variations in power is an uninterruptible power system (UPS). UPS protection does not come cheap. These are big-ticket systems that must be carefully matched to the requirements of the critical load to be protected. They can be very complex, and their selection for specific applications requires careful study of both performance and support capabilities.

The following checklist will be a reminder of the important, basic considerations that must be kept in mind when specifying and choosing a UPS. It will help to determine whether or not a UPS is really needed, serve as a guide in evaluating various systems, and provide information on what to look for in suppliers.

Need for a UPS

1. What is the possible loss (theft, fire, flood, sabotage) due to security-system alarm failure?

2. What is the possible in-process material loss due to power outage?

3. What is the potential manufac-

turing/process equipment damage due to power failure?

4. Are safety or life-support systems dependent on continuous electrical power?

5. How will erroneous data affect your operation?

6. What is the direct labor loss due to outage?

7. What are equipment downtime costs?

8. Is any data that may be lost irreplaceable?

9. What is the revenue loss during downtime?

10. How much revenue will be permanently lost to your competition as a result of downtime?

11. What are the costs of reprogramming? Rerunning?

12. What are the costs of outside reprogramming and processing necessary for catch-up and restart?

13. Is scrap account higher than normal (due to computer errors)?

14. Do thunderstorms often knock out power in your location? Icestorms? Windstorms?

15. Are your utility power lines above ground?

16. What is the frequency and duration of blackouts?

17. During power-company brownouts, what voltage is available at your distribution panels?

18. Do computer errors appear around the same time of day?

19. Are you experiencing media damage due to head crashes?

20. Is the failure rate of computer printed-circuit cards or power supplies excessive?

21. Has your computer supplier recommended a UPS?

22. Does your power meet the minimum requirements as specified by the computer manufacturer?

23. Is your critical load constant or variable and to what extent?

24. Can your critical load tolerate a break in power of 100 msec? 50 msec? 8 msec? None?

25. Is there heavy manufacturing in or near your facility?

26. Is downtime affecting your goodwill with customers?

27. What data-processing alternatives do you have?

Factors affecting choice of UPS

1. What is the apparent power of the protected load (kVA)? Real power (kW)?

2. What voltage is needed by each piece of equipment?

3. Is the computer input voltage convertible?

4. What is the power factor of the critical load?

5. Has the computer manufacturer identified power requirements?

6. Are the load figures calculated or measured (actual)?

7. Will the UPS be used to convert voltage or frequency?

8. Are essential lighting loads under UPS protection (high-pressure-sodium, safety lighting)?

9. Have all noncritical loads (air conditioning, general lighting, unnecessary wall outlets, etc.) been subtracted from UPS load requirements?

10. What are the computer start-up power requirements?

11. Are the critical loads powered up sequentially?

12. How severe will unbalanced loading be?

13. Will single-phase loads need to be closely matched?

14. If part of the critical load requires 415 Hz, how will this be provided?

15. Have requirements of existing and planned peripherals been determined for sizing UPS?

16. Can UPS be expanded to meet future load growth?

17. How easily can UPS be expanded?

18. Is UPS design inherently simple without unnecessary frills?

19. How conservative is the derating practice on key UPS components?

20. Are service-line voltage, amperage, frequency, and wiring configurations compatible with the load requirements?

21. Does UPS have AC walk-in feature in rectifier stage to gradually apply load (to prevent high current inrush)?

22. Is the recharge rate adequate for satisfactory UPS protection after an outage?

23. What is the minimum power factor that the rectifier will present to the AC service line?

24. How much voltage drop or frequency variation can the UPS rectifier tolerate and still charge the batteries?

25. Is the harmonic distortion in the UPS output within acceptable limits?

26. Can the UPS provide power to the loads with leading power factors as well as with lagging power factors?

27. What is the overall UPS efficiency?

28. Are isolation transformers provided to permit independent grounding of the UPS, battery bank and load?

29. Must the UPS be derated with altitude?

30. Is the UPS specified to 50°C or to only 40°C or less?

31. Will the UPS continue to operate with the battery bank momentarily disconnected?

32. What is the overload capability of the UPS?

33. If the UPS transfers the load to bypass due to an overload, will the load automatically retransfer upon removal of the overload?

34. Is the load transfer to bypass make-before-break? Retransfer?

35. Is the UPS static switch commutated by natural means or forced?

36. Are there any moving parts in the UPS? Motorized circuit breakers?

37. Does the UPS have fault-clearing and surge-current capabilities if the bypass source is out of limits or unavailable?

38. Is a maintenance bypass breaker provided by the UPS supplier?

39. Are all lugs and terminal blocks needed to connect the UPS provided?

40. Does the alarm system differentiate between minor problems and major problems (leading to shutdown)?

41. Does the alarm system indicate which minor problems will require action to preclude shutdown?

42. Is a remote alarm panel necessary?

43. Where is the best place for a remote alarm panel? Any distance limitations?

44. Should there be more than one remote alarm panel (computer room, security desk, control station, etc.)?

45. Before the alarm board is cleared, will a subsequent alarm reinstate the audible alarm?

46. Can the alarm system be summerized and extended to the central alarm system?

47. Does the UPS have an alarm test feature?

48. What battery reserve time is required?

49. How much battery reserve time is required for orderly shutdown?

50. What type of batteries is best for

your application?

51. Will the batteries be automatically recharged when required, or does the UPS require operator intervention (equalize switch, charge timer, etc.)?

52. Does the UPS contain automatic disconnect means to prevent irreversible battery damage due to excessively deep discharge?

53. Is a DC disconnect switch provided to protect battery cabling and to facilitate battery maintenance?

54. Is long-term emergency power (engine-generator) necessary in addition to reserve time from batteries?

55. What is the cost trade-off between a larger battery plant and an engine-generator?

56. Is the engine-generator sized to handle the UPS and other loads in the worst possible case?

57. What percentage of connected load on the engine-generator is represented by the UPS (check with engine-generator supplier)?

58. During a power outage, is the engine-generator adequate to provide the additional battery recharge requirement, or should recharge be delayed until AC power-line restoration?

Choosing the right supplier

1. Which UPS manufacturers have experience in applications similar to yours?

2. Is a complete user list available or a "selected" list?

3. Is a nearby installation available for inspection?

4. Can you readily check with users on supplier's service and maintenance capability and responsiveness?

5. Is the potential UPS supplier financially sound?

6. What is the design philosophy of the UPS supplier?

7. Has the UPS evolved as a collection of components or is the design an integral system?

8. What is the UPS supplier's reputation for technical competence?

9. Does the supplier have a strict quality-control program? Can it be documented?

10. How comprehensive is the potential UPS supplier's proposal?

11. Will the supplier take full system responsibility (UPS, battery, engine-generator, transfer switch, installation)?

12. Will the supplier arrange for you to witness testing of your UPS in the factory?

13. Is factory-certified test data available?

14. Is technical assistance available from supplier *after* installation?

15. What is covered under the warranty? What is excluded? Starting when? For what length of time?

16. How complete are instruction manuals provided with the equipment?

17. Does the UPS supplier maintain a dedicated spare-parts inventory?

18. What is the size and geographic distribution of service organizations?

19. What is the service availability (how many hours per day, how many days per week, covering what geographic territory)?

20. What are the qualifications of the supplier's service personnel?

21. Does the supplier directly assume responsibility for service, or is it farmed-out to another division or company?

22. Are your delivery deadlines critical?

23. Has the potential UPS supplier consistently met delivery commitments?

24. Does the potential supplier treat the UPS as a major product line or a sideline?

Installation decisions

1. Have wiring, protection systems, and disconnects been specified by UPS supplier?

2. Has proper cabling been specified to prevent excessive voltage drop between UPS and critical load and between UPS and battery plant?

3. Is building power adequate?

4. Is the AC input service sufficient to provide for battery recharge current in addition to the fully loaded UPS?

5. Have applicable local, state and federal codes been checked?

6. Is an engine-generator with adequate capacity available at the site?

7. Is the UPS electrically positioned between the engine-generator and the critical load?

8. Is the bypass fed from a separate circuit breaker?

9. Will a bypass transformer be required?

10. Is the load protection coordinated with the UPS protective devices?

11. Where will the UPS be located?

12. What floorspace is required for the UPS and related equipment?

13. Has sufficient floor space been alotted for future UPS expansion?

14. Will the UPS require side or rear access?

15. Where should the battery plant disconnect switch be mounted?

16. Is a good earth ground available in the UPS area?

17. Is the computer properly grounded?

18. Is the building capable of safely carrying both UPS and battery-bank floor loading?

19. Have applicable building codes been checked for seismic requirements affecting the battery racks?

20. Is air-handling equipment adequate for the UPS heat-load and temperature limits?

21. Is air conditioning sized to keep the battery plant within the recommended temperature range?

22. Is ventilation sufficient in the battery area to prevent an accumulation of hydrogen gas?

23. What is the maximum ambient temperature expected during a blackout in the UPS room? In the computer room?

24. What ceiling height is required for proper UPS ventilation?

25. Will overhead piping or wall-mounted equipment obstruct placement or operation of the UPS?

26. Is the UPS location subject to flooding or other water damage?

27. In the UPS and battery area, will improvements be completed (floor tile, windows, duct work, security) before equipment is installed?

28. Has an emergency eyewash been provided in the battery area?

29. Will a telephone be available in the UPS area?

30. Should access to the UPS area be limited?

31. Does the UPS require special packing, transportation and handling, or can a common carrier be used?

32. Is the unloading dock facility adequate to handle the size and weight of UPS and battery equipment?

33. At the site, can you use, rent or hire a forklift? Forklift operator? Pallet truck? Crane?

34. Is there proper access and clearance to move the UPS and batteries to their final installation location from the loading dock?

35. If an elevator is needed, is the door opening adequate and is load capacity sufficient?

36. Will anything obstruct the movement of the UPS (computer flooring, doorways, stairways)?

37. Who will install the UPS?

38. Does the supplier offer installation services?

39. How long does the complete installation take?

40. Should you have the manufacturer install the UPS?

41. Are there any local labor restrictions that could prevent the supplier's personnel from installing the unit?

42. Should the UPS supplier supervise the installation?

43. Will the UPS supplier fully inspect and test the installed UPS before turning it over?

44. How much on-site testing is involved?

45. Does the supplier have an installation testing procedure you can study?

46. During installation, does the supplier provide test loads?

47. Will a detailed, comprehensive test report be provided by the supplier after cutover?

48. If your project is delayed, what provisions should be made for storing the equipment until the site is ready?

It's up and running. Now what?

1. Does the UPS need a trained operator?

2. Is the UPS status regularly monitored?

3. Has an action/responsibility plan been developed in the event of a UPS alarm?

4. Does the UPS require maintenance?

5. Does the battery or other related power equipment require maintenance?

6. Can you maintain the UPS with your personnel?

7. Does the supplier offer user training programs? On-site? At the factory?

8. Is special test equipment needed? If so, how available is it?

9. What is the turnover rate of your maintenance personnel?

10. Are spare parts necessary?

11. Are various levels of spare parts available?

12. How long does it take to get parts to your facility?

13. Does the supplier offer preventive maintenance programs?

14. Does the supplier have sufficient qualified service personnel to service the equipment?

15. Can service contracts be tailored to meet particular maintenance needs?

16. Does the service contract identify *all* costs?

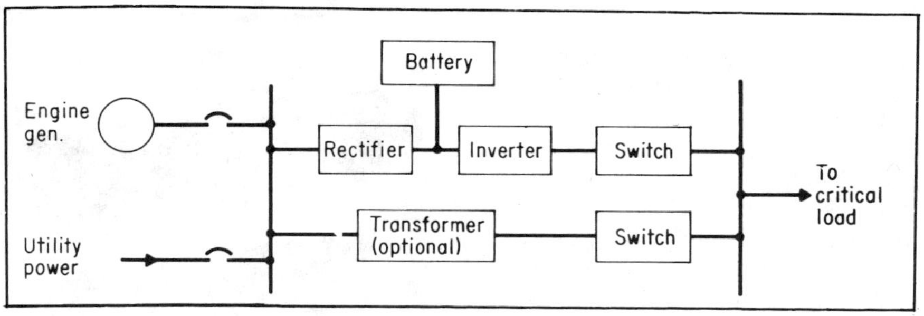

How to Specify Uninterruptible Power Supplies

Will the UPS solve your power problem? Here are the basic facts to consider.

By A. KUSKO, President, and T. KNUTRUD, Consulting Engineer
Alexander Kusko, Inc.
Needham Heights, Mass.

Uninterruptible power supplies (UPS) incorporate complex equipment that is expensive to install and maintain; therefore, they must be planned and specified carefully before purchase. These systems, consisting of solid-state rectifier-inverters (usually backed up by engine-generators), are being used to supply power to computers, on-line data processors, process controllers, and other critical loads, to prevent costly power interruptions.

The heart of the UPS is the rectifier-inverter unit, or module, which accepts ac line power and delivers transient-free ac power to the critical load. A battery supplies power to the inverter when the ac line power source is interrupted up to several minutes. The configuration of a typical UPS is shown in the accompanying diagram. Rectifier-inverter units are employed in many combinations to supply a critical load—singly, in parallel, with a bypass switch, backed up by engine-generator sets, etc. The particular combination selected is determined by the magnitude of the critical load power, the pattern of anticipated ac line interruption, and the sensitivity and critical nature of the load.

Equipment and costs

Commercial rectifier-inverter units are built from about 1 to 250 kva and have been arranged in systems up to 2000 kva. Units are designed to supply critical loads at typical commercial single-phase and 3-phase voltages and frequency, including 400 Hz. Batteries typically use lead-calcium cells and usually are selected to provide full-load power for 10 min. Battery ampere-hour ratings drop rapidly for short discharge times at high current, so that the cost saving resulting from selecting 5 min of capacity instead of 10 is minor. Except in the integral kva ratings, rectifier-inverter units are usually not "off-the-shelf" items; they are custom-made to variations of a standard design.

The cost of a rectifier-inverter unit with 10 min of battery ranges from about \$5000/kva at 1 kva to \$500/kva at 250 kva. Costs of engine-generator sets lie in the \$125-to-\$200/kva range. In large, several-hundred-kva, redundant-unit systems, the cost for units and engine-generators approaches \$1000 per load kw. To the component costs must be added the costs of installation of the UPS, the electrical distribution and supply system, and additional building space—all of which can double the original component costs.

UPS purpose and operating conditions

The customer (or his engineer) is faced with both high cost and a variety of UPS components and configurations. Considerable preparatory work must be done before purchase to insure that the UPS will meet his requirements. Above all, he must ask himself the question, "What do I hope to accomplish with the UPS?" To come to grips with this question, he must answer the following:

(1) Considering the critical load equipment that demands high-quality power—precisely how critical is it?

(2) What is the penalty in operating all or part of the critical load equipment from commercial power? What is its tolerance to line transients and outage?

(3) How frequently can failure be tolerated after the UPS is installed to supply the critical loads? How fast must the UPS be returned to service?

(4) What arrangement of UPS will satisfy the requirements at minimum cost? Can gas-turbine or engine-generator sets alone satisfy the requirements?

Assuming that the answers to these questions lead to the conclusion that a UPS is needed, the customer next must consider its anticipated operating conditions:

(1) Where will all components, such as rectifier-inverter units, batteries, engine-generators, and fuel tanks be located, and what will be their environment?

(2) Who will provide preventive maintenance and repairs of failed equipment; e.g., contract maintenance or own technicians? How rapidly can they respond?

(3) What spare parts and sub-

assemblies will be stocked? Will failed assemblies be repaired or returned to the manufacturer?

(4) The system must be exercised periodically by starting and running the engine-generator sets, by tripping the commercial power breaker, and by tripping the units of a redundant system one at a time. How often and by whom will this be done?

(5) How will the UPS be coordinated with the downstream distribution system for fuse-blowing, breaker-tripping and high reliability?

Answers to these questions will help determine how many spare (redundant) parts will be required to meet system objectives in the face of failures and repair time for equipment. (For example, solid-state bypass switches may be used around rectifier-inverter units to provide redundancy and improve downstream fuse clearing.)

Writing specifications

Equipment specs represent the translation of the customer's requirements and design into hardware. The customer should be responsible for his own specifications, rather than accept what a vendor offers him. In preparation, he can be guided by the vendor's specifications, or modify an existing set, or retain an experienced consultant to prepare specifications for him. The Federal Aviation Administration Specification FAA-E-2473 is an excellent basic document that can serve as a model and be modified for a customer's particular application.

Specification must contain the following items:

(1) *The steady-state and transient electric power requirements of the critical load equipment to be supplied by the UPS.* Most computer manufacturers will make this information available. The conditions under which the UPS must meet these load requirements must be given (commercial power failure, battery condition, engine-generator operation, and internal UPS faults).

(2) *The required reliability and maintainability of the UPS.* This is usually expressed as average hours between failure (MTBF) of the UPS to meet the electric power requirements of the critical load, and the average hours to repair (MTTR) any failed rectifier-inverter unit of the

LOW-POWER (2.5-kva), 90-min, single-phase rectifier-inverter supply with integrally mounted battery is used primarily for general-purpose building loads.

system. The specification must describe how these requirements are to be measured and how they are to be demonstrated by the vendor.

(3) *The terms of acceptance of the UPS by the customer.* These terms must include the tests to be conducted

MEDIUM-SIZE COMPUTER LOADS are served typically by 3-phase UPS equipment such as the 50-kva unit above and the 75-kva unit at right. Doors are opened to show rectifier at right, inverter transformers, switches, fuses and controls, in the mid-section, and output switchgear to the left.

FOUR RECTIFIER-INVERTERS in parallel make up this 750-kva UPS.

LARGE BATTERY BANK for a 200-kva, 10-min. UPS module.

within the vendor's plant, the tests of the complete installation at the site, and the equipment performance for acceptance.

In preparing his specifications, the customer must remind himself that the UPS must provide his critical loads with a more-reliable power supply than the commercial line itself; otherwise, there would be no reason to go through the trouble and expense of the procurement.

There are about 20 vendors of solid-state rectifier-inverter units and about 10 vendors of gas-turbine and engine-generator sets in the United States. Practically all vendors are well intentioned but not equally capable of meeting a UPS specification and delivery date. Each vendor should be required to submit as part of his proposal a list of installations of his equipment that are similar to the customer's. The customer should visit and/or telephone the vendor's previous customers to evaluate both the vendor's capability and the performance of the equipment. The most suitable vendor for the contract will generally be the most experienced; his quoted price and delivery schedule should be the most realistic.

Testing

The most important step in purchasing a UPS is the testing. The vendor and customer must agree upon and carry out a meaningful set of tests at the factory and at the site. The purpose of testing is twofold: to see that the equipment meets the requirements of the specifications, and to expose any design or manufacturing flaws.

At the factory, the complete UPS, if possible, sould be assembled and tested—including the rectifier-inverter units, the control console, bypass switches, and batteries. The switchgear and engine-generators may have to be tested independently and shipped to the site.

The first set of UPS tests is for steady-state performance; it should be conducted for variation of load, line voltage, power factor, frequency, and temperature and relative humidity (if adverse environment is expected). The second set of tests is for dynamic performance, such as for tripping the input power line, isolating and inserting of parallel units, transferring to and from bypass, switching load, short circuiting the output terminals, and fuse clearing. The third set of tests is to demonstrate the ability of a failed rectifier-inverter unit to isolate itself from the system while the UPS continues to supply critical load, under such conditions as loss of SCR gate drive, failure of inverter commutation, blower failure, and fault in the common frequency and load-division circuits. The final set of tests is to prove the reliability by operating the UPS without failure for a time and loading schedule to support the MTBF predictions.

At the site, the complete system should be tested to check on shipping damage and for operation with the actual switchgear and engine-generator. First, the factory steady-state performance test is repeated using both the commercial line and engine-generator ac power. Second, part of the factory dynamic tests are repeated to check on switching operations of the system. Finally, the system is run with a dummy load for an extended period as a final acceptance test.

Documentation and training

Coincidentally with the installation and testing of the equipment, the final documentation and training phase of the procurement should be complete. The following items must be included:

(1) Instruction manuals with complete diagrams, explanation of operation, and parts lists.

(2) Spare parts lists including provision of recommended quantity and naming sources of future requirements.

(3) Procedure for repair and/or purchase of special components, such as printed-circuit boards, special inductors, special SCR's.

(4) Maintenance procedure, recordkeeping forms.

(5) Training of customer's personnel to operate and maintain the equipment.

(6) Warranty agreements, maintenance contracts.

A proper vendor will recommend field changes in the future to improve the UPS, will provide special parts, and will provide special repair services for unusual failures and extended downtimes.

Design procedure for

Selecting a UPS system

Here's a discussion of the basic considerations in selecting an uninterruptible power supply to provide continuous power input to such critical loads as computers, communications equipment, control, and other essential services.

By J. F. McPARTLAND III
Northeast Area Sales Manager
Emerson Electric Co.
Industrial Controls Div.
Santa Ana, Calif.

The uninterruptible power supply (UPS) is an all-solid-state power conversion system designed to protect computers and other critical loads from blackouts, brownouts, and transients. It is usually connected in the feeder supplying the load, with bypass provisions to permit the load to be fed directly.

Typical UPS components

The diagram in Fig. A depicts a UPS and its associated components. Briefly, here is how the system operates. Incoming ac is rectified to dc, which is immune to ac input disturbances. The isolated, filtered dc bus, regulated by printed-circuit logic cards, delivers precise, regulated power to the inverter. A small amount of this power is also used to charge the battery, maintaining a fully charged condition. If ac power is lost for any reason, power is supplied to the inverter by the battery, which is connected or "floated" across the dc bus. The inverter delivers precise ac power to the critical load.

If the UPS module itself fails, a make-before-break transfer to a utility company bypass source is effected. Two interlocked circuit breakers and a static switch are used to accomplish the transfer. Normally, load current flows through one of the two CBs. However, due to its mechanical inertia, a circuit breaker takes 5 cycles to open or close and does not deliver quality power for this 5-cycle period. During this period, load current is conducted through the static switch that parallels the bypass circuit breaker. The net result is a completely uninterrupted transfer from UPS output power to utility bypass power.

Fig. A. Typical arrangement of UPS components

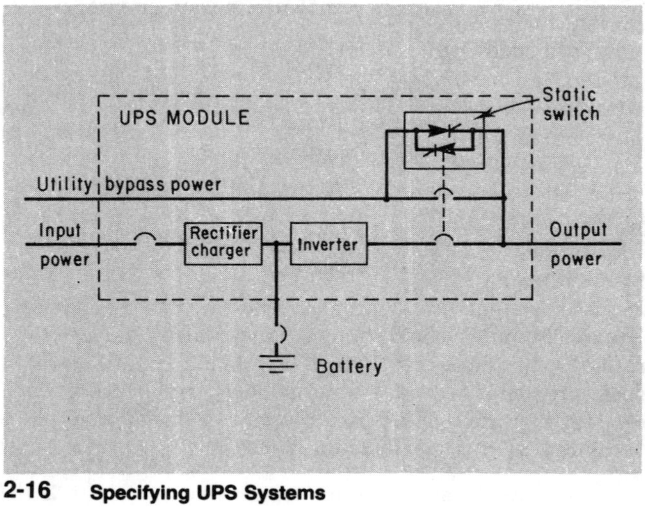

Ratings. Data-processing installations normally require medium-to-large 3-phase uninterruptible power systems ranging from 37.5 to over 2000 kva. Some of the typical ratings available from UPS manufacturers are 37.5 kva/30 kw, 75 kva/67.5 kw, 125 kva/112.5 kw, 200 kva/180 kw, 300 kva/270 kw, 400 kva/360 kw and 500 kva/450 kw. Larger systems are configured by paralleling two or more of these standard-size single modules. Paralleling is covered below.

The necessary rating is chosen based on the size of the critical load. If a power profile itemizing the power requirements is not available from the computer manufacturer, the load may be measured using a kilowatt meter and a power-factor meter. Since UPS modules are both kva-(apparent power) and kw-(real power) limited, a system should be specified with both a kva and kw rating. The required kva rating is obtained by dividing the actual load kw by the actual load power factor. For example, an actual 170-kw load with an actual 0.85 power factor would require a 200-kva-rated UPS. The system should be specified as 200 kva/170 kw, and the standard 200 kva/180 kw UPS module should be selected for the application.

Since many companies find the need to add various peripherals to their data centers as time goes on, some spare capacity should be included in the initial UPS sizing. Allowing 10 to 20% extra at the beginning is much more economical than attempting retrofitting at a later date.

Since UPS equipment costs between $300 and $800 per kva (the larger the UPS system, the lower the cost per kva), the air conditioning that is required to maintain a precise computer room environment should not be included as part of the UPS load. This air-conditioning equipment operates quite well on raw commercial power, and standby generators can be utilized to power this equipment during commercial power outages. If emergency lighting is not otherwise provided, then some of the computer room lighting should be powered by the UPS.

Hardware. The heart of the UPS is the power conversion module. It ranges in size from 3 ft deep, 5 ft long, and 6.5 ft high at 37.5 kva to 3 ft deep, 15.5 ft long, 6.5 ft high at 500 kva.

Economic and performance characteristics make the lead-calcium type battery the best suited for UPS applications.

The battery ratings are referred to in terms of critical load kw requirements for a specified amount of protection time—most commonly 5, 15 or 30 minutes. A 30-kw, 15-min battery would consist of about 120 series-connected cells mounted on a 2-tier rack measuring about 22 ft long, 18 in. deep, and 47 in. high. A 450-kw battery would consist of approximately 180 series-connected cells on four 2-tier racks, each about 16 ft long, 21 in. deep, and 66 in. high.

Other major hardware items are the remote monitor panel and the battery circuit breaker. The latter is used to deenergize the dc bus when maintenance of the power converter module is necessary.

The remote monitor panel, mounted in the computer room, should be as simple as possible with no more than four or five alarms. (Any more might confuse computer technicians unfamiliar with UPS operation.) Some suggested remote alarms are a green light indicating normal operation and red lights indicating battery operation, low battery, and load on bypass. A summary alarm should also be included to indicate a minor problem at the UPS that should be investigated immediately by UPS maintenance personnel, even though the problem does not represent an immediate threat to safe computer operation.

Location. Distances between UPS components should be kept short to minimize voltage drop. Distances of 50 ft between the module and the battery and 150 ft between the module and the critical load are considered safe. If measurements show excessive voltage drop, oversized conductors and/or step up, step-down transformers may be required.

Floor loading should be considered, since UPS equipment is quite heavy. A 75-kva module weighs 4400 lbs distributed over 18 sq ft (244 lbs/sq ft), and a 500-kva module weighs 15,000 lbs distributed over 46 sq ft (326 lbs/sq ft). The battery for the 500-kva module weighs 46,000 lbs distributed over a floor area (including 3-ft aisles) of 256 sq ft (180 lbs/sq ft).

The battery may be located in the same room as the UPS module. Lead-calcium batteries do not emit dangerous amounts of hydrogen in an adequately ventilated room, "adequate" being defined as one air change per hr. The battery warranty is based on a 77F ambient; thus this temperature represents the best environmental design goal.

Heat rejection of the UPS module ranges up to 134,000 Btuh for a 500-kva unit. Few rooms are large enough or sufficiently ventilated to disperse that much heat; therefore, air conditioning usually is required. A rule of thumb is one ton of air conditioning for each 12,000 Btuh, but the system should be professionally designed.

High-frequency UPS. Over the past several years, some computer manufacturers have developed equipment that requires 415-Hz power. Initially, this requirement was satisfied by motor-generator sets, but today the market is dominated by 415-Hz UPS systems.

There are several reasons. The solid-state frequency converter offers greater reliability and more precise output characteristics than rotating equipment. The solid-state UPS also runs much quieter—68 db vs 89 db. Annual power costs for operating a 75-kva rotary system run $3000 to $6000 more than a static system, since the solid-state UPS is more energy-efficient (approx. 90% vs 80%). Also, solid-state frequency converter systems are inherently expandable and easier to parallel than rotary systems.

The 415-Hz UPS has another great advantage where both 60-Hz and 415-Hz power are required by the computer, since it has the ability to reduce the size of a 60-Hz UPS by off-loading what otherwise would have to be part of the 60-Hz UPS load. Example: The power requirements of a data center are 175 kva at 60 HZ and 75 kva at 415 Hz. If the high-frequency power is furnished by an M-G set, the M-G set would become part of the critical load, and a 60-Hz UPS system would be chosen to handle 175 + 75 or 250 kva. The next standard size is 300 kva, leaving 50 kva of spare capacity. However, if a 415-Hz UPS were used for the 75-kva load, a standard 200-kva, 60-Hz UPS would be more than enough for the 175-kva load. This would be a considerable saving over the 300-kva UPS plus M-G set (see Fig. B).

Fig. B. Methods of supplying 415-Hz power

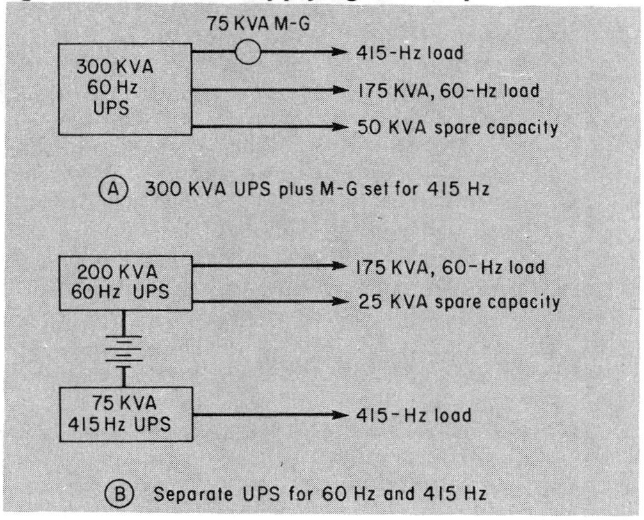

To minimize effects of the higher inductive reactance at 415 Hz, 415-Hz loads should be fed by either aluminum or plastic conduits, and the conductors should be twisted around each other.

Parallel and redundant operation. When one UPS module is not large enough to power the critical load, two or more modules are run in parallel, with their outputs feeding a common bus in a system control cabinet (Fig. C).

Fig. C. Parallel operation of two UPS systems

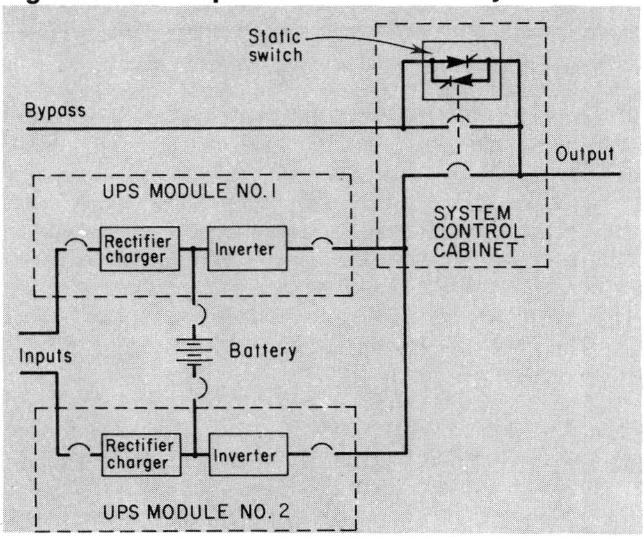

NONREDUNDANT 550-kva system, expandable to 1100 kva, includes two parallel UPS power conversion models (center), system control cabinet (right), and 15-minute battery (left).

In the parallel configuration, the static transfer switch and utility bypass input are located in the system control cabinet. Logic circuits are designed so that each module powers its fair share of the load (each of two modules in parallel carries one-half the load, etc.).

It is desirable to use as few modules as possible. It makes more sense to power a 600-kva load by two 300-kva modules than by three 200-kva modules. Two modules offer greater reliability by having fewer components that can fail and also cost less—$170,000 for the two modules vs $220,000 for the three.

A "redundant" UPS system employs one more module than is required to power the critical load. A 300-kva redundant UPS would thus consist of two 300-kva modules feeding a system control cabinet, one of which is required to power the load and the other of which is a backup. A 600-kva redundant UPS would consist of three 300-kva UPS modules feeding a system control cabinet, one being a backup.

As in normal parallel operation, the on-line modules share the critical load evenly. Where three 300-kva modules feed a common bus with a 600-kva load, each of the three modules would carry one-third of the load, or 200 kva. If one of the modules should malfunction, an isolating device called a "static interrupter" would disconnect the faulted module from the common bus without affecting the critical load. The two remaining 300-kva modules would share the 600-kva load until the faulted module was repaired and put back on line, making the system redundant once again.

With the perfection of the static-transfer switch, redundant operation of small 60-Hz systems no longer is recommended. The switch provides a reliable utility backup, and the single-module system (which does not require a parallel module or system cabinet) has fewer parts and fewer components that can fail. It can be said (somewhat qualifiedly) that the simpler the system, the better. In reality, a redundant 60-Hz system probably buys 10% more system reliability for 200% of the single-module system cost.

For very large installations, redundancy usually is a good idea. A 1500-kva system consisting of three 500-kva modules should have a fourth 500-kva module as backup. The rationale is that since three modules must be paralleled anyway, the extra module does not increase the system complexity but allows for maintenance and troubleshooting on any one module without shutting down the system.

Redundancy should always be provided for 415-Hz UPS systems, since there is no 415-Hz utility bypass available if the UPS breaks down.

Vendor selection. Experience of the vendor is an important criterion in the selection of any UPS equipment.* There are suppliers who have been in the medium-to-large UPS market for the past ten years. A prospective UPS supplier should arrange a visit to one of his nearby installations that is similar in size to the system under design.

Where an installation requires both 60-Hz and 415-Hz equipment, both systems should be procured from the same manufacturer. The installation will be more pleasing aesthetically; and should a malfunction occur, there is no possibility of two manufacturers both disclaiming responsibility. Standardization and unified responsibility are important advantages in selection and operation of UPS systems.

*It is in the best interest of the electrical designer to insist that any vendor of UPS systems provide a list of satisfied customers who may be contacted for their evaluation of the equipment and the company's service.

Specifying Power Supplies and Conditioning Equipment

Specifying power supply equipment for computers requires determining if the system to be protected is real-time and critical so that no downtime can be tolerated. Evaluating the site history of power outages and the cost per hr of computer downtime, against the cost of protective equipment is also necessary. In critical situations, a static or rotary uninterruptible power supply (UPS) will normally be cost-effective. It not only supplies uninterruptible power, but also provides line-side filtering and conditioning necessary to assure the power quality required by computers. In less critical situations, the type of less-costly power conditioning that can provide the required quality to the computer must be determined.

Other power conditioning devices and equipment to improve power quality, in approximate order of increasing cost and complexity, include transient suppressors, noise filters, voltage regulators, shielded isolation transformers, motor-generator (MG) sets, and magnetic synthesizers.

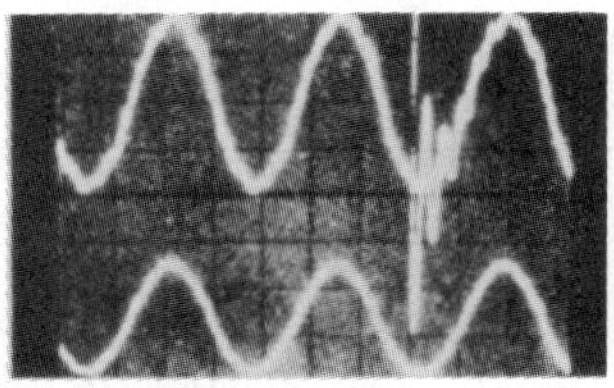
Oscillatory transient

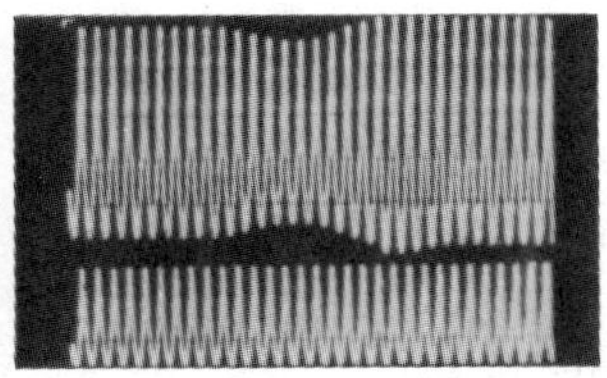
Sag

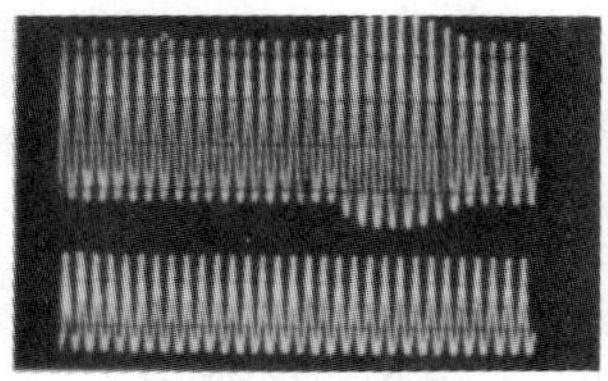
Surge

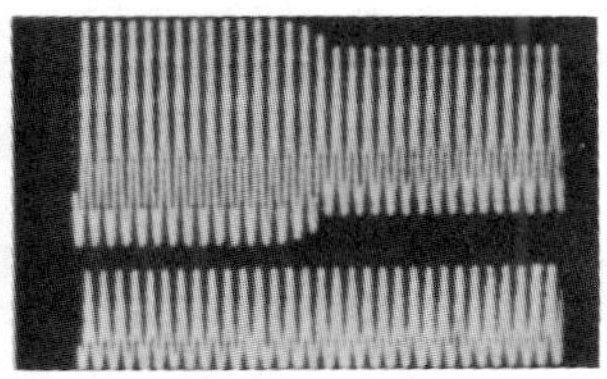
Brownout

Static UPS systems consist of a rectifier (AC-to-DC converter), a DC bus with a floating (continuously connected) battery, a DC-to-AC inverter, and a solid-state bypass switch. Static UPS units range from a few hundred VA for microcomputers to 750kVA and higher for large installations, with larger capacity achieved by using paralleled units.

A rectifier converts the raw input power to DC, which keeps the floating battery fully charged and supplies power to the inverter section. Rectifiers may use diodes, thyristors, or transistorized high-frequency switching power supply to provide the DC.

Inverters use thyristors to generate six- or twelve-step waves, pulse-width modulated waves, or a combination of the two to synthesize the output. Switching transistors are used in some inverters for high-frequency switching that requires less output filtering. Inverter output should be a stable, low-distortion sine wave.

On loss of input power or severe low input voltage, the batteries supply power to the inverter. Batteries are sized to provide power for a specific time at full load, usually between 5 min and 1 hr, with 15 min most common. Batteries for large installations are usually lead-calcium wet cells. Some smaller units use gelled-electrolyte cells or, more recently, immobilized-electrolyte cells. Both types are sealed, maintenance-free, and spillproof.

Since the batteries are connected at all times, there is no switching, and full, undisturbed power to the computers continues until either the computers are shut down in an orderly fashion, without loss of data, or until an alternate source of input power such as an engine generator is brought on line. When normal input power is restored or an alternate source provides the power, the rectifier again supplies the inverter and also recharges the battery. During these transfers, no disturbance appears in the output power to the computers.

A static bypass switch transfers power to the raw input source if the UPS fails. Also, the maximum output of the UPS under short-circuit conditions is limited and may not be enough to instantaneously actuate the fuse or circuit breaker protecting the faulted piece of equipment. Under these conditions, the bypass switch transfers the load back to the utility source until the protective device clears the fault, and then retransfers the load to the UPS. Each transfer ordinarily takes less than 4 msec, with the utility and UPS outputs overlapping, so there is no break in the continuity of power.

Rotary UPS systems vary more in concept and design than static UPS. Even "block diagrams" of rotary systems vary widely in basic components. The one thing most rotary UPS systems have in common is that the output to the load is supplied by a motor-generator (MG) set during normal operation and on loss of normal power. The output of an MG set is a clean, low-distortion sine wave, completely isolated from disturbances in the input power supply to the motor.

On loss of normal power, an MG set (sometimes with a flywheel) has enough inertia to continue supplying the load for a period of a few cycles to several sec. During "ride through," an auxiliary source of power must be brought on line to supply the motor that drives the generator in order to maintain uninterrupted power to the loads. In some designs, this auxiliary source "floats" on the line at all times. The auxiliary power can be supplied by batteries, with an engine-generator brought on line before the battery power is exhausted. In some cases, the system ride-through is long enough to permit the engine-generator to be brought on line directly (usually 10 sec or more), and intermediate power from batteries is not required. Some "hybrid" systems have backup power supplied by a rectifier-battery-inverter system similar to a static UPS, but with the inverter output driving the motor of the MG set. Since the generator puts out clean power even with poor-quality power input to the motor, the rectifier-inverter can be a simple, low-cost unit with little output conditioning. The motor driving the MG set may be DC, AC induction, or AC synchronous type, depending on the system design. Many rotary systems are quiet enough to be installed right in the computer room. Some available rotary UPS systems include:

- single-shaft synchronous MG, with a common rotor and stator winding, supplied by a rectifier-inverter system and battery backup. In normal operation, 95% of the motor power is direct from the utility source and only 5% from the inverter operating in parallel with the utility. On loss of utility power, the motor is supplied 100% from the on-line inverter, powered by the battery, with no transfer time required.
- DC-motor-driven MG, with the DC for the motor supplied from a solid-state rectifier, which also keeps a floating battery charged. On loss of normal power, the battery powers the motor directly, with no transfer time. This is a UPS with the MG set acting as a rotary inverter.
- 3600-rpm induction-motor with a flywheel, driving an 1800-rpm generator through belts and adjustable-speed pulleys. The ride-through on this system is more than 12 sec, allowing an engine-generator to be brought on line without intermediate battery power. Other units with 1800-rpm induction motors and adjustable-speed belt drives use batteries and inverters to drive the motor on loss of normal power.
- synchronous MG, with an additional DC machine on the same shaft. With normal power, the DC machine coasts, with controls driving it as a low-output generator to keep a backup battery charged. On loss of normal power, the control system switches so that the battery powers the DC machine as a motor, driving the generator while the AC motor coasts.
- an MG, vertically mounted, with a large flywheel providing enough ride-through to bring an engine generator on line. The engine generator is off until normal power is lost; then is started and on line within 2 sec. No batteries or rectifier-inverter are required.
- a variable-speed, constant-frequency synchronous MG. The machine varies the number of poles of both motor and generator, so that the frequency remains the same over a wide speed range. On loss of normal power, generator voltage and frequency output at full load remain within computer limits for 25 sec as the machine slows down, providing sufficient time to bring an ordinary engine-generator on line. No batteries or inverters are required.
- a European system that connects the load to the normal supply through a large inductor. Tapped to the inductor is a synchronous motor, overexcited in normal operation to act as a rotary condenser, which with the inductor acts as a resonant filter to provide clean power to the load. In normal operation, the synchronous motor rotates a magnetic clutch. On loss of normal power, the magnetic clutch provides enough inertia for the synchronous motor to act as a generator and supply the load. A diesel engine is brought up to speed in less than 1½ sec and connected by a separate mechanical clutch to drive the synchronous motor.

Transient suppressors protect against voltage spikes and oscillatory transient voltages. There are two types of suppressors: the gap type is relatively slow acting (microsecs), with large energy-handling ability; the clamping type is fast acting (nanosecs), with somewhat smaller energy-handling ability. Gap-type units are best near the power service entrance, and clamping-type units near the equipment being protected.

Noise filters are low-pass units connected to the power line supplying computers, passing the 60-Hz line frequency easily, but presenting a high impedance to block and attenuate unwanted signals and noise, usually of high frequency. Some filters can eliminate all voltages varying more than about ±2V from the correct instantaneous voltage at any point on the sine wave. Filters are often combined with transient suppressors, as well as with other power conditioning.

Voltage regulators maintain constant output voltage to computers despite a wide variation in input voltage. Each type has advantages and disadvantages in application.

Ferroresonant transformers can have separate input and output windings and can provide voltage transformation and common-mode noise isolation as well as voltage regulation. They provide excellent regulation, but have limited overload capacity and poor efficiency at low loads.

Magnetically controlled voltage regulators are a special autotransformer type, with buck-boost windings to control the voltage. The effect of the buck-boost windings is varied by a control winding with DC current that affects the saturation of the core. The control-winding current is produced by electronic sensing and control circuits. The response time is a relatively slow three to 10 cycles, but response is smooth.

Electronic tap-switching transformers have thyristor switching of a tapped coil winding that changes the turns ratio. The switching is controlled by electronic sensing circuits and can be very fast—in the order of one to three cycles. Thyristor switching at voltage zero is easier and less costly than at current zero, but can cause transient voltages in the system. Current-zero switching is preferred. Voltage change is in steps, not smooth, but the steps can be small. These regulators can have the high efficiency, low impedance, noise isolation, and overload capacity of a transformer.

Shielded isolation transformers provide good common-mode noise isolation and can be 1:1 ratio, for isolation only, or can also provide voltage transformation. The minimum shielding is usually a grounded Faraday shield between windings, but additional attenuation can be provided by also shielding either or both individual windings. Attenuation of 140 dB or more is available with maximum isolation for the most sensitive circuits, but for most systems isolation of 40 to 60 dB is usually adequate.

Power conditioners can include a combination of individual types of conditioning equipment. They can contain the voltage regulation, transient protection, noise filters, and shielded transformer as necessary to provide output power fully protected against all input power disturbances except sustained outages.

Computer power centers are usually desirable to distribute power and provide a common ground point close to the computer equipment. This power center will include a distribution panelboard, often with factory-fabricated cables to supply individual units of computer equipment. It will often contain a transformer to provide 120V to the computer equipment from a 480V supply. A power center is not a power conditioner, but provides an excellent location for the desired power-conditioning equipment.

MG sets provide excellent power conditioning, fully isolating the high-quality output power of the generator from all disturbances of the input power to the motor except sustained outages. The inertia of an MG set, especially if supplemented by a flywheel, can ride-through several seconds of input power interruption. The driving motor can be AC synchronous, AC induction, or DC, with speed control to maintain constant output frequency. The generator output can be of different voltage and frequency from the motor input. UPS-conversion backup power units for installation on existing MG sets are available.

Magnetic synthesizers use ferromagnetics (transformers and inductors) and capacitors to synthesize a high-quality 3-phase output, using the input only as a source of energy. The output is fully isolated from imperfections in the input power. A 3-phase output of reduced capacity can be maintained if one input phase is lost, although the angle between phases will not be 120°. A magnetic synthesizer has no moving parts or semiconductor devices, except in the controls; it has been called a static MG set. Enough energy is stored in the capacitors to ride-through about one cycle of outage. When a reliable, synchronized alternate power source such as a second utility line is available, the magnetic synthesizer has been combined with a static transfer switch that can transfer to the alternate source in $\frac{1}{4}$ cycle, providing uninterrupted power to the computer load and eliminating the need for a UPS.

Specifying Data System Components

Data transmission

A data-communications system within a facility is generally defined as a local area network (LAN) that moves data, voice, text, graphics, or images between devices located on the same premises. The system usually requires backup power to assure its reliability. LANs are classified as either baseband or broadband. Baseband means that only one signal can utilize the distribution medium at one time. In the broadband system, several signals can occupy the distribution medium simultaneously through the use of multiplexing, either frequency or time division. The channel or conductor that physically interconnects the transmitter and receiver in an LAN is called the medium. The three basic media are twisted-pair wires, coaxial cable, and fiber-optic cable.

Data cable characteristics

Characteristic impedance is determined by the cable's physical construction and the dielectric material surrounding the conductors. This rating must be matched to that of the connected equipment to prevent reflections and energy loss. When the line is not matched, energy is reflected from the end of the line with a delay between each reflection that is proportional to the length of the line. When sufficiently high in amplitude, this "noise" on the signal transmission line can cause data errors.

Velocity of propagation is a function of the dielectric constant of the insulating material and is a consideration in preventing waveform distortion. "Propagation delay" is expressed as a value representing the time required for a pulse to travel through a transmission line system, compared to the speed of light in a vacuum.

Capacitance of a cable is related to the dielectric constant (the ability to store an electrical charge) of the insulation on the individual conductors. The capacitance between two conductors of a cable pair is called mutual capacitance, and this value should be kept as low as possible to minimize loss of signal quality.

Attenuation is the reduction in amplitude (height) of a transmitted signal. This loss is expressed as decibels per 100 ft at a given frequency (dB/100 ft at 300 MHz). Losses due to heat generated by the transmission of signals (partially due to the resistance of the conductor) adds to the attenuation due to capacitance. Cable with the largest and highest conductivity conductors available for the application helps reduce attenuation. Insulation with a low dielectric constant and spacing conductors as far apart as possible helps reduce capacitive effect.

Crosstalk is the amount of signal pickup on one quiet line from lines in the same cable or from adjacent cables. These are usually termed line-to-line and cable-to-cable crosstalk, respectively. In addition to representing a loss of signal from the driven circuit, crosstalk contaminates adjacent lines and may cause false triggering, overloaded circuits, and interference.

Shielding. Four different types of metallic coverings or shields are used on data circuit conductors and cables to prevent or minimize crosstalk and electrostatic interference, thus improving transmission quality.

- Longitudinally applied metallic tape consists of flat aluminum tape bonded to the jacket or a corrugated aluminum, copper, or metal-alloy covering. It provides the best shielding effectiveness over a wide frequency range.

- Spirally applied metallic tape is of flat aluminum, copper, or metal alloy. It provides a good degree of shielding effectiveness at low (audio) frequencies but is unsuited for shielding against high-frequency interference.
- Braided shield consisting of copper wire or strips applied in a braid pattern is used for low- and high-frequency shielding. The braid effectiveness depends on how closely woven the braided metal is. The flexibility of the cable decreases with increasing coverage. To gain full effectiveness, more than one braid is used.
- Foil shield consists of metallic foil (aluminum) laminated to a plastic sheet to provide strength and flexibility. Although the foil-polyester-film, tape-wrapper shield provides 100% coverage, the degree of shielding effectiveness is less than longitudinally or spirally applied metallic shield because of the difference in DC resistance of the materials.

Cable types

Twisted-pair is an individually insulated pair of conductors, enclosed in a common protective sheath, that has its two conductors twisted together at various intervals to mimimize the coupling of noise and crosstalk into the circuit. The twisted pair may be shielded or unshielded, jacketed or unjacketed. If a cable assembly contains many such pairs, the twisting of the individual pairs within the cable sheath reduces the coupling between pairs; but as the frequency of transmission increases, this isolation decreases and crosstalk problems increase substantally, thus limiting the maximum practical data rate that can be transmitted over such cable pairs.

Type FCC cable. Flat conductor telephone/data cables generally are used in conjunction with Type FCC branch circuit power cable. The flat cable serves as a transition from the standard twisted-pair that carries the signals from the system terminal strips. By connecting to flat conductor cable, these signals are prepared for a horizontal run along a floor. The flat cable is available in 3, 4, and 25-pair.

Coaxial cable consists of a single-wire conductor centered within a cylindrical outer conductor or shield. The two conductors are insulated from each other by various dielectric materials including plastic or air (in which the concentric shape is maintained by insulating disks spaced at regular intervals). The inner conductor can be solid, stranded, or even tubular. The most common outer conductor is a single copper-braid covering, although a double-braid construction is also available. The outer conductor confines the signal within the cable, reduces interference, and serves as a return conductor, thereby completing the signal circuit. A jacket is normally extruded over the shield to isolate it from contact with other metal.

Coaxial cables are classified as follows.

RG# #/U coaxial is constructed to U.S. government specs. The R stands for radio-frequency; the G for government. Two digits (represented by # #) make up a government approval number and indicate a particular impedance value (RG11/U is 75 ohms, RG62/U is 93 ohms, etc.). The U stands for universal specification.

NonRG/U coaxial does not conform to government specs and may be applied where the federal government has no jurisdiction.

Military coaxial bears the designation MIL-C-17 indicating that it conforms to military requirements. A cable sample must pass a series of tests before it can be placed on a qualified products list (QPL).

Commercial coaxial is listed and meets requirements of electrical and building codes.

Cable accessories

A broadband coaxial LAN operates similarly to the CATV coaxial cable. However, while the CATV system is usually a 1-way signal transmission, a LAN system normally involves 2-way communication. The usual method for providing 2-way service is to divide the available bandwidth into two parts; one for downstream communication and one for upstream. The divisioning requires the use of a head-end remodulator at one end of the cable. Trunk amplifiers, and where necessary, distribution amplifiers, are placed at regular intervals to maintain the signal strength.

To tap into the coaxial cable, a CATV-type signal splitter is used. The splitter is inserted in the main cable and tuned for impedance and proper signal level. From the splitter, connection to the station outlet is made by a drop link and an RF modem.

Flat coaxial cable assemblies for serving computer workstations or video terminals are similar in construction to the flat telephone cable. The coaxial harnesses can be supplied preconnectorized up to 50-ft lengths, or the cable can be supplied, with the connectors being field installed. Directional changes along the floor are made by trimming the outer edges with a tool, and the cable radiused as required.

Optical-fiber cable

This type of cable is suited for high-speed, high-volume point-to-point transmission of data and for other specialized applications. An electrical signal is converted into light impulses by a transmitter and delivered through thin, flexible glass fibers to a receiver. At the receiving end, the light is reconverted back to electrical signals.

Optical-fiber cable is being used where data rates greater than 40 Mbits/sec are needed, and where immunity to electromagnetic interference and induced surges are required. The compositon of the fiber determines the transmission attenuation. Many types of optical fiber cables are available.

Multimode step index is normally used for short distance, low bandwidth systems. It is the easiest type to connect or splice because of its large core and high internal reflection properties. The core is surrounded by a silica glass cladding with a lower refractive index than the core.

Multimode graded index has higher bandwidth properties than the step-index type. The refractive index of the core gradually changes from the cladding to the core. Light pulses are bent back toward the center of the core as they travel down the fiber.

Single mode has performance, reliablity, wide bandwidth and design flexibility features. The light-carrying core is smaller than that of multimode fiber and only recently has the technique for economical manufacture of this type been perfected. It is superior for long distance use.

Flat cable construction utilizing optical-fiber cable can be run under carpet squares the same as the standard flat cable assemblies.

Splicing together two lengths of cable is done by fusion splicing or by a mechanical connection. The fusing technique involves aligning the two fiber ends very accurately and then joining them permanently by melting the fiber tips.

Chapter 3

Design of Computer Electrical Systems

A SYSTEMS APPROACH TO COMPUTER POWER

Alfred Berutti

Computers require good power and grounding systems and a clean voltage source.

To ASSURE proper coordination of building electrical distribution and computer power conditioning equipment and standby power, Affiliated Food Stores, Inc. called upon outside electrical engineering expertise to provide a fully integrated source of power for their new 3300-sq-ft computer center in Keller, TX, a suburb of Ft. Worth. The installation consists of a highly automated warehouse and an office building housing the computer center, which serves 135 on-line users.

Three distinct computer systems are involved. One handles the order entry, label generation for the warehouse order fulfillment system, and billing. A second computer and peripherals are dedicated to the gathering of historical data, providing customized pricing, strategy information, and inventory monitoring. These two systems are located in the office building. The third system, located in the warehouse, is used for controlling the operation of conveyor systems and other automatic equipment that transport items to trucks waiting at the loading docks.

Planning the facility

Letting of contracts followed the standard approach taken when developing any new location. The building shells and services, including electrical, were packaged together. Mechanical and electrical installations within the computerized warehouse were subcontracted out by the general contractor of the warehouse. The office building interior, minus the computer center, was let as a package. Complete responsibility for the electrical installation within the computer center was given to a consultant specializing in correction of computer power problems.

One of the principal reasons that Affiliated Food realized the potential pitfalls of a data center was that it had earlier installed and operated a computer facility at its warehouse, then located in Dallas, and had experienced many problems. They called a

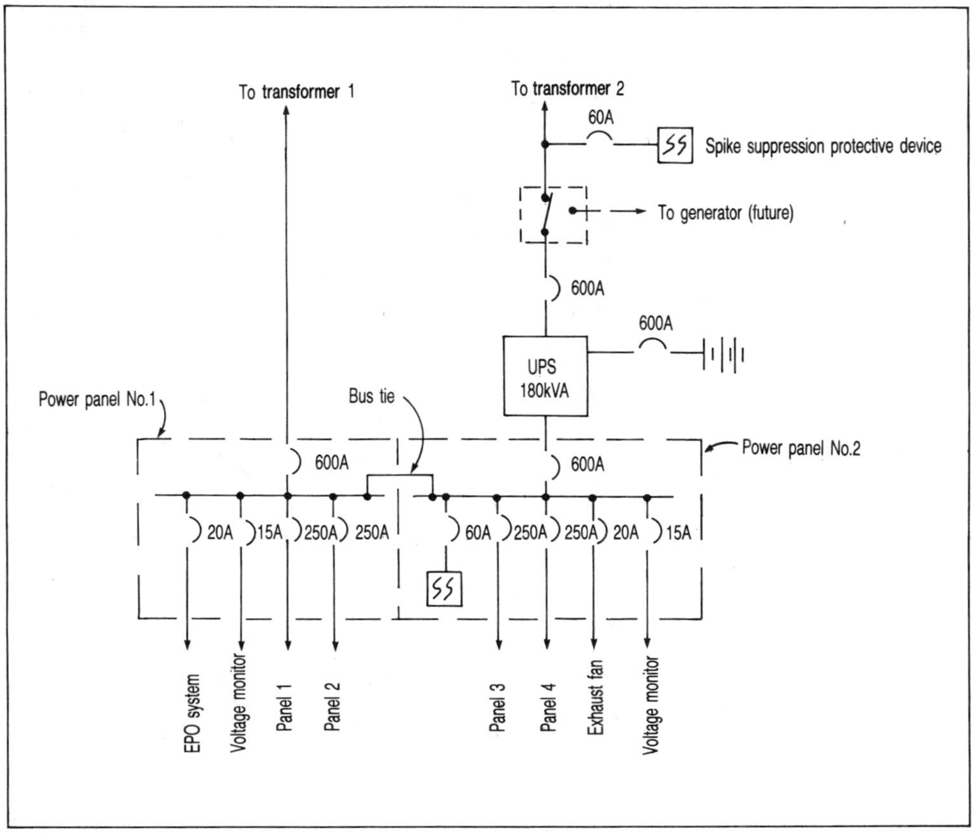

Fig. I. Simplified one-line diagram of the computer center.

Padmounted transformers are located outside the office building. Transformer No. 1 (to the right in photo at right) is a standby source of power for the computer center equipment. Transformer No. 2 (to the left in photo at right) is the main power source. Other transformers feed the remainder of the office building loads. Because of the importance of the grounding array system around the transformers, the dirt is kept moist by a sprinkler system (photo below) that is used during dry spells.

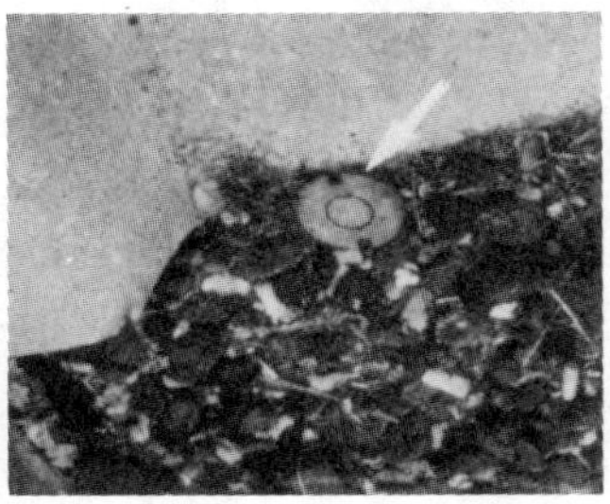

consultant to determine the cause and propose solutions. The results so pleased Affiliated that when the new Keller site was selected, management opted to break out the computer center electrical portion of the project and contract the same outfit to handle it.

Computer center electrical

Job-consultant Barry Epstein's guiding principle for any computer installation is that the incoming electric service voltage must be at the proper level, and that effective, single-point grounding must be provided. Only after these two conditions are met should the *quality* of the incoming power be addressed.

As part of the initial installation, primary power for the buildings was tapped from the serving utility's underground 13.8kV distribution system. Twin padmounted transformers at the warehouse, the office building, and the computer center were installed to reduce this voltage to that needed within the building. The two 150kVA transformers serving the computer center have a 208/120V secondary voltage level. Fig. 1 is a one-line diagram of the basic power-distribution system for the computer equipment.

The first step in meeting Epstein's criteria is to avoid skimping on wire size—this is the least- expensive item in an electrical system. To compensate fo voltage drop, feeder cables were increased one size for every 75 ft above that required to satisfy NEC rules. A similarly sized neutral should also be carried into the building from each transformer. These parameters were met in the Keller installation.

There is no reason why voltage at equipment should be low if voltage drop is compensated for by generous sizing of wire. If the problem exists even with oversized feeders, attention should be directed to raising the setting of the transformer taps. All too often these simple solutions are ignored in favor of installing expensive localized voltage-regulation equipment.

Flexibility of the installed system to meet the power requirements of the computer equipment once it is in operation is another prerequisite. Because the Keller site is in a rural area and the serving utility lines run overhead for most of their run, outages due to lightning had to be expected. While a UPS system was not part of the initial equipment purchased by Affiliated, provisions were made in the design layout so that one could be inserted into the system if necessary. This indeed proved to be the case, since 1- to 5-min outages occurred fairly often. A 180kVA unit was installed to allow the computers to ride through these power interruptions.

Because of the uncertainty of the length of the outages that could be expected, the potential for generating a block of power to maintain critical equipment in operation during long power interruptions also was worked into the equipment and electrical layout. A manual transfer switch was mounted in the electrical room in anticipation of the installation of a diesel generator. Thus far, the need has not developed.

Running load of the computer room equipment already installed totals only 80kVA out of the 300kVA capacity of the two transformers. For that reason, the electrical system was designed with transformer No. 1 acting as an alternate source of power in case of failure of transformer No. 2 or the UPS sys-

Power distribution panels in the office building utilities room direct power to subpanels that provide power to the computer-center equipment. The transfer switch to the right of the panels provides a means of tying-in a generator set if required in the future. The UPS system to the left of the power panels provides a means of riding out most expected power outages. It also serves to filter out some of the utility-line transients.

Computer center (photo below) houses the computers and peripherals that allow the company to offer customized financial and ordering services to member stores. The volume and complexity of the data to be handled require that the source of power to the equipment be reliable as well as clean. Such a result was achieved by addressing the total needs of the installation as a whole rather than just as an assortment of hardware.

tem. While the secondary cables from each of the transformers feed separate power panels, the buses of both panels are tied together. Power to the computers is thus easily restored with a minimum of labor and lost time.

The grounding system

All electrical installations require a good grounding system for personnel and equipment protection. Computer systems are particularly sensitive to transient signals that can be fed back through their digital circuitry, and errors can result. Because of the delicate physical makeup of circuit boards and the logic elements they contain, computers are also subject to internal damage and component failure. A good grounding system minimizes these hazards.

Much of the soil in the area of the Affiliated Food facility is very dry, making a driven electrode a poor ground reference. To increase grounding effectiveness, a number of ground rods were driven around the transformer pad location and connected together to provide the required transformer ground. To counter the dry conditions, a water sprinkler system was installed to keep the ground moist. A grounding conductor (sized the same as the power conductors) was then run from the transformers to the ground bus of the power panel fed by each transformer.

Many installations involving office buildings go wrong in grounding the computer equipment. It is essential that neutrals and grounds be segregated in orange "isolated ground" receptacles, which are permitted by NEC Sec. 250-74 Ex. 3. There should be only *one* path of lowest impedance to ground via

Manager of data processing David Berry (photo left) checks the tape of the automatic monitoring system installed in the computer center. A log of the voltage available to the equipment, any downtime experienced, and other factors that are important to the successful operation of the units is registered. Because careful attention was paid to proper grounding and voltage for the installation, only very minor disturbances have been noted.

Computers located in a separate room (photo above) near the computer center are used to receive orders from stores and to print labels that are used to route the tote bins containing the filled orders to the correct truck assigned to serve the participating store location. The conveyor systems in the highly automated warehouse (photos above right and far right) are controlled by a separate computer system. Recurring failure of equipment had in the past been experienced among the almost 135 terminals scattered throughout the huge facility.

the most direct path to the transformer array ground. Any alternative paths—including routes through the communications ports, circuit boards, and computer power supplies—can cause the mysterious burnout failures that are often blamed on poor equipment.

Because wire is cheap in relation to power-conditioning equipment, it is worth the added cost of running independent insulated ground wires from the system ground to the various isolated ground receptacles. Where these receptacles are physically located close to each other, a single conductor can be tapped at several points to provide a ground for more than one receptacle (daisy chaining). However, the use of the insulated ground wire does not eliminate the need to ground the boxes and raceways containing this conductor by some other independent means.

Avoiding power-line transients

High-voltage spikes reaching 6000 to 10,000V due to lightning strikes, switching in the utility's system, and similar events often appear on power lines feeding installations containing computer equipment. Frequently flicker or short power outages are accompanied by high-intensity spikes that are generated *within* the facility's electrical system. For instance, reactive spiking is generated by transformers during collapse and buildup of their magnetic fields. Random noise picked up from sources such as SCR equipment, capacitive discharge, cycling of motors, etc. is also embedded in the power distribution system. High-frequency noise picked up by conductors that act as antennas can further complicate the picture.

The intensity of the impulse or noise causes different problems to computer-based equipment. *High-intensity spikes* due to lightning, failure or return of power, and similar events cause damage that is readily visible, such as burned-out or failed equipment. *Medium-intensity spikes* weaken components, causing them to fail days or weeks after the original event and making it difficult to identify a particular storm or outage as the cause of a specific parts failure. *Low-level impulses* or random noise generate system errors, stoppages, wrong counts, erratic operation, loss of control, etc. All three types of spikes—high, medium, and low— exist in every power system. Protection against impulses and noise is usually justified.

Sags or dips in voltage levels are less serious to the proper operation of computers because of the voltage regulation circuitry of power supplies contained within most equipment. Required output voltage will be maintained within acceptable tolerances except under the most extreme conditions approaching a total outage.

Spike-suppression protective devices that provide a shunt discharge path for transient overvoltages were installed at Affiliated Food to protect the equipment in the new computer center. The heart of these units uses a specially processed selenium cell having a zener-type characteristic; the current passing through the suppressor varies with the applied voltage. The suppressor imposes a ceiling above which the surge voltage cannot rise. Because the units are non-polarized, they can discharge in either direction. For instance, they can protect against outside surges or from reactive transformer spikes. Internal high-bypass filters eliminate the problems caused by high-frequency noise.

Improving warehouse computer reliability

Affiliated Food unintentionally provided a comparison between good (the computer center) and bad (the warehouse) computer power installation techniques. After commissioning the warehouse system, many repetitive failures occurred on the approximately 135

VDTs scattered throughout the area. Additionally, order-taking and shipment programs that were terminated almost daily because of power-related problems had to be rerun, resulting in delayed shipments. The computer-room consultant was assigned to track down the source of the problem and recommend corrective action.

Many cables connect the computers in the office complex and warehouse terminals. One of the first things noted was that no consistent grounding scheme had been followed in the warehouse. While isolated ground receptacles had been installed, ground wires on the receptacles were either nonexistent, bonded to the box, or tied to the neutral terminal; or the ground wire was run in conduit up to and connected to the roof-truss steel. The multiplicity of ground-return paths encouraged the circulation of ground-loop currents. Lack of a single-point ground reference allowed surges induced by lightning to seek the lowest-impedance path to ground, which often was through computer peripheral equipment.

The only columns in the warehouse are those internal ones that support the roof trusses. Adequate grounding of roof steel thus becomes important. Because of the huge size of the warehouse, the data cables running from the computers in the office building to the terminals in the warehouse were strung along the roof-truss steel. Without a solid grounding system, the roof steel acted as a giant lightning rod and antenna. Sneakpaths for induced currents within the cable shielding were causing much of the erratic computer operation.

Initial recommendations for improving the situation primarily focused on the grounding system and installing protective devices similar to those used in the computer center. Because this is an operating facility, correcting the receptacle ground-wire situation is proceding on an "as available" basis. Similarly, cable strapped to roof steel is being rerouted as time permits. To answer the immediate need, small individual surge-suppression protective devices were installed at all major pieces of equipment.

Among the problems (photo left) that were uncovered while searching for the cause of the failure of equipment within the warehouse and garbled data in the order-processing computer was that data cable (arrow) was strung along roof-truss steel. Unreliable grounding methods that had been employed during building construction had created a huge antenna. Sneak circuits offered by the data cable shields allowed circulating ground currents to flow and spikes to be introduced into the logic circuits. Improving the grounding system and the installation of individual surge-suppression protective devices at sensitive equipment greatly reduced the error rate.

Dramatic improvements in the reliability and life of the warehouse equipment resulted from the relatively simple approach taken. The failure rate of the equipment is now virtually identical with that of equipment located in the computer center.

The major lesson learned from this comparison of installation techniques is that more is not necessarily better. While it may seem that there can never be too much grounding, this may not be true where sensitive electronic equipment is involved. *Better* grounding rather than *more* grounding is what is needed.

Computer power, grounding, and surge protection should be treated as an entity. Best results are obtained when the installation is not thought of as consisting of a room containing a conglomeration of hardware. Unfortunately, just as people in the electrical industry may not understand digital techniques, those well-versed in computers often do not understand power systems. Only a good-faith effort to achieve a coordinated system will produce an efficient, dependable installation.

DESIGNING A POWER SYSTEM FOR A LARGE COMPUTER CENTER

Robert J. Lawrie

UPS modules, huge engine generator, and transfer and bypass switching assure maximum reliability.

POWER SYSTEM reliability and "clean" computer power are vital to effective operation of the modern data center at Wakefern Food Corp., a large, cooperative food retailer in central NJ. Each day, the 50,000-sq-ft complex handles hundreds of orders, stock inventory, pricing data, and related information for nearly 200 Shop Rite supermarkets located in five northeastern states.

At the headquarters location are two large warehouses and the data center, which includes a 10,000-sq-ft computer area. Most stores are equipped with minicomputers that place orders on line. The data-processing center communicates with each store, obtains these orders, processes them on two IBM 3083 B-16 mainframes, and sends the information to warehouse facilities where orders are filled and shipped to the supermarkets.

The computer network also incorporates 500 on-line terminals in the data center to aid dispatching, invoicing, accounting, and other paperwork, as well as computerized bar-scanning systems at stores for pricing, inventory, and accounting.

Because accuracy is necessary for all functions, transient-free power as well as total reliability was considered essential. To assure this high level of performance, Won Kim, P.E., project engineer for Wakefern, worked with design consultants and the installing electrical contractor in specifying and developing a total power plan designed to provide an extremely high degree of power-supply quality and dependability.

Dependable power system

The elements of the reliable power system for the data center are shown by a modified one-line diagram. Accompanying photos, which are keyed to the diagram, permit a "photo-tour" of the installation.

Major elements include the normal incoming utility power; an alternate supply consisting of a 1000kW diesel engine-generator set; a lead-acid battery bank; transfer switching; UPS distribution panel; one 400kVA, 60-Hz

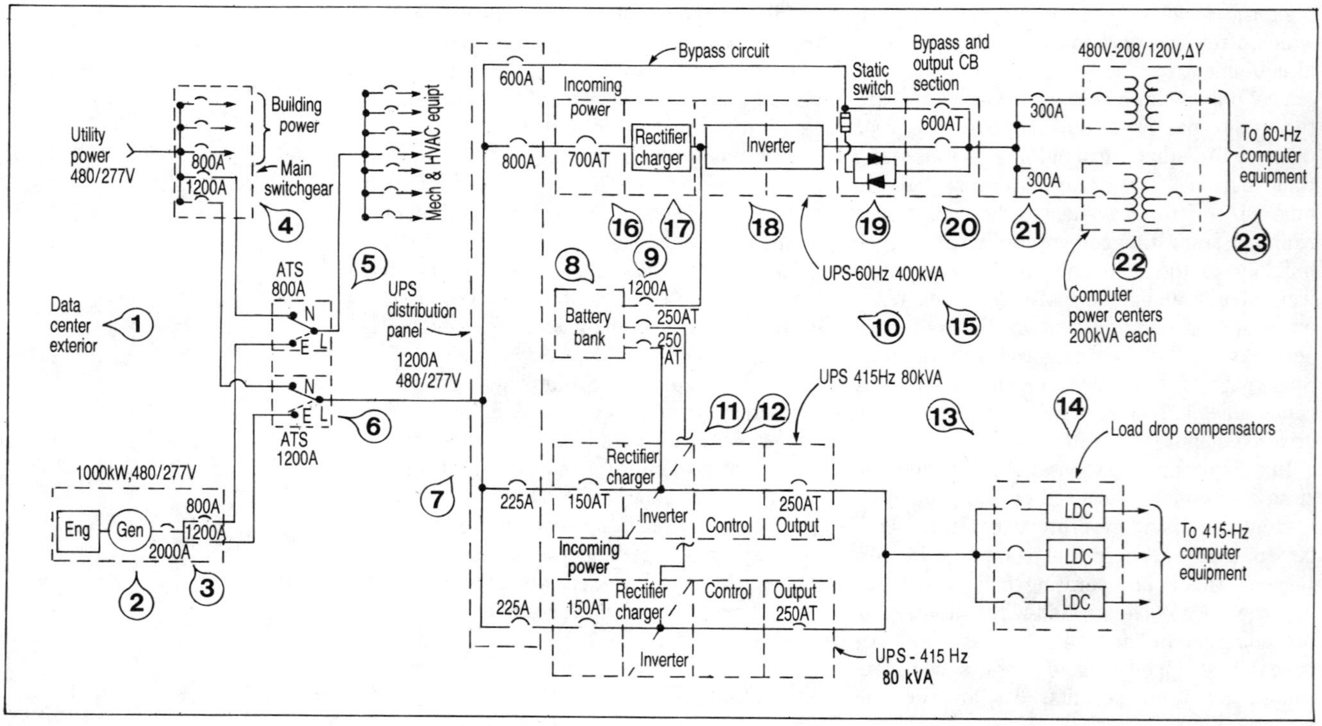

Power distribution for data-processing center is shown in modified one-line diagram with major components arranged to correspond with actual physical layout. Numbered arrows are keyed to accompanying photos.

Photo 1. Data-processing center is hub of computer communications and operations for a 200-store retail food cooperative. The entire 50,000-sq-ft structure is furnished with raised floors to simplify cabling and other services to computer area and several hundred on-line terminals.

Photo 2. 1000kW diesel engine generator, protected by a weatherproof enclosure, dwarfs engineer at right. Muffler can be seen on the roof of the enclosure. The generator supplies 480/277V to the computer center in the event of utility power failure.

Photo 3. Inside diesel-generator housing, switchgear and controls are at right, generator at left. On wall in background are special filters to remove water from fuel supply. Control panel provides for local operation, protection, and monitoring of the diesel-generator set.

Photo 4. Main switchboard consists of two sections. Molded-case circuit breakers for building power and light are at left; a 1200A CB that protects the circuit supplying 480/277V to the UPS transfer switch and an 800A CB for computer-area HVAC equipment are in the right compartment.

Photo 5. Two automatic transfer switches and emergency-power panels are installed close to the main switchboard and UPS room. Supplying the UPS room is an automatic transfer switch (ATS) at left, which is rated 1200A; at right is an 800A ATS supplying computer HVAC equipment.

UPS module; two 80kVA, 415-Hz UPS modules (one redundant); a bypass circuit; a solid-state UPS bypass switch; load drop compensators; and two 200kVA computer power centers.

Both UPS systems draw all power from the utility line during normal periods; the diesel engine-generator set supplies power whenever utility power is not available. The UPS systems supply regulated output voltages with precise 60-Hz and 415-Hz frequency stability. The systems must operate unaffected by any transients in the utility supply and must continue to supply filtered, uninterrupted power in the event of a utility power failure. The battery bank assures absolute uninterrupted power to critical loads during transfer of the load from the utility to the generator set. The UPS systems isolate the computer equipment from the utility line, keeping all transients, via inherent filtering systems, from reaching these critical loads. The UPS also supplies power during momentary power dips, surges, or internal faults,

Photo 6. Normal-power conductors come into 1200A automatic transfer switch at the bottom, generator supply conductors are at top, and conductors to load are at right. All are 500MCM THW.

Photo 7. UPS main distribution panel, which supplies 480/277V power to the UPS equipment, contains two 225A CBs at top that protect circuits to the two 80kVA, 415-Hz UPS modules. At bottom is a 600A CB protecting the AC bypass circuit and an 800A CB protecting the 400kVA, 60-Hz UPS module.

Photo 8. Battery bank, (photo above) consisting of 186 lead-acid batteries, is installed in a separate room adjacent to the UPS room. In the event of normal utility power failure, the battery bank will carry the entire UPS load for at least 15 min.

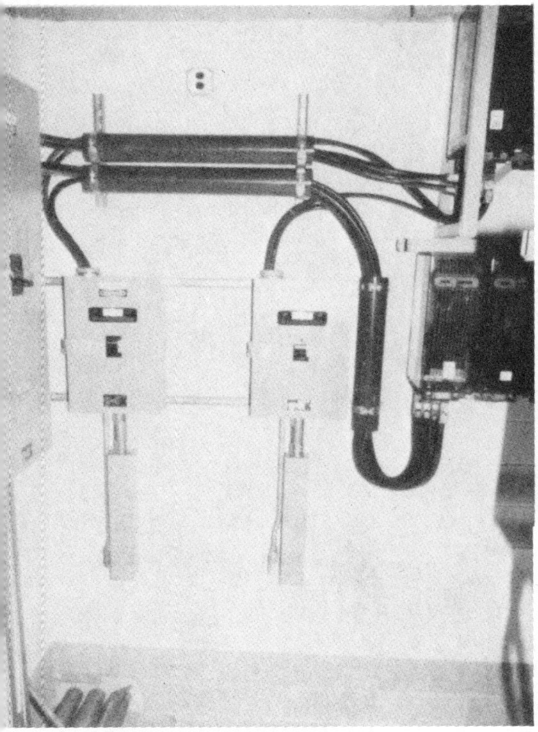

Photo 9. Battery output arrangement includes eight 500MCM THW conductors and 1/0 ground extending through protective sleeves to 1200A CB at right, where they continue in 4-in. rigid steel conduit through wall to 400kVA UPS module. The two 400A CBs mounted on wall in background supply 80kVA units.

while permitting only conditioned, regulated voltage and frequency to be supplied to the computer loads.

The UPS modules are installed in a room adjacent to the computer area. This UPS room, a nearby battery room, power distribution equipment, and pertinent components are all located in the rear of the building close to the diesel engine-generator set installed outdoors in a weatherproof enclosure.

Computer power centers are compact units specifically designed to simplify circuiting to computer equipment. Two 200kVA, 480/208-120V power centers are installed in the computer area. Units come with flexible, multiconductor cables of the proper size and length with appropriate connectors for the equipment supplied. Shielded isolation transformers minimize transmission of spikes or other transients to the computers. This feature is especially valuable when computers are operating on "bypass" utility power. Both units have two main sections and an add-on section to provide for additional CB

Photo 10. UPS room contains two 80kVA, 415-Hz modules at left, load compensator panel in background, and a 400kVA, 60-Hz module at right. Battery bank is on other side of wall at left; computer room is on other side of wall at right. Inspecting metering section is Tom Noto, project engineer, Datasphere, Inc. (left), and Won Kim, project engineer, Wakefern Corp.

Photo 12. Metering and control section on 80kVA, 415-Hz module is furnished with input meters for 60-Hz voltage and current at top left; meters for DC output voltage and current are at center: output meters for 415-Hz voltage and current are at right. Center section is mimic panel of system, with indicating lights. At bottom is a visual and audible alarm panel and reset button.

Photo 13. Load drop compensator (LDC) panel (photo below) is furnished with control panel (left), which synchronizes outputs of the two 415-Hz UPS modules, and compensator section at right. Each of the top three sections contains an LDC; bottom three sections are spares.

Photo 11. Rectifier/inverter section of 80kVA, 415-Hz UPS module (photo above) includes 480/277V, 60-Hz input to rectifier (A); rectifier section (B); inverter section (C); capacitor and choke filter for rectifier DC output (D); and output filter trap that removes 5th and 7th harmonics (E). Both rectifier and inverter sections consist of thyristors mounted on heat sink.

Photo 14. LDC detail consists of common vertical 415-Hz input bus (A); circuit breaker (B); conductors to tap transformer (C); capacitors (D); and output terminal plate (E). The LDC provides for a tuned circuit that compensates for various values of inductance and voltage drop on different lengths of 415-Hz circuits extending to computers.

spaces. Each unit has 168 CB spaces and main CB protection. The computer power centers are provided with monitor lights, alarms, metering, and controls.

UPS details

A typical UPS module includes a main control panel, a bypass CB and static switch, input and output CBs, a rectifier-charger, an inverter, and filters. The rectifier-charger consists primarily of an input isolation transformer, thyristor phase-controlled rectifier bridges, and a DC output filter. Firing of the thyristors is controlled continuously to provide regulated DC at the bridge output. The output ripple frequency is six times the input frequency. This pulsed DC output is smoothed by a series choke and further filtered by a large capacitor bank connected across the DC positive and negative buses. The gate drive to the thyristors is controlled by logic circuitry that senses the DC bus voltage and current.

The inverter consists of twin 3-phase bridges made up of thyristors controlled by digital logic. The arrangement results in a 6-phase circuit with 12 switching elements. The square-wave output is filtered to obtain optimum wave shape and lowest harmonic distortion.

A static bypass switch operated in conjunction with a paralleled power circuit breaker allows instantaneous load transfer from the 60-Hz UPS output to the utility power bypass circuit in the event of a system failure. Normal fault-isolating devices such as circuit breakers are not fast enough to effect a transfer without causing adverse transients on the computer load supply.

Photo 15. 400kVA, 60-Hz UPS module (photo above) consists of six sections: incoming/CB (A); rectifier/filter (B); inverter/filter (C); additional inverter (D); meter/control/static bypass switch (E); and output/CB (F).

Photo 16. Input section (photo left) contains 60-Hz power input CB (arrow) and terminals for six 500MCM conductors that are carried overhead in two 3-in. rigid steel conduits. DC supply from battery bank, carried in three 4-in. conduits, also enters at this section.

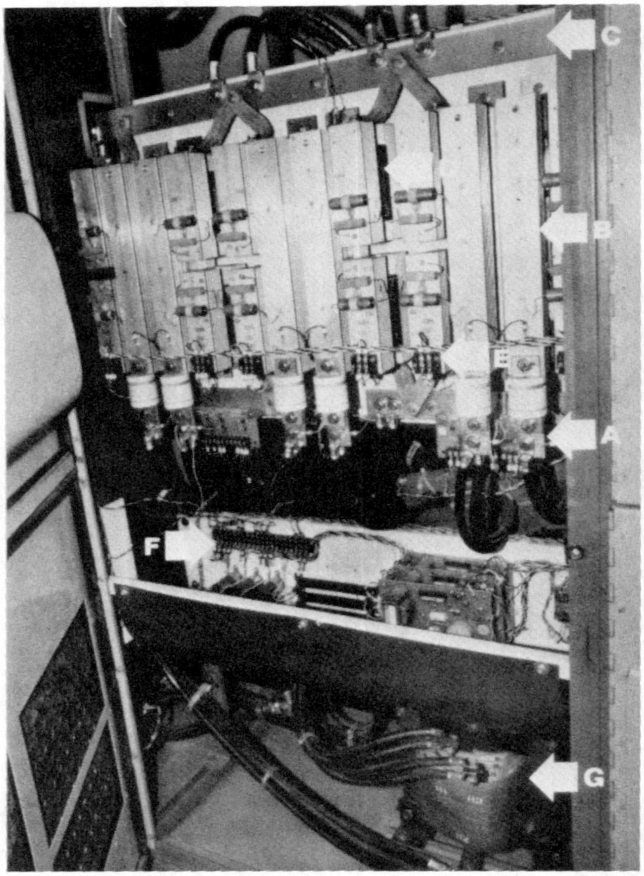

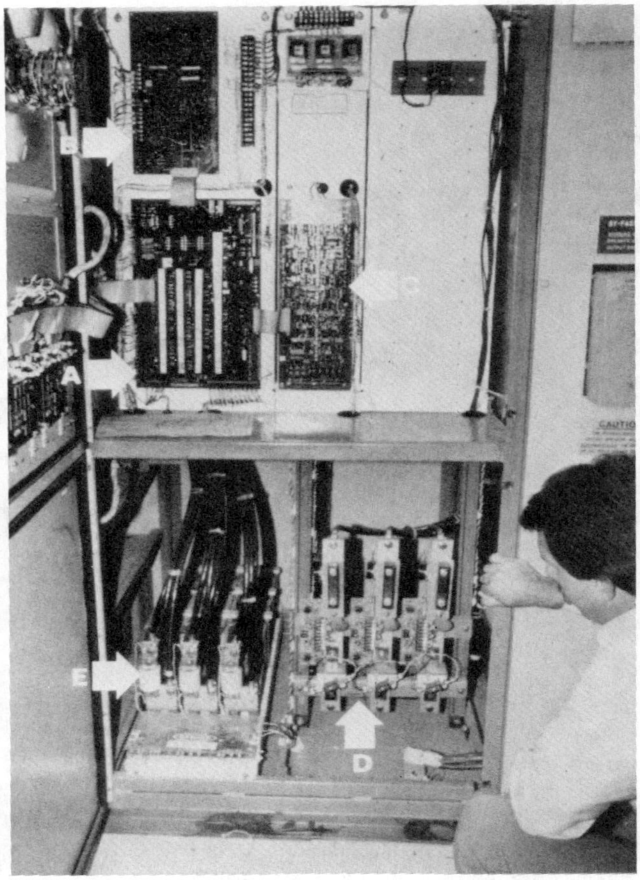

Photo 17. Six-step rectifier input (A) (photo above) is 60 Hz, which extends to rectifier assembly (B) and is brought to DC bus (C). Thyristor (D) is typical of several in assembly. Firing termination boards for thyristors are mounted on heat sinks (E). Terminal board (F) is for control circuits extending to main control. Inductor (G) smooths DC output.

Photo 18. Inverter input bus (A) (photo right) extends into next section, providing for two sections of inverter components. Inverter assembly (B) consists of thyristors compressed between heat-sink plates. Terminal boards for firing thyristors (C) simplify circuiting back to main control board. Capacitors (D) and inductors (out of sight) act as an output filter trap.

Photo 19. Interior of meter and control section (photo above) includes main logic board (A) with 4 plug-in PC boards (edges of PC boards appear as vertical white lines). These boards provide for AC and DC power regulation, firing control of thyristors in the rectifier and inverter, and digital logic control of multiple sensing—temperature, frequency, voltages, etc. At the top of the panel is the high-voltage isolation and signal-conditioning board (B). A synchronizing board is provided for line-sync operation of the bypass switch, solid-state switch, and output CB (C). A high-speed static bypass switch is at lower right (D) and high-speed fuses that operate in conjunction with the switch are at lower left (E).

Photo 20. Metering and output sections of 60-Hz UPS module include meters for incoming AC voltage and current (left-top); UPS volts and current frequency Hz (right-top), and DC voltage and current (center). Mimic diagram of UPS system includes an inverter voltage adjust control (arrow). At lower left are alarm diagnostic lights and reset and control switches. Static bypass disconnect switch control is at right top of output power section; bypass CB is at left; UPS output CB is at right.

The static bypass switch consists of thyristor circuitry controlled by fast-acting sensing devices and electronic logic circuitry. The switch is a *time-rated* device, rated at 250A for 150 ms. When an automatic high-speed transfer is required or an internal UPS fault occurs, the static switch carries current for approximately 40 ms while the paralleled bypass circuit breaker is closing. When the bypass CB has assumed the load current, the static switch will automatically open.

The data center and computer facility were designed and built by a team consisting of Datasphere, Inc., computer facility engineers, Clark, NJ; Won Kim, project engineer for Wakefern Food Corp., Edison, NJ; and Lessner Electric, electrical contractor, Elizabeth, NJ. Developer and general contractor was DiGirolamo Construction Co., Newark, NJ.

Photo 21. UPS output distribution panel contains two 300A CBs and space for additional CBs. From each circuit breaker, three 350MCM and ground in a 3-in. conduit extend under the raised floor in the computer room carrying precisely regulated 480/277V power to a 200kVA computer power center. Rigid steel conduits overhead carry input power, battery-supply power, and bypass circuit conductors to the UPS module.

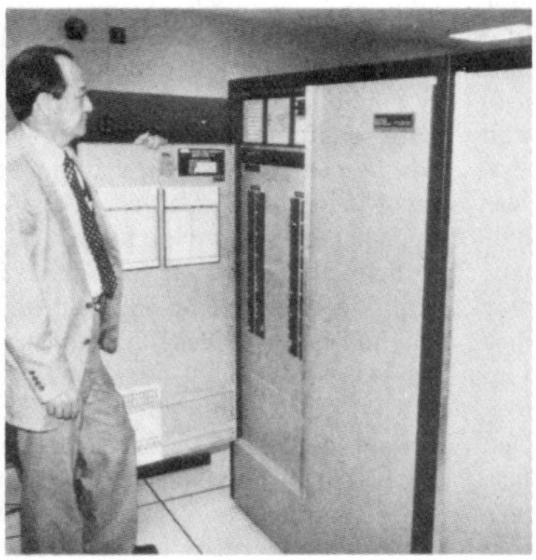

Photo 22. 200kVA computer power center, one of two installed in the computer area, is inspected by Won Kim, project engineer. Each center consists of two main sections and an add-on panelboard section, providing a total of 168 CB spaces in each center. The unit is furnished with a shielded transformer that steps primary voltage to 208/120V for reliable operation of computers, particularly in the event that they must be operated on bypass utility power.

Photo 23. Computer area is equipped with computer systems, including two 3083 mainframes, 500 on-line terminals, tape drives, tape library, printers, and communications equipment. Extensive HVAC and fire/security protection systems are installed.

Rotary equipment supplies clean computer power

New developments in motor-generator technology make rotary power systems an excellent solution to computer power problems.

A MODERN motor-generator set can provide a very cost-effective solution to all power problems other than sustained outages. Where a real-time computer *must* have continuous power during sustained outages, only a true uninterruptible power supply (UPS), supplied by batteries and backed up by an engine-generator, will do. However, sustained outages are a very small percentage of the total number of utility power aberrations—about 0.5%—and a UPS/generator system is very costly.

Motor-generator sets for computer power must, of course, be high-quality, dependable units. Other than that, they often differ little from MG sets used for other applications, although some are manufactured with special characteristics for computer application. The motor can be either a synchronous or induction type. Some manufacturers call the induction motor "asynchronous." The generator, since its output is AC, is sometimes called an "alternator" to differentiate it from the DC generator; and the combined unit is referred to as a motor-alternator set. However, the dictionary defines a generator as a device for converting mechanical energy to electrical energy, and so we will use the term motor-generator, or MG set for equipment with an AC output. The motor and generator can be in one housing, with a single shaft for both units, or they can be separate pieces of equipment.

One of the best ways to supply power to computer equipment is by means of a computer power center, which provides power distribution, grounding facilities, and often many electrical and other monitoring and alarm functions for the computer power and equipment and computer-room environment. By designing the MG set for low noise and putting it in a sound-absorbing enclosure, it can be incorporated directly into the power center. This provides a combined power-conditioning, monitoring, and distribution center supplied with raw utility power and supplying the computers with clean power. Locating the entire system right in the computer room has many advantages. Short power cables, and short, single-point grounding connections minimize noise pickup that can affect computers. Not the least important advantage, by far, is the tax benefit. A computer power center can be depreciated rapidly for tax purposes, like any other computer equipment. Usually, depreciation can be over 5 to 8 years. If the equipment is installed as a part of the building's permanent wiring system, it must be depreciated over 22 to 30 years if the building is owned by the user, or over the life of the lease if the space is leased.

The most common utility supply voltage used for the motor of the MG set is 480 V. The generator output can be either 480 V or 208Y/120 V in smaller units rated about 75 kVA or less. Usually in larger units the most economical generator output is 480 V. For 480-V generators, a 480-208Y/120-V 3-phase isolating transformer is supplied to provide 208Y/120-V power to the computer equipment.

Single-shaft vs separate machines

There are two basic types of MG sets used for computer power: single-shaft, single-housing machines, and those with a separate motor and generator unit. Either design can use either a synchronous or asynchronous (induction) motor. In each case, there are advantages and disadvantages; manufacturers are able to provide successful operation by using good design to maximize the advantages and minimize the disadvantages of the arrangement they select.

The single-shaft, single-housing design is compact and has fewer bearings than separate machines—often only two bearings, one at each end of the shaft. The motor windings are at one end of the housing and the generator windings are at the other. Single-shaft, common-housing machines have inherent inductive and capacitive coupling between the motor, housing, shaft, and generator that result in shaft currents. The motor input power can also have transient voltages that cause currents to flow in the common shaft. Bearing pitting (see photo) is evidence of the reality of these shaft currents. These shaft currents appear as voltage disturbances in the generator output and therefore in the power to the computer. However, since most large-generator output is at 480 V and must be transformed down to 208Y/120 V, most of these disturbances can effectively be prevented from reaching the computer by using a shielded isolating step-down transformer.

Separate machine design uses a motor, either induction or synchronous, driving the physically separate generator by such means as an insulated coupling or belts and pulleys. This eliminates any electrical coupling and transmission of shaft electrical noise to the output power. However, each machine has two bearings, giving a total of four lightly loaded bearings. This provides greater shaft rigidity, but there are four bearings to be maintained. In the case of belt drive, the belts must also be maintained. Here, also, good design can minimize mainte-

By ARTHUR FREUND

nance and problems by using large bearings and redundant long-life belts. Separate machines lose a small amount of power in the belt drive, so their efficiency may be slightly lower than a similar motor and generator on a single shaft.

Using separate machines, it is possible to oversize the motor for the generator application, which can have several advantages. For example, a 75-kVA generator requires, at full load with no coupling losses, a motor capacity of 89 hp, assuming a typical 80% power factor and 90% generator efficiency. Usually, this generator would be driven by the next standard-size motor, 100 hp; but with separate units, a 125-hp or even 150-hp motor could be used to drive the generator. There are a number of benefits from using the larger motor. The MG combination will withstand larger voltage input variations without a deterioration of output performance. The larger bearings, operating at less than rated load, will have longer life. On loss of one input power phase, the motor can continue to run on single phase for a considerable time without damage, since the oversized motor is only partially loaded and can withstand the single-phase current without overheating. The slip range with voltage or load variation is reduced, helping to maintain output frequency stability. Also, the increased rotational mass will provide more power outage ride-through.

The case for induction vs synchronous motors

Both synchronous and induction motors have been used successfully to drive generators for computer power. Each has advantages and disadvantages. The major advantage of the synchronous motor is that while running normally, it is synchronized with, or locked into, a particular speed—depending on the supply frequency. There is no "slip." An 1800-RPM motor rotates at exactly 1800 rpm for a supply frequency of exactly 60 Hz, and the generator output will be exactly 60 Hz. Utility frequencies *average* 60 Hz, although utilities do vary the frequency slowly to maintain this average value under load changes. An exhaustive study by Bell Laboratories at 24 sites in the U.S. over two years found utility frequency ranges from 58.7 to 60.7 Hz. As the utility supply frequency varies, the generator output with synchronous motor drive will vary to match it. Although frequency tolerances permitted by most computer manufacturers are typically given as ±½ Hz, or from 59.5 to 60.5 Hz on a nominal 60-Hz system, these utility variations are spread over a 24-hr period or longer and do not seem to cause problems. Nevertheless, staying within computer manufacturers' recommended tolerances will result in the most reliable computer operation.

The major disadvantage of a synchronous motor is that it is difficult to start. In fact, it cannot be started as a synchronous motor. It must be started and brought up to "pull-in" speed by some other means. "Pull-in" speed is that minimum speed, close to synchronous speed, at which the motor will pull into synchronization if the excitation is applied to the field. Most synchronous motors have an auxiliary winding on the armature, known as the *amortisseur* winding, usually of the squirrel-cage design, although it may be of the wound-rotor type in some cases. This winding allows the synchronous motor to start and come up to speed as an induction motor. When pull-in speed is reached, automatic sensing equipment applies field excitation, and the motor locks in and runs as a synchronous motor. (Some large synchronous motors are brought up to speed by an auxiliary, or "pony," motor, but these have little application for computer power systems.)

The amortisseur winding can produce only limited torque, so synchronous motors are usually brought up to speed without a load. This is not a problem in computer rooms for initial startup. However, if there is a momentary power outage, complications develop. When power fails, the synchronous motor must be disconnected from the input line immediately, or it will act as a generator and feed power back into the line, thus rapidly using its stored (kinetic) rotational energy, which it needs to ride through a momentary power outage. Automatic controls can do this disconnecting, but control operating time reduces ride-through ability, and the additional circuitry adds complexity.

On power failure, the speed of the unit under load quickly drops below pull-in speed, and when supply power returns, the amortisseur winding must reaccelerate the motor, under load, until the field can again be applied. This is a difficult problem, requiring a large winding and sophisticated controls. While speed is below synchronous, the generator output frequency may be too low for proper computer operation.

BALL BEARING from a motor shows pitting of both inner and outer races. This provides evidence that shaft currents exist. These disturbances can be transmitted to the generator output on single-shaft machines.

The induction motor has no synchronization problems, but instead has "slip." To produce torque, the rotor must rotate at slightly lower speed than the stator field. For a nominal 1800-RPM standard motor, actual speed will be about 1750 RPM, varying slightly with load and applied voltage. This represents a slip of about 2.8%. The generator, if directly driven or on a common shaft, will have an output frequency 2.8% low, or about 58.3 Hz. This is below the minimum permissible 59.5 Hz of most computer equipment, so such an MG set cannot be used. Special precision-built low-slip induction motors are available, at a considerable premium in price, with a NEMA standard slip of 0.7% at the NEMA nominal motor voltage of 460 V. At the usual actual motor voltage of 480 V in a computer room, the slip is only about 0.5%, not a NEMA standard value. With 0.5% slip, speed at full load will be about 1791 RPM, and the directly driven or common-shaft generator will have an output frequency of 59.7 Hz. This is within tolerance, but close to the minimum permissible frequency, and leaves very little for additional slip as a result of voltage sags or increased load. Therefore, single-shaft or direct-coupled MG sets are best driven by a synchronous motor.

A simple, low-cost solution to this problem of slip is available in the MG set with adjustable-speed drive. One low-cost, adjustable-speed device is a belt-and-pulley system. By making the pulley on the motor slightly larger in diameter than the pulley on the generator, with the actual diameters adjustable, the generator can be driven at synchronous speed, or even higher, while the motor rotates with normal slip. Other types of adjustable-speed drives may also be used. No premium motor is required.

Induction motors have several other advantages over synchronous motors. They are simpler in construction and lower in cost, do not have a need for field application systems during start or after a momentary power outage, and tend to be quieter (which is important in a power center installed in the computer room).

Voltage sags and output frequency

Voltage sags have no effect on the output frequency of a synchronous-motor-driven MG set until the voltage gets so low that the torque is reduced to a point where the machine pulls out of synchronization; then resynchronizing becomes a problem. When voltage sags, induction motor slip increases, and the motor slows down, driving the generator more slowly. The result is a drop in generator output frequency. Since even precision-built, 0.5%-slip, direct-coupled induction-motor-driven generators produce close to minimum acceptable frequency output (59.7 Hz) at full voltage, a voltage sag or load

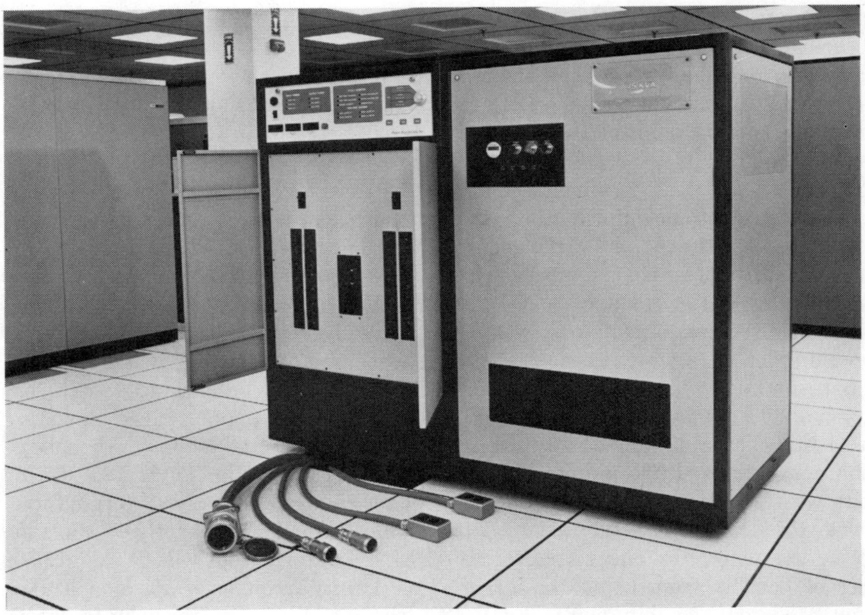

COMPUTER POWER CENTER is located in the computer room. It combines motor-generator set with flywheel, for continuous power of computer quality, with power input, power output, and environmental monitoring (top left), and power distribution (lower left). Flexible cables with receptacles or connectors as required by computer equipment are supplied by the power center.

Frequency at full load with voltage variation

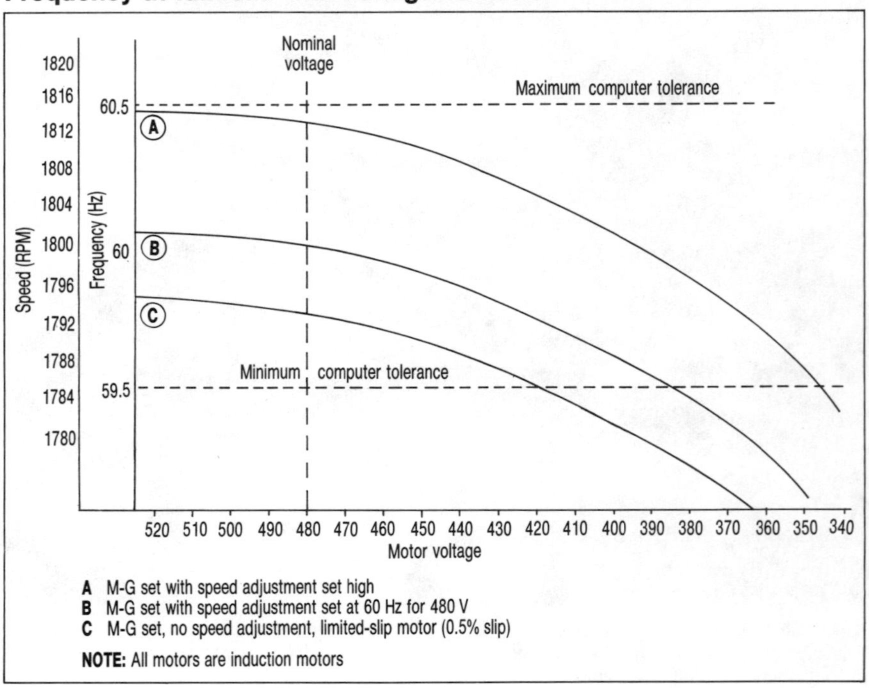

increase will, through increased motor slip, cause the generator output to fall below the minimum standard of 59.5 Hz. The Bell Laboratories study showed that voltage sags make up the greatest portion (about 87%) of incoming power aberrations, and similar indications have been obtained by other investigators.

The adjustable-speed drive between induction motor and generator solves the problem for separate machines, as described earlier. With the motor at the speed for normal slip, say 1750 RPM, the generator can be rotated at 60-Hz synchronous speed, 1800 RPM. As voltage sags, slip will increase, but the voltage will have to sag considerably before output frequency falls below the permissible minimum of 59.5 Hz. If severe sags are anticipated, the drive can be adjusted so that at maximum expected voltage the output is on the high side of frequency tolerance, close to 60.5 Hz. This allows large drops in voltage, with larger slips, before frequency drops below minimum tolerance of 59.5 Hz. As seen in the accompanying curves, the costly limited-slip motor, with 0.5% slip, will drop below 59.5 Hz at about 420 V (Curve C). A standard 1750-RPM induction motor, with drive adjusted to drive the generator to 60 Hz at 480 V, will not fall below 59.5 Hz until the voltage sags to about 385 V (Curve B), and if it is set high, close to the upper tolerance limit of 60.5 Hz at maximum voltage, it will not drop below 59.5 Hz until the voltage drops to about 345 V (Curve A). An oversized motor will have less slip and maintain frequency even better.

The adjustable drive permits good speed and frequency control during voltage sags, permitting the use of a reliable, low-cost, easily maintained standard induction motor to drive the generator. One such drive is the adjustable pulley with belts. For reliability and long life, not one pulley and belt, but sheaves of four or more pulleys and belts are used. One manufacturer has made the sheaves field-adjustable, so the speed can be tailored to the voltage actually available in the computer room and the actual computer load. That manufacturer also uses oversized motors, further enhancing speed performance. Once the speed is set using a frequency meter on the generator output, it rarely needs to be reset, since the computer load is relatively constant. If additional computer equipment is added, the speed can be readjusted for the new load, if necessary. Using four or more high-quality belts, belt life should exceed bearing life. When the bearings are replaced, new belts should be installed as part of normal maintenance, while computer operation is uninterrupted in the bypass mode.

Ride-through on outages

The ability of an MG set to continue to provide computer-quality power during a voltage sag or extended brownout is excellent. This ability during complete outages—blackouts, or total loss of incoming power—is also impressive. Here, the duration of continued satisfactory power comes from the speed and mass of the rotating parts, which store large amounts of kinetic mechanical energy as rotary motion. When the input power fails, the momentum of these rotating parts keeps the generator turning and producing power for the computer equipment. The power produced uses up the rotational stored or kinetic energy and continues to be satisfactory until the kinetic energy falls below a certain minimum level. Therefore, the duration of continued acceptable computer power is a function of the *change* in kinetic energy.

Kinetic energy in ft-lbs of a moving object can be calculated from the formula

$$KE = Wv^2/2g,$$

where

W = weight of the moving object (lb)

g = acceleration of gravity (32.2 ft per sec per sec)

v = velocity (ft per sec).

For a rotating object, the velocity is measured at the radius of gyration—the distance from the axis at which the rotating mass can be assumed to be concentrated. Velocity in ft per sec can be calculated from the formula

$$v = 2\pi K \times RPM/60$$

where

K = radius of gyration (ft)

$2\pi K$ = circumference of the circle at radius K (ft)

RPM/60 = speed (revolutions per sec).

This is why a flywheel of as large a diameter as is convenient, with as much of its weight as possible concentrated in the outer rim, will give a large mass and radius of gyration to increase the kinetic energy of the rotating machine.

For a given MG set, weight and radius of gyration are constant. Therefore velocity (v) is proportional to speed (RPM), and kinetic energy (KE) is proportional to v^2, or $(RPM)^2$. This is expressed as $KE = k \times (RPM)^2$, where k is a constant. It is important to note that doubling the RPM of a given machine will quadruple its kinetic energy.

When input power is lost, ride-through is provided by the kinetic energy of the rotating machinery, which continues to turn the generator. As the machine gives up its KE, it slows down. The change in KE can be expressed as the change from initial KE_1 to final KE_2, or

$$\Delta KE = KE_1 - KE_2.$$

ΔKE can be expressed as being proportional to initial $(RPM_1)^2$ less final $(RPM_2)^2$, or

$$\Delta KE = k[(RPM_1)^2 - (RPM_2)^2]$$

for a given machine.

Let us examine the significance of this for computer power application. At synchronous speed of 1800 RPM, generator output will be 60 Hz. For the minimum acceptable frequency for most computer equipment, 59.5 Hz, the minimum permissible generator speed is 1785 RPM. A generator driven by a direct-coupled motor, or a single-shaft machine, with limited motor slip of 0.5%, will have a speed of 1791 RPM with an input at 480 V. A unit with adjustable-speed control, set to the maximum frequency tolerance limit of 60.5 Hz, will rotate at a speed of 1815 RPM. The direct-coupled machine can slow from 1791 RPM to 1785 RPM before dropping below minimum frequency specification. The adjustable-speed machine can drop from 1815 RPM to 1785 RPM under the same frequency requirement. Thus,

$$\Delta KE_1 = k[(1791)^2 - (1785)^2] = 21{,}456k$$

$$\Delta KE_2 = k[(1815)^2 - (1785)^2] = 108{,}000k$$

where

ΔKE_1 is the change in kinetic energy for a direct-coupled or common shaft MG set with 0.5% slip motor drive

ΔKE_2 is the change in kinetic energy for an adjustable-speed MG set.

Note that for machines of the same constant k, the adjustable-speed separate-unit MG set has *more than five times as much kinetic energy available for power-outage ride-through* than the similar 0.5%-slip motor-driv-

THE ALTERNATIVES

There are two fundamental types of power-supply equipment for computers: power conditioners and power synthesizers. Power conditioners take the incoming power, condition or process it to remove as many aberrations as possible, and then supply that power to the computer. Power conditioners include, singly or in combination, such equipment as voltage regulators, isolation transformers, and transient suppressors. Power synthesizers use incoming power only as an energy source for a rotary or static synthesizer, which creates a completely new power supply to the computers. Rotary synthesizers are MG sets. Static synthesizers are ferromagnetic or semiconductor types. The power conditioners cost less but solve fewer problems and provide less isolation than the power synthesizers.

Among the power synthesizers, static (solid-state) or rotary UPS systems are considerably more costly than others, but they can be used to overcome sustained outages. Rotary MG equipment costs slightly more than ferromagnetic synthesizers but can maintain power for several seconds or more during a momentary outage, while ferromagnetic synthesizers cannot ride through an outage exceeding about one cycle. Power synthesizers are effective against all other types of input power disturbances, since the input power is never fed directly to the computers.

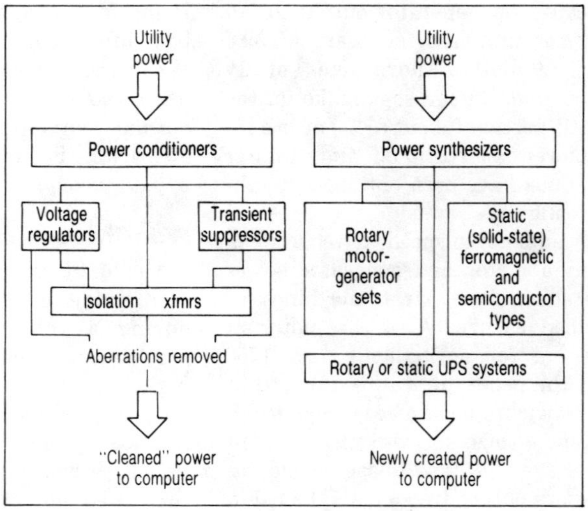

The ride-through capability of MG equipment is a result of the large amount of kinetic energy stored in the rotating parts, such as the rotor, shafts, and armature. Sometimes a flywheel is included to increase this stored rotational energy. When power input to the motor is lost, the inertia of these rotating parts continues to drive the generator. With a large flywheel, power could continue for as much as 20 seconds, according to one manufacturer. A more typical design is for a ride-through (sustained normal power output) of several seconds of total input power outage.

en single-shaft or direct-coupled MG set.

Thus, by applying an adjustable-speed mechanism to a standard 480-V induction motor driving the generator, it is possible to get complete isolation of input power transients from the generator output, to keep the frequency within the acceptable range down to a motor input as low as 345 V, and to have considerably more kinetic energy available for momentary blackout, or input power loss, ride-through.

One manufacturer of an adjustable-speed drive system has taken his design even further from the conventional to obtain even greater ride-through at minimal increase in cost. Remember that, as shown above, kinetic energy is proportional to the square of the speed, but that output frequency requirement of a maximum of 60.5 Hz limits the maximum generator speed to 1815 RPM. With these two facts in mind, this manufacturer's designers simply moved the flywheel onto the motor shaft and replaced the standard 1800-RPM induction motor with a 3600-RPM induction motor. By using a 1-to-2-ratio, adjustable-speed controller instead of a 1-to-1 device, the generator is still driven at 1800 RPM and can still be set by speed adjustment of the controller to precisely 1815 RPM. Now, by doubling the motor and flywheel speed, the kinetic energy of both is increased to four times what it was previously!

Let us assume for simplicity that our MG set, with the flywheel on the nominal 3600-RPM motor, has the same constant k as the previously discussed MG sets. Let us also assume a standard-grade induction motor with a typical slip of 3% rotating, with the flywheel, at 3492 RPM and driving the generator, through the adjustable-speed controller, at 1815 RPM. The generator could slow down to 1785 RPM before the output dropped below 59.5 Hz. This represents a minimum permissible motor speed of (1785/1815) × 3492, or 3434 RPM. Calculating the permissible change in kinetic energy as before,

$$\Delta KE_3 = k[(3492)^2 - (3434)^2] = 401{,}708k.$$

This represents, for the 3600-RPM motor and flywheel, almost 19 times as much energy for ride-through as the low-slip, single-shaft MG set (ΔKE_1), and almost four times as much as the separate units driven by an 1800-RPM motor (ΔKE_2), for the motor-flywheel part of the system. The separate generator contribution to ride-through will still be more than 5 times as much as the single-unit generator because of the lower acceptable speed limit.

This is very impressive theory. How it actually describes the real-world operation of MG sets depends upon many variables that differ from one manufacturer to another, such as windage losses and bearing friction. However, these variations are small compared to the effect of the flywheel weight, diameter, and shape. As a rule-of-thumb for most computer-room MG sets, the rotating mass of the generator contributes about 15% of the total ride-through, the motor about 15%, and the flywheel 70%.

It is, of course, possible to operate both the motor and generator at a nominal 3600 RPM. However, since the generator contributes only about 15% of the total kinetic energy for ride-through, the higher-priced 3600-RPM generator is not cost-effective when compared with a 3600-RPM motor and flywheel driving an 1800-RPM generator. Using these estimates, we calculate:

$$\begin{aligned}\Delta KE &= k[0.85\ (\text{motor and flywheel}) \\ &\quad + 0.15\ (\text{generator})] \\ &= k[(0.85 \times 401{,}708) \\ &\quad + 0.15(108{,}000)] \\ &= 357{,}652k\end{aligned}$$

With a 3600-RPM motor and flywheel driving an 1800-RPM generator, we have 357,652 k/21,456 k, or 16.7 times as much total input power loss ride-through energy as a similar single-shaft machine.

Conclusions

The adjustable-speed controller between the motor and generator solves many of the problems of using a standard induction motor to drive a standard generator and provides numerous benefits. First, it provides excellent brownout and sag protection, maintaining frequency within computer requirements over a wide range of input voltage. Second, it separates the motor shaft and stator housing from the generator shaft and stator housing, eliminating any capacitive and inductive coupling and providing total isolation of input power transients from the generator output. Third, it allows increased motor and flywheel rotating speed, which produces significantly greater kinetic energy for total power blackout and low-voltage sag and brownout ride-through capabilities. Fourth, it permits the use of induction motors rather than synchronous motors. This provides greater simplification of motor control, eliminating field-application problems. It eliminates the difficulty of accelerating the synchronous motor back to pull-in speed before applying the field after a momentary power outage and increases reliability through simplicity. Fifth, it permits the use of an oversize motor to further increase these benefits. Such an application can provide economical, quality power for computers.

Rotary UPS equipment

Advances in the design and capabilities of nonstatic uninterruptible power supply systems have resulted in greater maintainability and higher efficiencies.

QUALITY-CONDITIONED, ultra-reliable power supply is in very great demand today for all kinds of computer operations, for critical communications, and for an ever-increasing variety of other commercial, industrial and institutional applications. The broad, intense demand for such clean and/or fail-proof power supplies has spurred rapid design development of the many types of equipment involved with power conditioning and maximum continuity of power feed—electrical noise suppressors, voltage regulators, isolation transformers, motor-alternator sets, magnetic synthesizers, and UPS (uninterruptible power supply) systems.

Although the static types of such equipment—regulators, transformers, static UPS and synthesizers—have found increasingly wider and more-popular application because they have no moving parts, the rotary type of conditioner/UPS—the motor-generator, or motor-alternator set—was the earliest type and has been achieving greater visibility due to rapid advances in the design and capabilities of such equipment. The following review of the generic characteristics of rotary power conditioners looks at the historical evolution of one line of such units up to its most recent design as a combination conditioner/UPS.

One of the first rotary types of UPS consisted of a nonphase-controlled diode rectifier, DC motor, flywheel and AC synchronous generator (Fig. 1). The rectifier converted AC power to DC to drive the DC motor. On loss of input AC to the rectifier, a battery was connected to the DC motor through an electromechanical contactor. Output voltage was maintained by a solid-state voltage regulator and static exciter. Output frequency was maintained by changing the excitation current in the DC motor. If the excitation of a DC motor is increased, the motor delivers less power; and if the excitation is decreased, it delivers more power. The closed-loop system for the frequency regulation (changing the excitation on the DC motor) had a response time that could vary from 100 milliseconds to 1½ seconds. The flywheel was used to maintain frequency under changing loads until the excitation could be adjusted.

This UPS system was highly reliable and had an efficiency in the 84% range; however, monitoring and adjusting the brushes to compensate for varying loads was required.

The early rotary UPS was upgraded by replacing the DC motor and large

Fig. 1. Early rotary UPS using DC motor

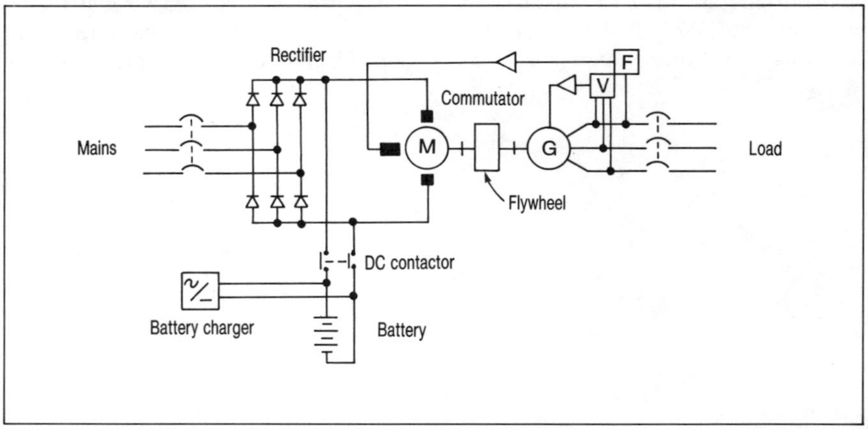

Fig. 2. Use of AC motor and SCR inverter

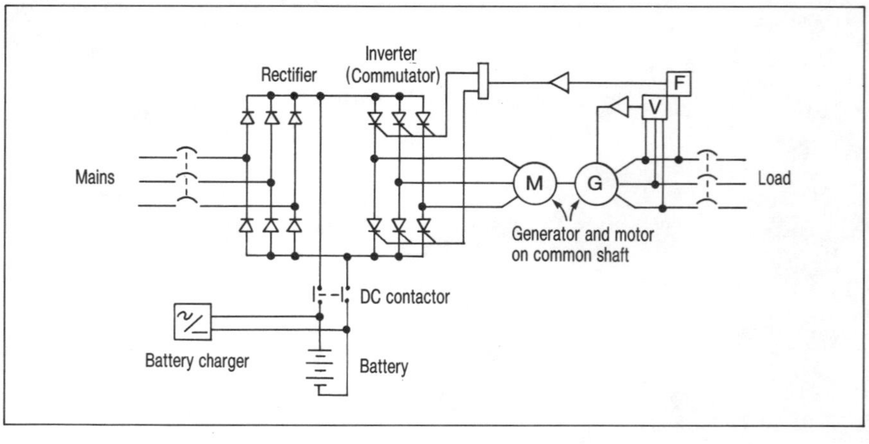

By J. F. McPARTLAND

flywheel with an AC synchronous motor and SCR phase-controlled commutator or inverter (Fig. 2). The frequency was then controlled by changing the firing angle of six SCRs and could be adjusted within three milliseconds, thereby eliminating the need for the large, separate flywheel required with the slow-responding DC motor. Also, because an AC motor does not require brushes or sliprings, brush maintenance was eliminated. The resulting UPS had an efficiency increase of about 3%, raising it to the 87% range. The increased efficiency was due to the removal of the flywheel and elimination of four of the six original bearings in the first UPS. The rotating section, in this case, is composed of both a synchronous generator and a synchronous motor combined on one common shaft with a bearing at each end.

Next, it was decided to increase the efficiency of the UPS above the 87% range by use of a new type of rotating section. This particular design combines the individual generator and motor stators into one common stator (Fig. 3). Efficiency of the overall system is in the 91% range. With the addition of a bypass circuit around the rectifier-commutator sections so that, under normal operating conditions, most of the power to the motor is shunted around the inverter by the more-efficient bypass switch (Fig. 4), the efficiency of the system is increased to 93%.

The common-stator rotating power conditioner substantially reduces the mechanical energy losses associated with traditional motor-generator designs using a separate motor and generator connected by a common shaft. The result is a rotary-based UPS system that offers significantly higher efficiency than other rotary-based UPS systems relying upon conventional motor-generator designs, and equal or better efficiency than static UPS systems.

Developments that led to the single-stator design (Figs. 3 and 4) were motivated by maintainability and efficiency and were not made to improve reliability. The reliability of the original system was in the 300,000-hour range. The calculated "mean time to fail" (MTTF) or "mean time before failure" (MTBF) of the single-stator design exceeds 500,000 hours. When used as a power stabilizer or UPS, electrical power is fed via a static switch and choke to the rotary converter and then out to the load. The converter consists essentially of a stator in which alternate slots are wound with input and output windings. When fed with a 3-phase supply, a rotating magnetic field is created, causing the DC-excited rotor to spin at synchronous speed. By controlling the electrical characteristics of the rotor, control of the output of the secondary stator windings is accomplished.

Relatively large mains-voltage variations on the input windings are not passed onto the secondary winding, and in practice even complete loss of the AC mains does not cause loss of AC output. High-voltage spikes and frequency variations are similarly attenuated. The rotary converter also inhibits the transmission of harmonics. A higher harmonic applied to the input winding will induce a rotating field, but of differing speed to the rotor. A damper winding acts as a narrow band-pass filter for the fundamental mains frequency.

Because this particular rotary converter has two completely separate sets of windings on its stator, a typical 450-kVA machine has approximately a 900-kVA frame size. The larger frame size raises efficiency. In addition, the

Fig. 3. Increased efficiency with common stator

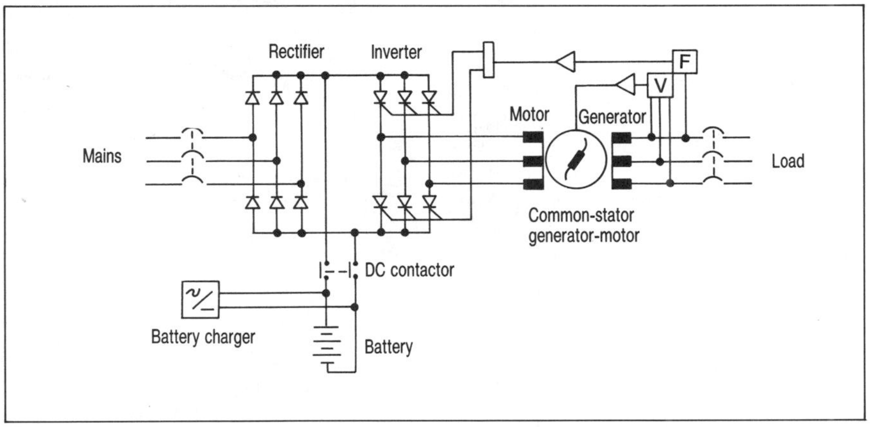

Fig. 4. Addition of bypass switch

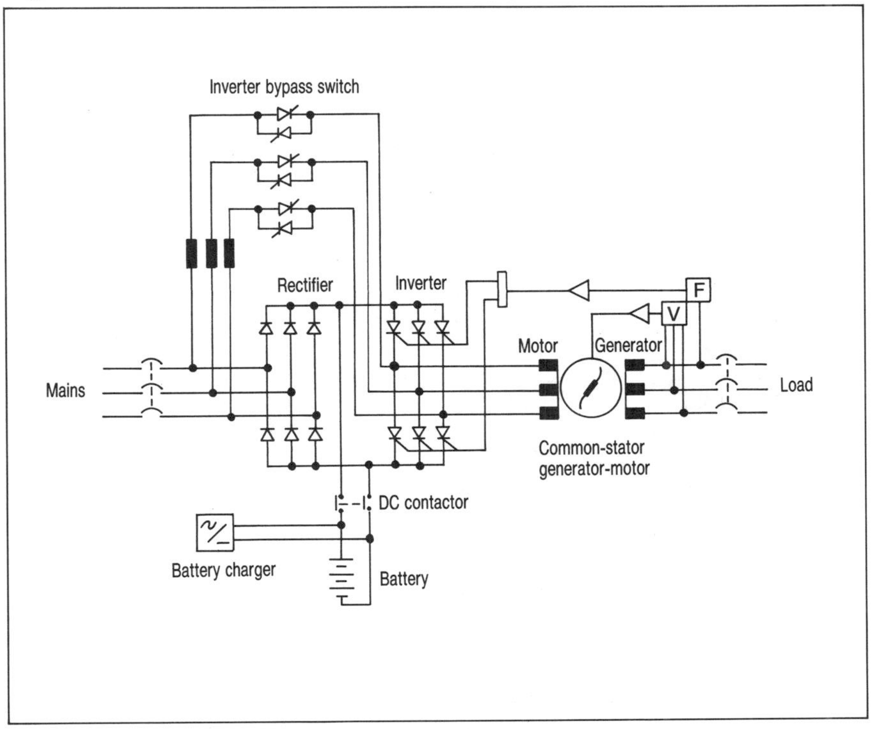

large frame size creates a relatively low-impedance power source capable of clearing subcircuit fuses under fault conditions much more readily than other systems. Because the output of the unit can supply up to seven times full-load current under fault conditions, a short circuit within a single computer device can be cleared from the system without the need to take the UPS system off line, and without affecting output to the remaining computer load.

Despite the increase in frame size, the set is smaller and lighter than comparable systems, due largely to a reduction in the number of bearings used and obtaining adequate flywheel action from the motor's ventilator rather than from a separate flywheel. In operation it is quiet and, when enclosed in a sound-proofed controlgear cabinet, it provides almost silent running.

THIS 360-KVA UPS system was installed recently in Albany, NY. The total system includes the common stator rotating section (with a separate battery and battery charger), a bypass transfer switch, two 60/415-Hz frequency converters, a paralleling system, a main computer-room power-distribution panel, and two 60-Hz power-distribution centers.

In the Fig. 5 diagram of the overall power-supply system, the utility input is split into two parallel current paths through the system: the main path via the static switch and choke, and the other path via a simple diode rectifier, standby battery, and static inverter. The static inverter section is an inherently simple design: commutation is achieved via the windings. Only six thyristors are used, which afford high reliability in comparison with a forced-commutation static system using up to 48 such devices.

Under normal operating conditions, the two circuit paths through the unit are used together, but only 95% of power passes through the static switch. The second path is used for 5% of the normal-condition power, and this is deliberately done to maintain the rectifier/inverter in an active state. That approach increases system efficiency and allows the more complex thyristor inverter system to virtually idle while lack of system stress enhances reliability.

For normal operation, as described, the conditioner is powered simultaneously by both the transient-protected inverter bypass-switch path and the full-wave rectifier/inverter path. However, because each path is fully capable of supplying full power to the rotating power conditioner, should the inverter, rectifier, or inverter bypass switch independently fail, the rotary portion of the system will still provide protected power to the computer load via the alternate path. The system therefore has built-in redundancy without the need for redundant UPS modules, as is often necessary with the other UPS systems.

In the event of utility power failure, the off-line battery is automatically connected to the inverter, assuring complete continuity of power to the computer load. However, unlike static UPS systems, the battery is not used to "filter" the utility power under normal

Fig. 5. Complete power-supply system

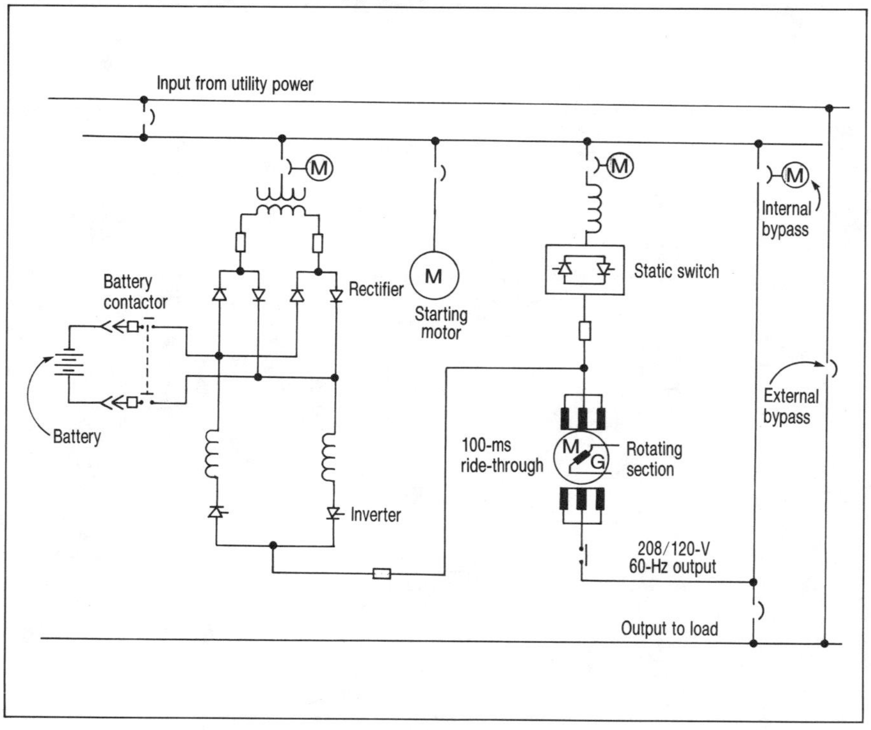

operating conditions and remains off-line until it is needed. This results in longer battery life and lower maintenance for the battery portion of the system.

If desired, the system shown in Fig. 5 may be installed as only a power conditioner *without* a battery (at significantly lower cost) to provide protection from surges, sags, transients, brownouts, and even complete loss of power in excess of 100 milliseconds. Later, if reliability of utility power should deteriorate or the computer load become more critical to the extent that the cost of a battery is justified, a battery system can be added without modification of the original system. And if future systems growth requires additional output capability, multiple modules of the basic system can be paralleled to provide whatever level of power is needed, without modification of the original unit.

The accompanying photograph shows a typical rotary UPS assembly of the type described for Fig. 5.

Operating modes

Refer to Fig. 5.

1. An auxiliary start system is utilized to start the rotating section. The start system is an integral part of the rotating section and is not coupled to the main motor by belts or gears. The starting system allows for reduced-current starting by limiting the inrush current normally associated with starting a motor. It also enables the inverter control electronics to be tested before applying full power to the system.

2. Once the rotating section is running at 1800 RPM, the rectifier is connected to the main power source. The rectifier consists of an input transformer with two secondary windings displaced ±15° from the primary winding, thus allowing more even distribution of current from the incoming power. The power from the input transformer secondary is converted to DC power by two full-wave diode bridge circuits. This eliminates controls required by a normal phase-controlled rectifier utilizing SCRs (silicon-controlled rectifiers), as found in normal static UPS systems.

The system operates in this mode until the total power output is more than 20% of its rated load or if the frequency of the input power is not within a .5 Hz tolerance.

3. When power is within specifications, the static switch is turned on and operates in parallel with the rectifier inverter path. When the static switch is operating, the power through the inverter is reduced to 5% of the rated load. This permits monitoring of the rectifier and inverter operation, reduces heat buildup in the inverter, and avoids subjecting the inverter to thermal shock when it is required to supply full load.

The static switch allows for fast disconnect of the input circuit when input power fails. It also allows for operation of the rotating section as a stabilizer if the inverter fails. During inverter failure, the system has the same operating characteristics as a 60/60-Hz stabilizer with complete brownout protection, transient protection, and 100-ms ride-through.

4. Upon loss of input power, the static switch isolates the rotating section from the input power. The stored energy in the rotating section supplies power until the battery is connected to the inverter. The battery supplies power to the inverter to power the rotating section.

When input power is returned within tolerance, the rectifier supplies power to the inverter and the battery is disconnected. The phase relationship across the static switch is monitored, and at the appropriate time, the static switch is gated ON to allow the bulk of the power to be supplied from the input power source to the rotating section.

5. If the inverter/rectifier system should fail, the load is maintained by the static switch and rotating section. In this mode of operation, the load is not subject to raw utility power as with a static UPS system but is still supplied isolated, regulated power from a power system with the same operating characteristics as a 60/60-Hz power stabilizer. In this operating configuration, the load is protected from transients, brownouts, and total loss of power for at least 100 ms.

6. When the rotating section has to be serviced, an automatic bypass around the rotating section is utilized. Since this is an electromechanical device and not a static switch, complete isolation between input power and output power is maintained when it is in the open position. An external bypass is also provided.

Transistorized UPS for "Sure" Power

Here's a look at the design, layout, and installation considerations involved in using UPS equipment incorporating power-transistorized electronics rather than thyristor types, along with the economic factors involved.

Fig. 1. Cost comparison: thyristor vs transistor UPS*

Equipment and tasks	Thyristor/wet-cell system (dollars)		Transistor/gel-cell system (dollars)	
UPS module	72,803		72,803	
Battery	18,490		23,112	
Equipment subtotal		**91,293**		**95,915**
Input wiring	5,721		4,229	
Module installation	4,551		3,109	
Battery installation	12,457		4,430	
Output wiring	14,074		2,702	
Installation subtotal		**36,803**		**14,470**
Installed cost		**128,096**		**110,385**

* Cost comparison figures include all material and labor plus sales tax, overhead, and profit.

When a Manhattan, New York City financial company thought they had been completely frustrated in their attempts to secure UPS protection for their computer, a state-of-the-art UPS technology offered a solution. An uninterruptible power supply utilizing transistors for power conversion and sealed gel-cell batteries for standby power could be installed on the computer room floor—the only place where space was available. Conventional UPS systems using power thyristors and lead-calcium wet-cell batteries had been considered initially, but they were too bulky and noisy for the computer room. What's more, the transistorized approach pared the cost of the installed UPS from $128,096 to $110,841 (Fig. 1). Let's go through this pre-installation study step-by-step and see why site conditions at many existing facilities could make the selection of a transistorized UPS located in the data-processing room a good idea.

Existing conditions

The firm involved in this installation specializes in financial analysis and management and leases space on the 14th and 15th floor of a midtown high-rise for offices and data processing. Because a computer failure could cost tens of thousands of dollars in lost commodities transactions, they decided to protect their critical computer mainframe and peripheral equipment with a 150-kVA UPS and 15-min battery backup. Then the problems began. The electric closet (8 × 8 ft) was not large enough to house even the smallest conventional UPS. All the offices were occupied, so there was no opportunity to block off office space, soundproof and ventilate it, and convert it to a UPS room. The computer room itself was large, having been designed so that additional pieces of data-processing equipment could be added. But the space available (15 × 8 ft) was not large enough to fit a conventional UPS and wet-cell battery system (Fig. 2).

In addition, the thyristor UPS was not suitable for the computer-room environment because it was noisy at 75 dB, and lead-acid batteries require special venting for the hydrogen gas they emit. It seemed that the computer power-protection equipment would have to be located in the basement, some 15 floors below. What was worse, this area would have to be leased from a reluctant building owner (Fig. 3).

Transistorized UPS

It was only after the UPS system had been priced as a basement installation that this tenant became aware of the inherent design features of the transistor technology that would solve his problems by allowing installation on the computer-room floor. The transistorized approach featured compactness, light weight, no special venting requirement (because sealed gel-cell batteries do not emit gas), and a low noise level of 60 dB, which is the ambient noise level in an average office area. Because each 3-dB increase represents a doubling of power and an audible increase in noise detected by the human ear, the 75-dB noise level of a thyristor UPS would sound five times louder—the noise level of a machine room.)

Compactness can be a critical consid-

By J. F. MCPARTLAND, III, and CHARLES M. HOLT

J. F. McPARTLAND III is the Vice-President of Electrical Information Services, Piermont, NY; **"MIKE" HOLT** is the President of Contractors Computer Systems, Inc., Pompano, FL.

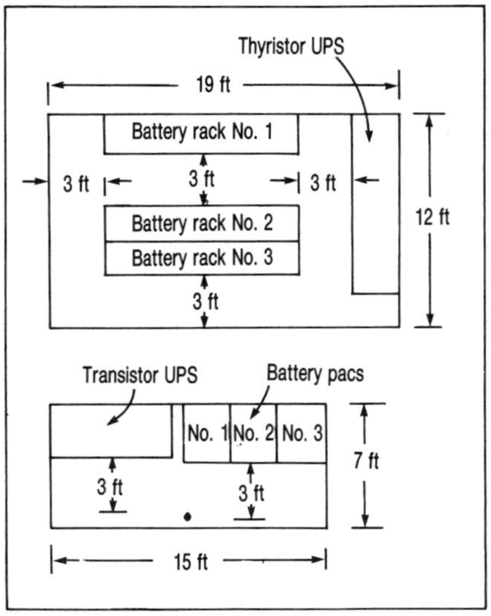

FIG. 2. After the required 3-ft clearances around the UPS and batteries are provided, it is evident that the transistor UPS is far more compact, requiring 105 sq ft of floor space compared to 228 sq ft for the thyristor and wet-cell system.

Fig. 3. Alternative equipment and locations

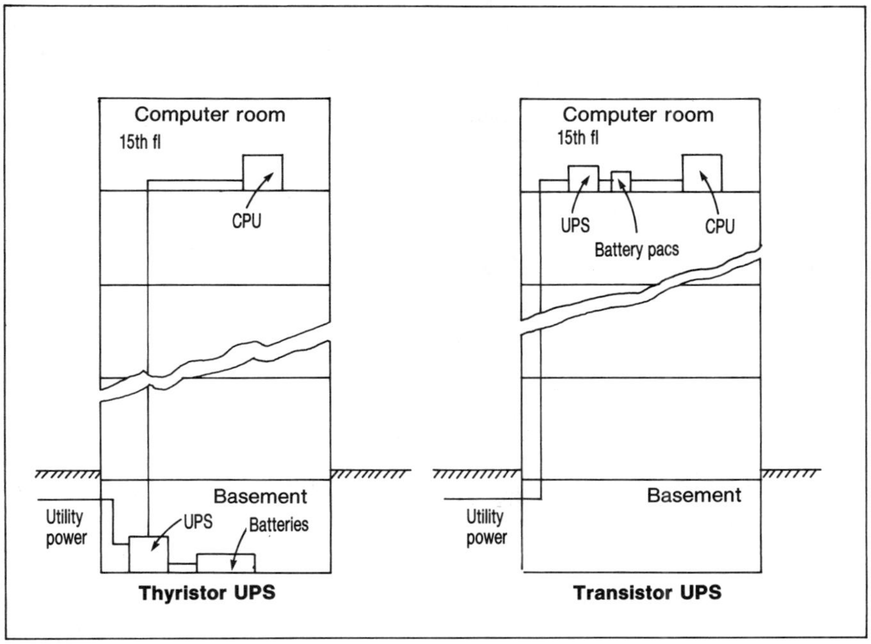

eration with respect to available space. For equal kVA rating, a transistor UPS may take up as little as 65% of the floorspace required for a thyristor UPS. Another benefit of the transistorized UPS in this application was that it qualified as a piece of computer equipment and could be depreciated as such, resulting in additional savings (Fig. 4).

One of the interesting aspects of the study was the use of a computer-assisted estimating system to figure the cost-effectiveness of the two UPS technologies quickly and accurately. While the thyristor UPS equipment costs about the same, it would be far more expensive to install. A review of the work required for each job will show how the installation savings were realized (Fig. 5).

FIG. 4. Except for its small size (79 in. wide, 32 in. deep, 86 in. high for 150-kVA size), the transistorized UPS module looks very much like the conventional thyristor UPS.

Installation requirements

The transistorized UPS and gel-cell batteries require only 105 sq ft of floor space, far less than the 228-sq-ft requirement of the other approach. The transistorized power-conversion module weighs 4400 lbs (compared to 7700 lbs for the thyristor UPS), and each of the three factory-wired battery packs weighs 1750 lbs (compared to 14,000 lbs for the wet-cell batteries, whose 180 individual cells must be interconnected in the field). For these reasons, as well as low noise and heat generation, the transistor system can be located in the data-processing room adjacent to the existing 600-A computer distribution panel. The feeder to this panel can be routed through a junction box and spliced to feed both the UPS input and bypass terminals (Fig. 6). The UPS is 90% efficient, so it will draw 456 A to feed the maximum continuous computer equipment load of 410 A. Since the input voltage is 208 V, 3 phase, 3 wire with a continuous input load of 456 A, two 300MCM THW copper conductors per phase will satisfy NEC Sec. 220-10(b), which requires conductors to be rated at 125% of continuous load.

Per Table 3A at the back of the NEC, two 2½-in. rigid conduits may serve as the raceway for the input power conductors. However, it was decided to allow some margin, and 3½-in. rigid conduit was used in the estimate. The bypass power requirement was 3 phase, 4 wire, with a fully rated 4/0 neutral. The bypass feed does not have to account for the power losses in the UPS module, so its continuous loading is 410 A, the same as the data-processing load. For the sake of simplicity and uniformity, this feed was figured using one 300MCM conductor per phase plus the 4/0 neutral in each of two 3½-in. conduits.

The output power was identical to the bypass power. It was handled the same way and routed directly back to the computer panel. Since the junction box is located right next to the panel and the UPS module, all input, bypass, output, and DC feeds are less than 10 ft long. The input and bypass conductors are protected by an upstream 600-A breaker, and the 10-ft tap exception of NEC Sec. 240-21 allows the splice and feed at the junction box. The output and DC conductors are protected by overcurrent devices in the UPS module. The maximum DC current is 500 A, so two 250MCM THW copper conductors

Fig. 5. Typical bid summary for lead-battery UPS system

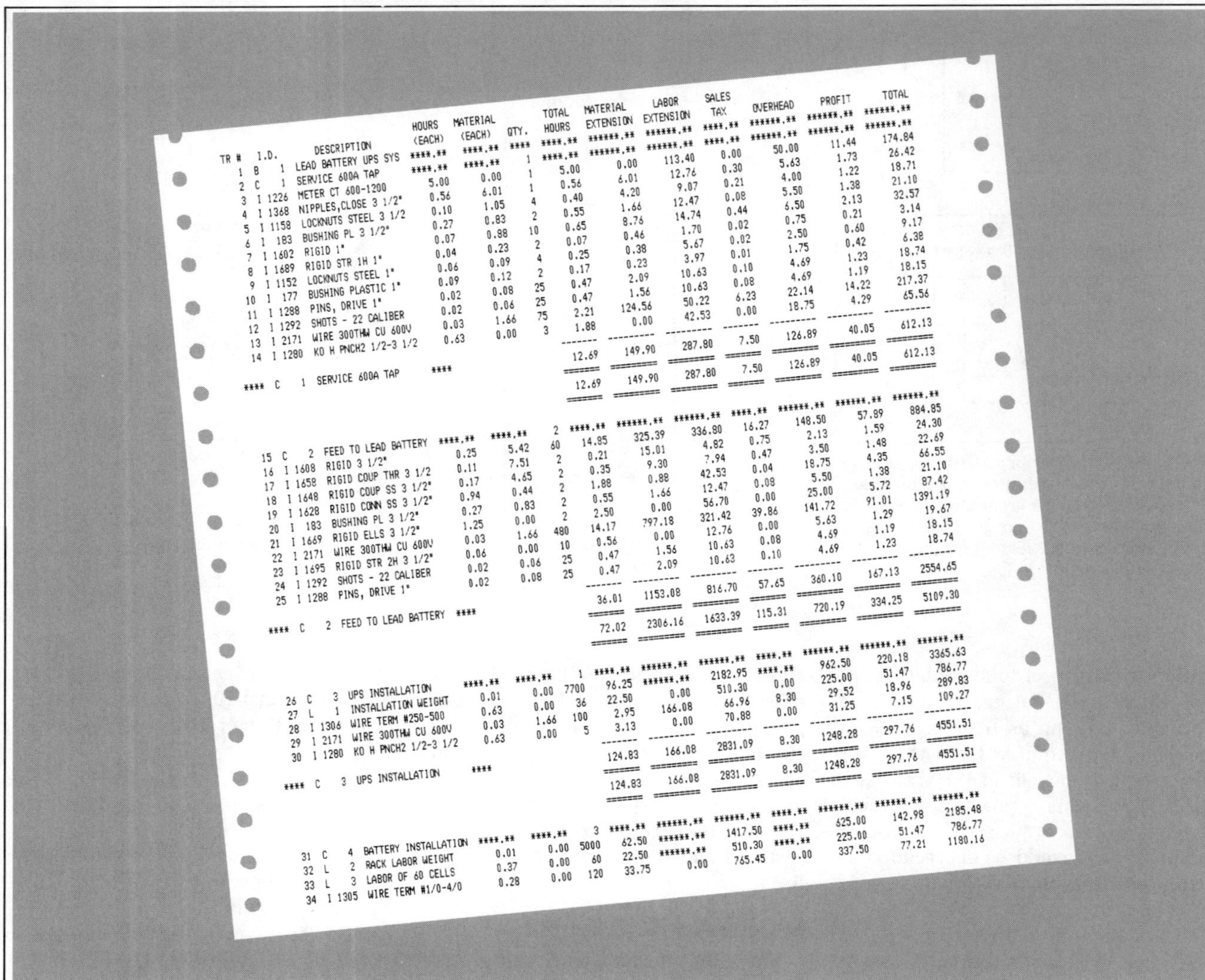

TR #	I.D.		DESCRIPTION	HOURS (EACH)	MATERIAL (EACH)	QTY.	TOTAL HOURS	MATERIAL EXTENSION	LABOR EXTENSION	SALES TAX	OVERHEAD	PROFIT	TOTAL
1	B	1	LEAD BATTERY UPS SYS	****.**	****.**	****	****.**	******.**	******.**	****.**	******.**	******.**	******.**
2	C	1	SERVICE 600A TAP	****.**	****.**	1	****.**	******.**	******.**	****.**	******.**	******.**	******.**
3	I	1226	METER CT 600-1200	5.00	0.00	1	5.00	0.00	113.40	0.00	50.00	11.44	174.84
4	I	1368	NIPPLES,CLOSE 3 1/2"	0.56	6.01	1	0.56	6.01	12.76	0.30	5.63	1.73	26.42
5	I	1158	LOCKNUTS STEEL 3 1/2	0.10	1.05	4	0.40	4.20	9.07	0.21	4.00	1.22	18.71
6	I	183	BUSHING PL 3 1/2"	0.27	0.83	2	0.55	1.66	12.47	0.08	5.50	1.38	21.10
7	I	1602	RIGID 1"	0.07	0.88	10	0.65	8.76	14.74	0.44	6.50	2.13	32.57
8	I	1689	RIGID STR 1H 1"	0.04	0.23	2	0.07	0.46	1.70	0.02	0.75	0.21	3.14
9	I	1152	LOCKNUTS STEEL 1"	0.06	0.09	4	0.25	0.38	5.67	0.02	2.50	0.60	9.17
10	I	177	BUSHING PLASTIC 1"	0.09	0.12	2	0.17	0.23	3.97	0.01	1.75	0.42	6.38
11	I	1288	PINS, DRIVE 1"	0.02	0.08	25	0.47	2.09	10.63	0.10	4.69	1.23	18.74
12	I	1292	SHOTS - 22 CALIBER	0.02	0.06	25	0.47	1.56	10.63	0.08	4.69	1.19	18.15
13	I	2171	WIRE 300THW CU 600V	0.03	1.66	75	2.21	124.56	50.22	6.23	22.14	14.22	217.37
14	I	1280	KO H PNCH2 1/2-3 1/2	0.63	0.00	3	1.88	0.00	42.53	0.00	18.75	4.29	65.56
							12.69	149.90	287.80	7.50	126.89	40.05	612.13
****	C	1	SERVICE 600A TAP ****				12.69	149.90	287.80	7.50	126.89	40.05	612.13
15	C	2	FEED TO LEAD BATTERY	****.**	****.**	2	****.**	******.**	******.**	****.**	******.**	******.**	******.**
16	I	1608	RIGID 3 1/2"	0.25	5.42	60	14.85	325.39	336.80	16.27	148.50	57.89	884.85
17	I	1658	RIGID COUP THR 3 1/2	0.11	7.51	2	0.21	15.01	4.82	0.75	2.13	1.59	24.30
18	I	1648	RIGID COUP SS 3 1/2"	0.17	4.65	2	0.35	9.30	7.94	0.47	3.50	1.48	22.69
19	I	1628	RIGID CONN SS 3 1/2"	0.94	0.44	2	1.88	0.88	42.53	0.04	18.75	4.35	66.55
20	I	183	BUSHING PL 3 1/2"	0.27	0.83	2	0.55	1.66	12.47	0.08	5.50	1.38	21.10
21	I	1669	RIGID ELLS 3 1/2"	1.25	0.00	2	2.50	0.00	56.70	0.00	25.00	5.72	87.42
22	I	2171	WIRE 300THW CU 600V	0.03	1.66	480	14.17	797.18	321.42	39.86	141.72	91.01	1391.19
23	I	1695	RIGID STR 2H 3 1/2"	0.06	0.00	10	0.56	0.00	12.76	0.00	5.63	1.29	19.67
24	I	1292	SHOTS - 22 CALIBER	0.02	0.06	25	0.47	1.56	10.63	0.08	4.69	1.19	18.15
25	I	1288	PINS, DRIVE 1"	0.02	0.08	25	0.47	2.09	10.63	0.10	4.69	1.23	18.74
							36.01	1153.08	816.70	57.65	360.10	167.13	2554.65
****	C	2	FEED TO LEAD BATTERY ****				72.02	2306.16	1633.39	115.31	720.19	334.25	5109.30
26	C	3	UPS INSTALLATION	****.**	****.**	1	****.**	******.**	******.**	****.**	******.**	******.**	******.**
27	L	1	INSTALLATION WEIGHT	0.01	0.00	7700	96.25	******.**	2182.95	****.**	962.50	220.18	3365.63
28	I	1306	WIRE TERM #250-500	0.63	0.00	36	22.50	0.00	510.30	0.00	225.00	51.47	786.77
29	I	2171	WIRE 300THW CU 600V	0.03	1.66	100	2.95	166.08	66.96	8.30	29.52	18.96	289.83
30	I	1280	KO H PNCH2 1/2-3 1/2	0.63	0.00	5	3.13	0.00	70.88	0.00	31.25	7.15	109.27
							124.83	166.08	2831.09	8.30	1248.28	297.76	4551.51
****	C	3	UPS INSTALLATION ****				124.83	166.08	2831.09	8.30	1248.28	297.76	4551.51
31	C	4	BATTERY INSTALLATION	****.**	****.**	3	****.**	******.**	******.**	****.**	******.**	******.**	******.**
32	L	2	RACK LABOR WEIGHT	0.01	0.00	5000	62.50	******.**	1417.50	****.**	625.00	142.98	2185.48
33	L	3	LABOR OF 60 CELLS	0.37	0.00	60	22.50	******.**	510.30	****.**	225.00	51.47	786.77
34	I	1305	WIRE TERM #1/0-4/0	0.28	0.00	120	33.75	0.00	765.45	0.00	337.50	77.21	1180.16

are run in each of two 2-in. rigid conduits, making up a single-phase (2-wire) circuit with two 250MCM per circuit leg. The UPS output is a separately derived system and must be grounded per Secs. 250-26 and 250-94 with a 1/0 copper conductor.

The alternate installation utilizing thyristor UPS and lead-calcium battery would require the installation of a CT and metering device in the basement adjacent to a spare 600-A breaker in the service switchboard. The input, output, bypass, DC and grounding would be sized the same as above. However, the runs would be much longer—60 ft versus 10 ft on the input and 180 ft versus 10 ft on the output. The greater distance substantially increases the total weight, and the resulting additional hours of labor are reflected in the module installation costs of Fig. 1. The battery installation cost is increased by both the greater weight and the necessity to assemble racks and to torque the intercell connectors on some 180 cells.

Fig. 6. Single-module UPS configuration

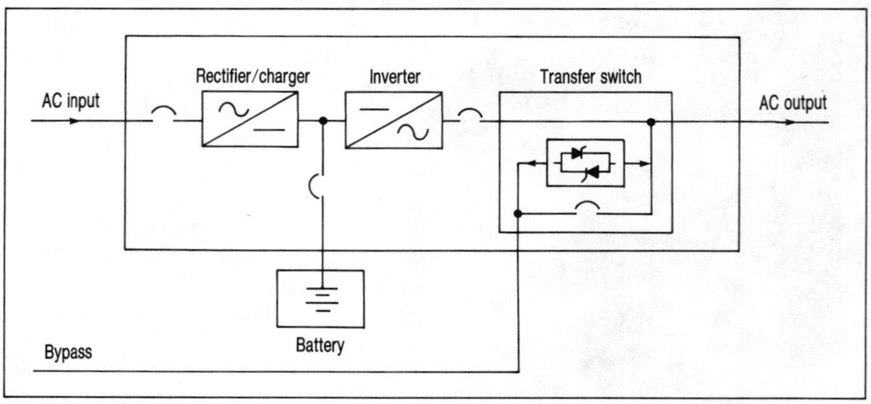

Continuous power for communications satellites

Dependable, uninterruptible sources of power were prime considerations in the design of Westar and Comsat, two systems of commercial communications satellites currently circling the earth. Sophisticated ground-based UPS systems and backup generating sets play important roles in both installations.

COMSAT

By FRANK R. BOYD, Professional Electrical Engineer
Santa Clara, Calif.

THE COMSAT system, owned and operated by Comsat Corp., Washington, D.C., includes two earth stations, one in California and one in Connecticut. Each station houses transmitters and receivers that "talk" and "listen" to miniature radio terminals orbiting the earth 23,000 miles out in space. The stations provide communications links to telephone companies on both coasts of the United States handling commercial business, telephone calls, computer data transmissions, and television signals 24 hours a day.

The three satellites in space are designed to hover above a definite spot on the earth's surface; however, the natural "wobble" of the earth causes the satellites to appear to move from the intended spot. The communicating antennas at the earth stations must thus be capable of following this wobble so that they point exactly at the satellites.

THREE REFLECTORS are in operation at the COMSAT earth station in Santa Paula, Calif. At left and right are 42-ft-dia steerable units; 34-ft limited-motion (hour angle-declination) antenna is in center.

Each of the two stations has three antennas. Two are 42-ft-dia, steerable reflectors that can be maneuvered to look at the entire sky; the other is a 34-ft-dia limited-motion type. Each parabolic-dish antenna "looks" at its own satellite.

All of the antennas, controlled by servo-motors, are capable of collecting very-low-power signals with maximum efficiency. If the servos (controlled by signals from the satellites) do not operate, or if the radio receivers lose power, the antennas will not point directly at the satellites, and the signals received will fall off drastically. If the signal strength becomes too weak, it cannot be separated from noise, and the station output is translated directly into a loss of revenue.

This makes continuous power absolutely vital. Consequently, the stations are served by a sophisticated power system consisting of an uninterruptible power supply and backup generator. The radio and computer equipment in

CONTROL CENTER (foreground) serves microwave horn heaters on antennas at Southbury, Conn., to keep rain and ice off their openings. The three cubicles in background house (left to right) rectifiers and inverters, circuit breakers and control electronics, and a solid-state static switch and manual bypass switch.

Information exchange

The authors of these articles have become specialists in power supply for critical communications equipment. As such, they are vitally interested in the reactions of readers to the details set forth. They believe that a continuing interchange of information among all electrical people involved in UPS work can do much to advance the best interests of all involved in such work—designers, installers, maintenance personnel, and manufacturers. The authors would welcome any questions, evaluations or comments about the experiences of others.

Please send correspondence to:
Stan Plato, Mgr. of Maintenance
Western Union
One Lake Street
Upper Saddle River, N.J. 07648

or

Frank Boyd
2183 Augusta Place
Santa Clara, Calif. 95051

each station receives 208/120-volt, 3-phase power from a 100-kva UPS at all times. The UPS is a modern, solid-state electronics unit that produces 208/120-volt, 3-phase, 4-wire, 60-Hertz power. In case of low line voltage, overvoltage, or low frequency, the UPS can maintain power to the radio equipment via 240-volt batteries. Normally, the UPS takes in 480/277-volt, 3-phase, 60-Hertz power, rectifies it to 240 volts dc, and then inverts the dc to 208/120 volts ac to power the technical equipment. The batteries, connected in parallel with the rectifier, are kept charged. If the ac input is interrupted for any reason, the batteries furnish the dc the inverter needs to power the load. The batteries are always connected to the inverter input and can supply the full output power of the UPS for 10 minutes.

The UPS is composed of passive elements; there are no moving parts. If the ac output becomes deficient in frequency or voltage due to internal circuit troubles, an electronic static switch can be operated to connect the load directly to the incoming ac power, bypassing the failing UPS. Thus, the UPS can also protect against its own failure to a limited degree.

The station is also served by a diesel-engine-driven generator with a capacity of 350 kva—sufficiently large to drive the UPS (and the radios), the station air conditioning (two 50-ton machines), station lights, and antenna drive motors. The accompanying abbreviated one-line diagram shows the power system.

The emergency-power generator has a built-in 3-phase voltage-sensing relay connected directly to the station incoming power bus. If the voltage drops or rises beyond established limits for more than a minute or so, the engine-generator will start up and automatically assume the station load. An interlock system prevents the generator from coming on-line while normal ac power is being received. The engine can be exercised and switched manually to assume the station load as desired.

The station in Connecticut presented a particular problem, since it was subject to ice formation on the "dishes" of the antenna reflectors. Ice would affect reflector shape and radio reception, so an ultrasonic detector was placed on the bottom lip of one of the 42-ft antennas where icicles would first form. An outside-air-temperature meter was also installed at the control building. If the air is cold enough and moisture is present, the station operators can prevent ice formation by turning on the antenna reflector electric heaters. Resistance heater strips were cemented to the back of each aluminum panel forming the surface of the reflector. Connected to contactors

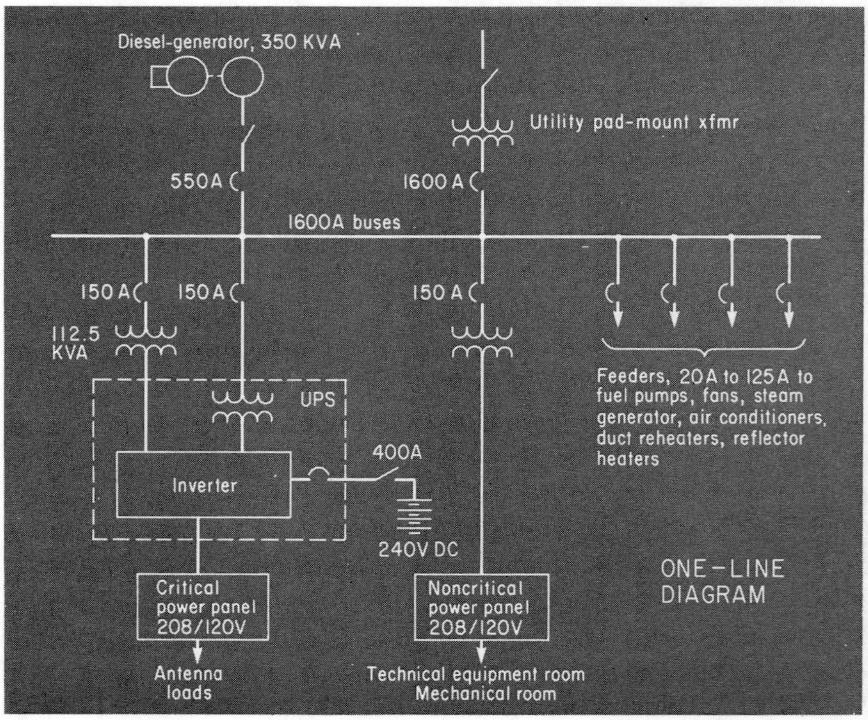

in a motor control center at the base of the antenna, the heaters receive 277-volt, single-phase power. The contactors cycle on and off under control of thermostats embedded in the reflector panels.

Each of the two earth stations is small but neatly laid out and houses the radio equipment (20% of the space) plus the UPS, switchgear, generator, air blowers, electronic maintenance workspace, and offices for station operators. Maintained lighting levels are 200 fc in the office area for the manager and his secretary, 100 fc in the radio rooms, and 50 fc in the corridor. Exit lights above the doors have built-in, miniature UPS systems. If the building power fails, battery-powered inverters light the exit paths.

Each building is protected from fire by combustion product detectors in the radio equipment rooms and air temperature detectors in other rooms, plus manual pull-alarm stations and alarm bells.

The facilities were designed by John Knorpp, David Sharpe, Frank Boyd, and Alex Bien of Aeronutronic-Ford, Palo Alto, Calif., who also oversaw construction and tested the completed installation. Foothill Electric of Santa Paula, Calif., was responsible for the electrical system on the West Coast; Busher Electrical Services, Danbury, handled the Connecticut station.

CONTROL of antennas in Santa Paula is handled from this console.

BATTERY BANK consisting of 120 cells provides 240 volts to UPS when ac power fails.

WESTAR

By STAN PLATO, Manager, and LOU FORTE, Senior Engineer
Maintenance Engineering, Western Union, Upper Saddle River, N.J.

THE WESTAR system, owned and operated by Western Union, consists of two earth-orbiting satellites and five earth stations located close to major cities—New York, Atlanta, Chicago, Dallas, and Los Angeles. The satellite communications system is tied into a coast-to-coast microwave ratio network, forming an overall satellite-landline system.

The New York area is served by a facility in Glenwood, N.J. Its power-supply arrangement was carefully designed at the outset and has been altered and improved periodically to achieve maximum reliability with a minimum of transients. This completely automatic system consists of two diesel generators, uninterruptible power supply (UPS) equipment, and a utility service.

Three types of loads are served.

1. Critical communications loads are connected to the UPS output, which is backed by a large bank of batteries and either of two diesel generators, in addition to the utility service. As long as one of the four possible inputs to the UPS is operable, the critical loads are assured of no-break power. Or, if the UPS fails, the loads can be picked up directly by the utility service or by either of the two generators.

2. Essential loads are important to the continuing operations of the facility but can stand a power lapse of short duration—emergency lighting, ventilation, pumps, and lighting and air conditioning in certain important areas. These loads are fed by any one of the three ac sources—the utility service or one of the generators.

3. Nonessential loads are all loads not classified as critical or essential. They include lighting and air conditioning in office areas and other locations not directly involved with the communications operations, as well as electric de-icing equipment on the giant antennas that is used during ice and snow storms. The de-icing load, normally fed directly by the utility service when needed, can also be handled by the two generators operating in parallel.

The accompanying one-line diagram shows the overall electrical system; the

MAIN SWITCHBOARD houses the main service CB, feeder CBs for essential and nonessential loads, and feeder CBs for ac input to each of the four UPS modules. The two diesel control panels (arrow)—one for generator No. 1 and the other for generator No. 2—include the synchronization and paralleling controls and are here at the end of the switchboard being checked-out by (L-R) Stan Plato, Lou Forte (authors of this article), Al Meliere, president, United Control, Paterson, N.J. and George Straub, maintenance engineer for Western Union at the Glenwood facility.

TRANSFER CUBICLE at one end of main switchboard houses the two power CBs through which the ac output of either of the two diesel-generators is fed to the 800-amp ac load bus that is normally supplied from the inverters of the four UPS units.

following schedule of conditions indicates how critical loads are assured of continuous power.

A. Normal status: utility supplies power to main bus; UPS functioning.

1. Battery chargers supply dc to UPS inverters, which in turn supply 208/120-volt, 3-phase, 4-wire, ac to critical loads through static switches. Output of UPS (voltage, frequency and phase angle) is constantly synchronized with ac on the main 4000-amp switchboard bus.

2. Essential and nonessential loads, including de-icing, are served through feeder CBs on service switchboard.

3. Generators 1 and 2 are in standby mode, with CB1 and CB2 open.

B. Fault status—loss of utility; UPS functioning from battery supply.

1. Output of UPS to critical loads continues right through utility inter-

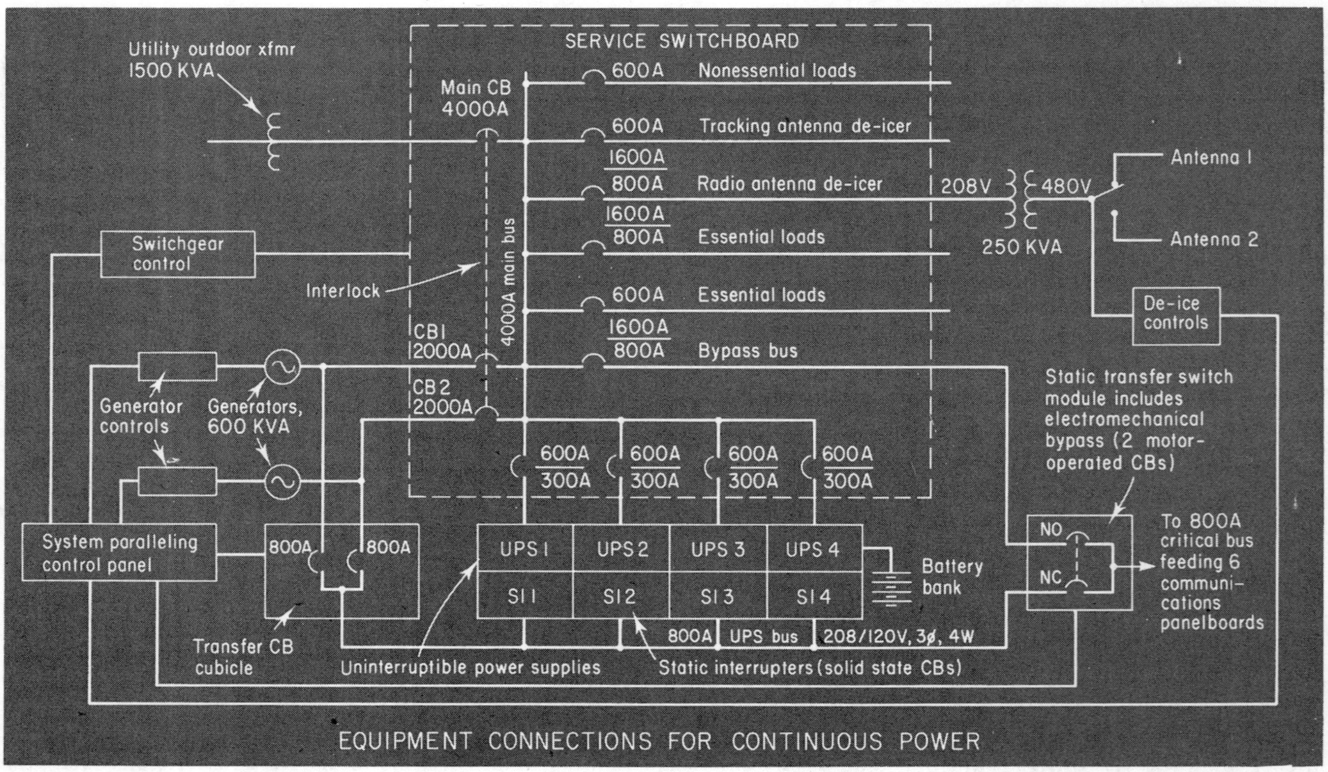

EQUIPMENT CONNECTIONS FOR CONTINUOUS POWER

ruption because of battery bank feeding dc to UPS inverters.

2. Main CB opens, disconnecting service from outdoor transformer.

3. Nonessential load CBs and de-icing CBs open.

4. Diesel-generator No. 1 starts; CB1 closes, supplying power to main bus and UPS units.

5. Critical communications loads continue to be supplied by UPS system but with ac input to battery rectifiers from main bus.

6. Essential loads and battery chargers are now supplied by generator No. 1.

7. Generator No. 2 is standby for generator No. 1 or UPS failure, or for use when de-icing is needed.

C. Fault status—utility fails, UPS operating; de-icing required

1. Generator No. 1 assumes main bus load through CB1 as in B above, with UPS continually feeding critical loads.

2. Generator No. 2 is started and synchronized with generator No. 1.

3. CB2 closes, paralleling the two generators on the main bus.

4. Nonessential load and de-icing CBs close to supply de-icing load.

5. When de-icing is no longer needed, generator No. 2 is shut down and assumes standby for generator No. 1.

D. Fault status—utility available, UPS inverter fails

1. Static switch transfers critical load from failed UPS to main 4000-amp switchboard bus via the bypass CB.

2. Generator No. 1 is started and synchronized (voltage, frequency, phase angle) to utility on main bus.

3. CB1 is closed, paralleling generator No. 1 and utility as dual supplies to main switchboard bus, with generator No. 1 controlled to supply exact amount of power drawn by critical loads.

4. Failed inverters are isolated from switchboard bus by individual 300-amp CBs and static interrupters.

5. Transfer-breaker CB1 is closed, connecting generator No. 1 to 800-amp UPS bus.

6. UPS bus is isolated from main bus by return of static switch to UPS position; CB1 opens.

7. Essential and nonessential loads will be served by utility on main bus. Generator No. 1 is supplying the critical loads in the communications room directly via CB1A and the static switch.

8. Generator No. 2 is standby in support of generator No. 1 and utility. If generator No. 1 is lost, no-break transfer is made to the main bus via the static switch. Generator No. 2 will go through the same procedure to assume the critical load in a no-break fashion. Δ

BATTERY BANK assembly (above right) for dc input to the inverters of the UPS units when ac input to the UPS chargers fails is in room adjoining main switchboard area and is arranged in two racks (left and right) with an aisle between. Output circuit conductors are run in conduit (upper left) to the dc bus of the UPS units and provide for keeping the batteries on trickle charge when the normal ac input to the UPS rectifiers is present. The voltage of the battery bank is maintained at 270 volts dc and has a supply capability of 15 minutes under full load for supplying the inverters and enabling them to handle a 150-kw load—when ac input to the inverters fails.

DIESEL-GENERATOR SETS (right) are located side-by-side in room adjacent to main switchboard. Each unit has a preheated engine block to assure fast starting. Each diesel-generator has its own set of cranking batteries (one set and its charger is shown at left rear). In addition, a third set of cranking batteries is installed and is diode-coupled to both sets of normal cranking batteries. This third set provides a backup, or emergency, cranking source that is in parallel with each battery set dedicated to one of the diesels.

REACTANCE CHOKE (arrow) is installed in the feeder circuit from each of four CBs in the main switchboard to the input of each UPS unit to prevent feedback of harmonic frequencies that are produced by the switching characteristics of the SCR rectifier circuits in the UPS input sections. Feedback of those frequencies ("noise" or "hash") into the ac bus of the main switchboard can disrupt and/or disable effective synchronization of the generators. Because the synchronizing controls sense the frequency characteristics of the ac on the main bus, presence of the harmonics causes the generator output to "hunt"—that is, to keep changing its adjustment trying to stabilize. Such high-frequency feedback can also adversely affect other loads fed from the main bus. Determination of the correct inductance (in microhenries) and resistance involves establishing the proper number of turns, the adequate size of conductor to minimize resistive drop, and the correct magnetic core structure. Such calculations are generally empirical and are based on the amplitudes and frequencies of the harmonics of the particular number and hookup of rectifiers in the UPS input. Such a choke must have low line loss and must not produce regulation problems.

MIMIC ONE-LINE DIAGRAM on UPS main control panel in static inverter room simplifies operation of system. Indicating pushbuttons and pilot lights show status of major components.

UPS supplies continuous power for computers

Here are design and installation details for an uninterruptible power source incorporating solid-state rectifiers and inverters, nickel-cadmium batteries and a well-planned protective scheme to assure reliability.

By PAUL KATZAROFF, Project Engineer
Jaros, Baum & Bolles, Consulting Engineers, New York City

AT THE New York City computer center of Merrill, Lynch, Pierce, Fenner & Smith, Inc., a 250-kva static uninterruptible power source supplies precise, unvarying power to computers and other sensitive electronic equipment. This computer equipment is connected to all of the brokerage firm's 220 domestic and international offices and major exchanges. It is of paramount importance that voltage and frequency applied to the computer be maintained within specified limits to eliminate any possibility of lost or distorted market information.

In the event of a total power outage, a nickel-cadmium battery bank feeds dc power to an inverter system which, in turn, supplies ac power for an orderly shutdown of the computer.

The primary purpose of the UPS is not to maintain computer operation during a normal power failure, but to eliminate voltage and frequency variations—which may be of comparatively small magnitude and last only a few milliseconds—but which can be a source of computer error. Any computer

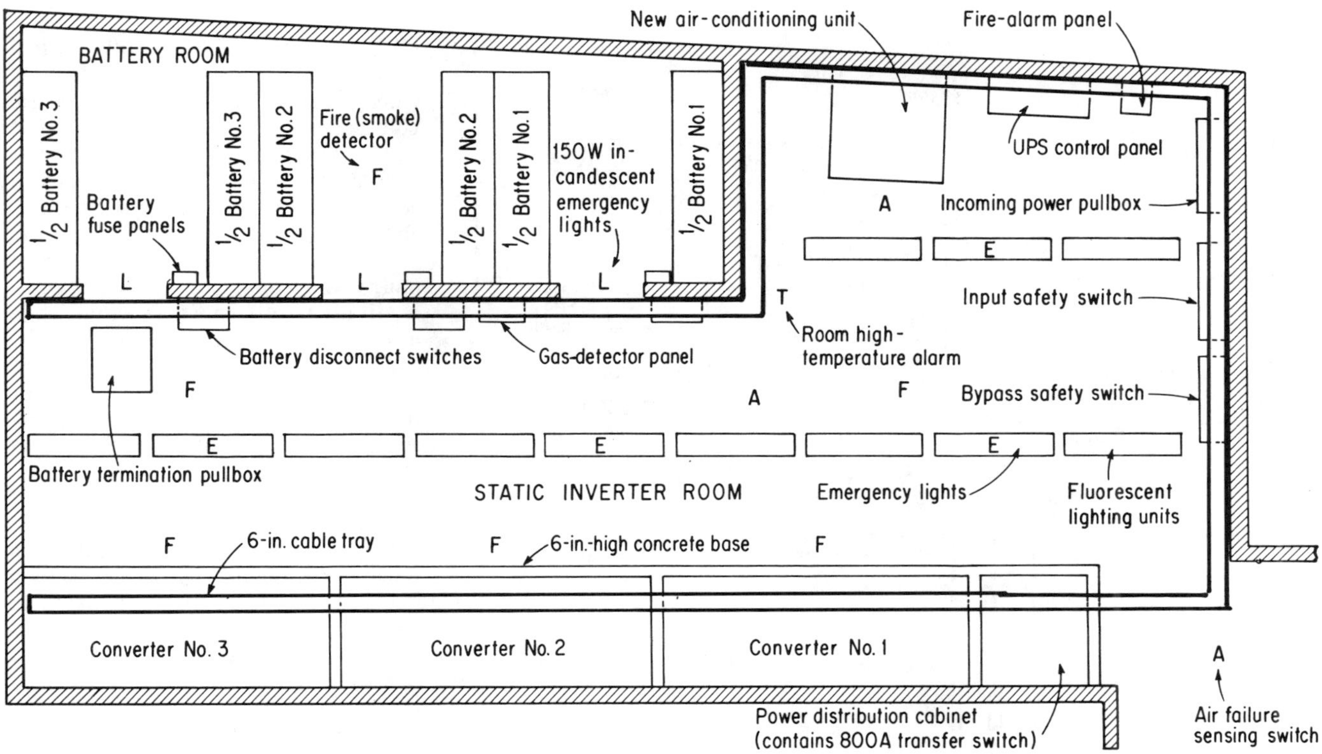

FLOORPLAN shows physical arrangement of UPS components and installation details.

error can be costly; however, in this instance, the problem is more acute because the computer is a "real-time" facility. A real-time computer receives, stores, processes and dispenses information all at the same time. Therefore, the firm called for a UPS at their New York facility and additional (but smaller) UPS systems at offices in major cities.

The installation in New York consists of a 250-kva solid-state rectifier-inverter system, operating with three 125-kva power converter modules, any two of which can carry the full critical load of 250 kva. In addition, a storage battery is provided to sustain UPS system operation for a 15-minute interval in case of total failure of the commercial power source.

The UPS system supplies power whenever the critical load is in operation; however, if the UPS system itself should fail, the critical loads may be powered directly from the normal power source. This is accomplished through a bypass circuit which permits transfer of supply without interruption of power to the critical computer load. This also provides for maintenance of the equipment, if required.

The UPS system, as installed, consists of three major components: power converter modules, battery bank, and logic and control circuitry.

Power converters

The power converter modules are each rated 100 kw, 125 kva at 0.8 power factor continuously, and 150 kw, 187.5 kva intermittently (for 10 seconds only). Each converter incorporates a regulated power supply section which rectifies the incoming 208-volt, 3-phase power providing regulated and filtered dc voltage and current for battery charging and for supply to the inverters. Each converter also includes a solid-state frequency inverter which changes the dc energy to precise 208-volt, 3-phase ac for the critical load.

Also included in the power converter modules are function-monitoring sensors and solid-state switches which enable the converters to disconnect rapidly in the event of a component failure. High reliability of the UPS system is achieved by a redundant power converter. Should one of the three power converters fail, the remaining two continue to maintain the combined full-rated power output requirements of the load.

THREE POWER CONVERTERS are installed in cabinet lineup on 6-in.-high base. Near ceiling are cooling-air ducts and 6-in.-wide cable tray that extends around room. Battery room is out of sight at right.

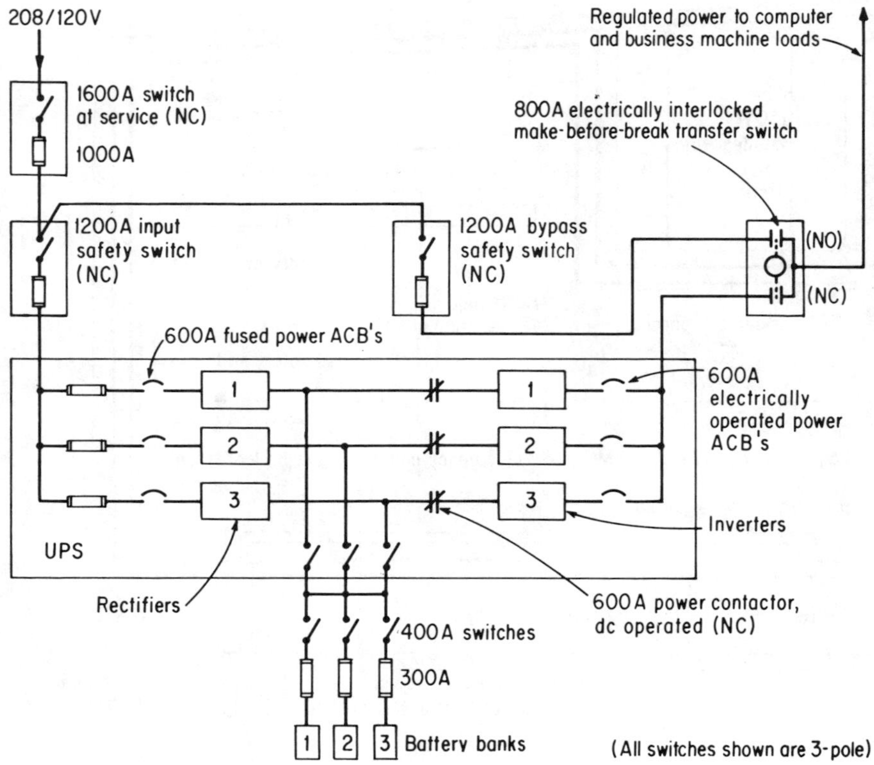

ONE-LINE DIAGRAM illustrates operation of UPS system. Normally, rectifiers provide dc to the inverters which, in turn, provide a regulated ac output to the critical computer load. Batteries absorb voltage transients or surges, helping to provide a stable output to the load.

NICKEL-CADMIUM BATTERY, below, consists of 1200 cells on six racks. Each rack supports 200 batteries, 100 per shelf, 25 per tier. Fuse panel is at upper right.

All three converters are always on-line and automatically share the load. Dropping off-line or bringing on-line of a power converter does not cause voltage transients greater than + or − 8% maximum with recovery to 100% voltage within 50 milliseconds (3 Hertz), which is within limits of input power requirements of the computer equipment.

Battery

When the normal power source is interrupted or is not within acceptable tolerances, the full critical load is powered through the UPS by the battery. The battery has sufficient capacity to provide proper operation for up to 15 minutes. In the future, a diesel-engine generator set may be installed to supply power through the UPS system in the event of an extended power failure. Upon return of the normal power source, the batteries are returned to a fully charged state by the system.

The batteries are nickel-cadmium type of pocket-plate construction. Under normal operation with ac input power, the battery is held at the float potential of 1.40 dc volts per series cell. On failure of input power, the batteries sustain UPS operation under full critical load and are considered discharged when terminal voltage drops to 1.0 volt per cell. On return of ac input power, the dc voltage applied to the battery is 1.60 volts dc per series cell until the battery attains full charge.

The battery, rated 100 ampere-hours and consisting of 1200 cells, is divided into three sections with associated wiring and disconnecting means so that any one section may be disconnected for maintenance while the two remaining sections maintain the battery standby function.

Logic and controls

The UPS system includes a power "walk-in" feature which prevents sudden load application to the primary power mains at the instant of power return. When input power is restored after failure of the commercial power source, the control limits input power at the UPS terminals to 25% of rated load current and causes the current to gradually increase to 100% of full rating over a 15-second time interval. Also included is a "power-limiting" circuit whereby a maximum limit is established on input power for all operating conditions regardless of the magnitude of the critical load. This feature establishes a maximum load which the UPS system will demand from the power mains. Thus, regardless of any transitory high load on the UPS system, and regardless of the demand which might be presented by way of a discharged battery, the input line current is limited to a maximum preset value. The battery takes up any differences between the demands of the load circuit and the available input current. The maximum input current per phase is limited to 1150 amps ac for the 250-kva UPS regardless of the magnitude of total demands.

The inverters are each furnished with dual master oscillators so that the failure of one oscillator results in an immediate transfer to the alternate with no interruption of power to the critical load. As a result, failure due to oscillator malfunction is highly improbable.

A central control panel in the sub-basement contains all metering, switching, control and alarm apparatus of the UPS system. Signal lights indicate the status of various connections, and various meters provide numerous monitoring functions. A remote panel on the computer floor provides operating personnel with information regarding the status of the system.

Installation

The UPS system is installed in a sub-basement area of the building; the computer equipment is on the 3rd floor. The UPS area is divided into two sections—a static inverter room and a battery room. Installed in the static inverter room are the three converter modules with an associated power-output distribution cabinet, the system control panel, a protective system control panel, two 1200-amp, fused safety switches, air-conditioning equipment and cooling-air ducts, incoming power pullbox, and protective devices.

Each converter module incorporates a dc power supply section, or rectifier, which consists of solid-state diodes protected by individual fuses. Main protection for the power converters includes 600-amp fused power air CBs on the line side and 600-amp power air CBs on the load side of each unit. In addition, a 300-amp normally closed, electrically held and electrically operated dc contactor is connected to the line side of each inverter, providing operational flexibility.

Three banks of batteries are installed in the battery room on six racks provided by the battery manufacturer. Constructed of steel channel, each rack supports 200 battery cells, 100 on each of two main shelves. Both shelves on each rack are tiered so that all cells can be easily reached for inspection or maintenance. However, manufacturers of the nickel-cadmium batteries point out that the batteries are designed for maintenance-free operation and may require no attention for years.

Each bank of batteries is protected by 300-amp fuses mounted at a fuse panel installed adjacent to each bank in the battery room. The main disconnect for each bank is a 3-pole switch rated 400 amps, 600 volts, dc, installed on the wall outside the battery room.

The 3-pole switches were used to divide the 500-volt battery bank into two parallel groups with only 250 volts across them, providing greater safety during maintenance.

Installed around the perimeter of the static-inverter room is a run of 6-in.-wide cable tray. Rod-suspended on trapeze hangers, the tray is galvanized-steel ladder type and was installed in 12-ft sections. Conductors carried by the tray are easily picked up or dropped off at equipment or pullboxes as required.

REMOTE MONITORING PANEL informs operater in computer room of status of UPS system. Alarm signals and intercom from UPS area to computer room help assure proper operation of all equipment and systems.

All conductors are either THW or RHW, copper. Battery interconnections and leads were made up with multiple runs of 200-amp welding cable to maintain flexibility and reliability. From each battery disconnect switch, two 350MCM conductors in 3-in. conduit extend to the battery terminal pullbox. (See floor plan.) From here four 500MCM conductors in 3½-in. conduit carry power to the converters. Incoming power at 208/120 volts is carried by three sets of four 500MCM conductors in 3½-in. conduit. UPS output power, at 208 volts, 3-phase, is carried by six 500MCM conductors in 3-in. conduit.

The inverter and its sub-assemblies (transformer, reactor, capacitor and power switch) were installed by the electrical contractor's men using a forklift truck, and all necessary wiring was completed. Components were installed in the three cabinets, which were mounted on a 6-in.-high concrete base at one side of the static inverter room. This elevated base will protect the converters against damage from water should any seep into the sub-basement. Each converter, which weighs approximately 6500 lbs, is 108 in. long, 38 in. deep, and 84 in. high.

A number of special features were included to assure proper environment and protection for the UPS system. As an example, ambient temperature in the static inverter room is kept at approximately 72F by two 7½-ton air-conditioning units. Cooling air is distributed through ducts; the converters are convection cooled.

A variety of devices and systems sounds an alarm if environmental conditions or other malfunction threatens proper operation of the UPS. Air-flow switches sound an alarm if normal air flow in cooling ducts is lost; if ambient temperature exceeds a preset level, a temperature-sensing device initiates the alarm. A smoke detection system incorporates ionization detector heads to obtain early warning of fire. In the battery room, a sensitive gas detector will sound an alarm if hydrogen gas is detected.

The electrical contractor on the project was Lorson Electrical Contracting Corp.; consulting engineers were Jaros, Baum & Bolles—both of New York.

Reliable Computer Power

Dual feeds plus new technology give UPS results at a fraction of UPS costs.

WHEN Michigan's third-largest bank, the Detroit Bank & Trust Co., undertook a major expansion of its computer room, management wanted assurance that the computer equipment would be protected against power outages, transients and fluctuations. Serious consideration was given to an uninterruptible power supply (UPS) with a backup diesel-powered generator—and a price tag of over a million dollars. However, thanks to an existing dual utility service and an ingenious design approach, the bank now has the equivalent of a UPS system at a quarter of the cost.

Part of the solution to DB & T's problem hinged on the fact that its downtown Detroit headquarters building is served from two entirely separate utility substations. In the entire history of this dual-feed situation, power has never been lost from both sources at the same time. Hence DB & T could be virtually certain of having power from one source or the other. The two power sources were fed into a 4800-V primary auto/manual transfer switch designed to throw the load from one substation to the other in the event of a power failure. Unfortunately, this auto/manual switch takes from two to three seconds to transfer, too slow-acting for the computer. The computer-room's 208-V transfer switch could throw over in about 3 cycles, but this was also too long for the computer. There was no equipment on line capable of bridging the power gap caused by the transfer. Consequently, a power failure would cause a break in power continuity that inevitably would cause a computer outage and serious damage to any programs then in progress (Fig. 1).

By ARTHUR FREUND

Defining the problem

The grim fact of computer outage was very much on the mind of DB & T's properties officer Richard B. Hamlin when he began to study the problem. The study was precipitated by a decision of the Computer Department to upgrade its system of two Burroughs B 4700s to a system of three B 4800s. To meet the increased power demand and to provide for further expansion through 1985, additional power equipment had to be purchased. But Hamlin was not concerned solely with the amount of power needed but rather about power quality and continuity. He was particularly alarmed about the consequences of more frequent blackouts and brownouts. They'd

Fig. 1. Original installation

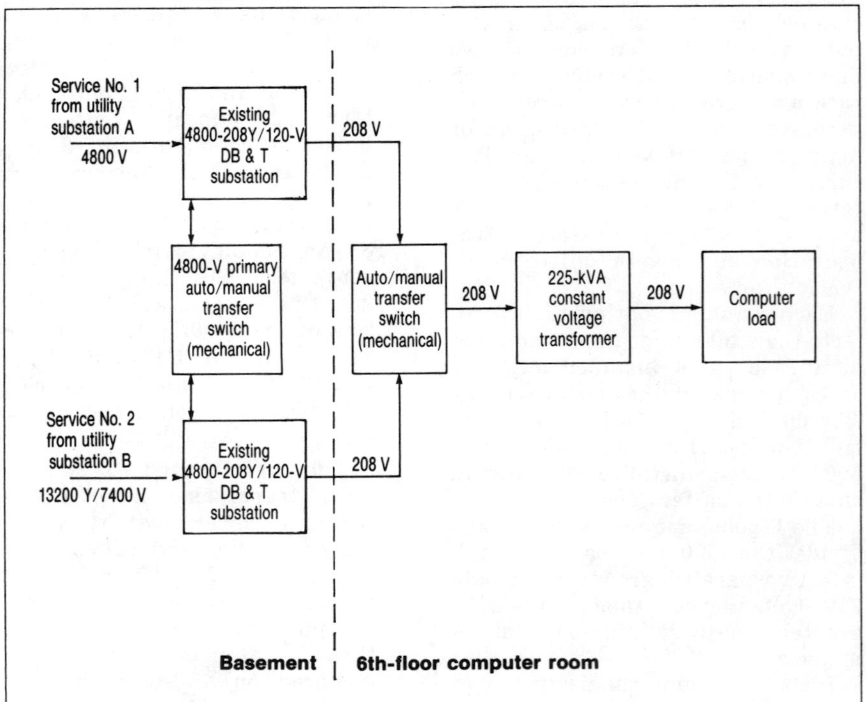

The power management center

The power management center consists of two major parts—a power-conditioner section and a power-monitoring and distribution section. The power-conditioner section is unique.

In a UPS system, incoming ac power is rectified, and the resulting dc power is used to feed an inverter that synthesizes an ac voltage, single- or 3-phase, of the desired frequency (usually 60 Hz, but occasionally 400 Hz). The dc rectifier output also keeps a battery connected across it charged, and on loss of incoming ac power, the battery powers the inverter, which continues to supply power without interruption or disturbance as long as the battery lasts. If ac power returns before the battery is discharged below its minimum acceptable voltage, the computer power is never interrupted. The inverter in a UPS synthesizes the ac by one of several methods, including the 6-step wave, the 12-step wave, and a pulse-width modulated wave, all of which are filtered to produce a low-distortion ac sine wave.

The power conditioner portion of the PMC consists of an ac synthesizer that uses a combination of transformers, inductors, and capacitors to provide an output similar to that of a UPS without ever going through a dc stage. Six 3-phase pulse transformers are connected in parallel with phase-shifting inductors so that the output is a 12-step synthesized sine wave, which is then filtered to produce a clean 60-Hz, 3-phase ac sine-wave output with low distortion. The output is fully isolated from the input and can be at the same or a different voltage. The 3-phase output will continue even on the loss of one incoming phase, with only a small drop in output voltage at loads up to 60% of full load. The circuitry that accomplishes this is patented and proprietary information.

Although the method of operation is different, the quality of the output voltage and the complete independence of the output from any and all disturbances on the input voltage are similar to those of a UPS. There is one major difference. On loss of input power, the UPS will continue to provide undisturbed output power as long as the battery voltage lasts. The PMC, with no dc stage or battery, can continue to supply power for only about one cycle.

SINGLE-UNIT PMC contains power conditioner and power monitoring and distribution together. This construction is available up to 30 kVA. From 50 kVA to 125 kVA, power conditioner is in one unit and monitoring and distribution in a separate unit, which can be adjacent to or remote from the conditioning unit. Jeff Powell, engineer for the PMC manufacturer, is indicating the location of CTs and circuit breakers in the monitoring and distribution section. The power monitor readouts are at the top, and the transformers, capacitors, and synthesizing circuitry in the rear.

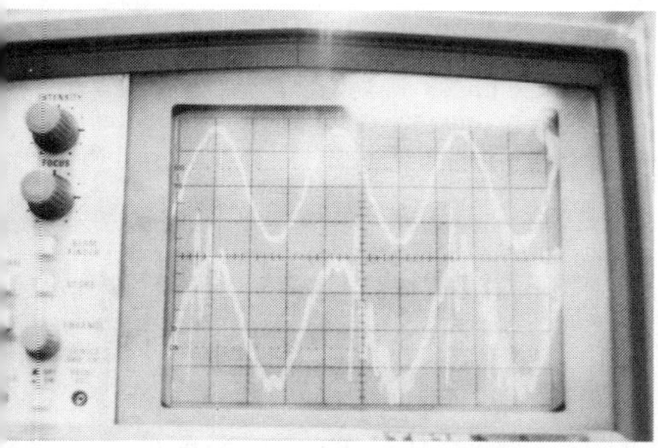

OSCILLOSCOPE TRACINGS clearly show multiple transients on the input (lower) trace, with no visible disturbance on the PMC output (upper) trace. Glare is reflection of overhead lights. Long-persistence oscilloscope permitted traces of many transients injected from test equipment over a period of several seconds to be seen together.

TEST AND DEMONSTRATION EQUIPMENT (photo, right) include a control board, on left, which can provide undervoltages, overvoltages, transients, single-phasing and other incoming line disturbances from adjustable transformers. Load can be varied from no load to 100% overload from the same control board. The demonstration unit PMC is to the right of the control board and is subjected to any desired line and load disturbance. On the top shelf to the right of the PMC is a dual-trace oscilloscope that can be used to examine the PMC input and output waveforms simultaneously.

In function, the PMC is a UPS without battery backup.

The specifications of the ac synthesizer are impressive.

- Output voltage: ±2% for input voltage variation from +15% to −30%, at all loads from no load to full load, steady or varying.
- Output voltage: ±4% for input voltage variation of ±40% from no load to full load, steady or varying.
- Output voltage: −4% for complete loss of one phase of input voltage for loads up to 60% of full load.
- Output voltage: ±2% for any unbalance in loads, up to 100% on one phase.
- Harmonic distortion: less than 4% for up to 40% input voltage distortion at all specified operating conditions.
- Automatic current limiting: maximum of 300% of full load. A short circuit can be sustained for 30 minutes without equipment damage.
- Efficiency: 91% at full load; 87% at ½ load; 79% at ¼ load.
- Power factor: 99% or better.
- Line noise and transients: output down 60 db below input for both single-phase noise (normal mode) and 3-phase noise (common mode). Single-phase output noise down 140 db for 3-phase input noise.

Since the conditioner has no moving parts and no semiconductors, it requires no maintenance except cleaning. It has no control feedback loops and no destructive failure modes. It is convection-cooled, requiring no fans.

The power-monitoring section has three digital readouts to monitor input and output voltages, line-to-line or line-to-neutral, line-and neutral-output currents, ground current, kVA load, and output frequency. Any reading can be shown, or the unit will automatically scan all readings continuously. Line output current and output voltage deviation are indicated on color-coded LED scales, with automatic scanning. An alarm panel monitors temperatures, output voltages and current, ground integrity and current, frequency, and other conditions (smoke, water, unlocked doors, and the like) as desired.

The output, protected by individual circuit breakers, is delivered to the computer through flexible cables with connectors to match the computer cables.

CONNECTING CABLES of PMC unit that is being readied for shipment are wired to PMC unit, tested, and left connected. Cables are wound around cylinders for shipment and entire assembly is shipped as shown, for easy and errorproof installation, since each cable can have a different connector as specified by the user. The connectors vary from single-phase duplex receptacles to special connectors as required for individual units of computer equipment.

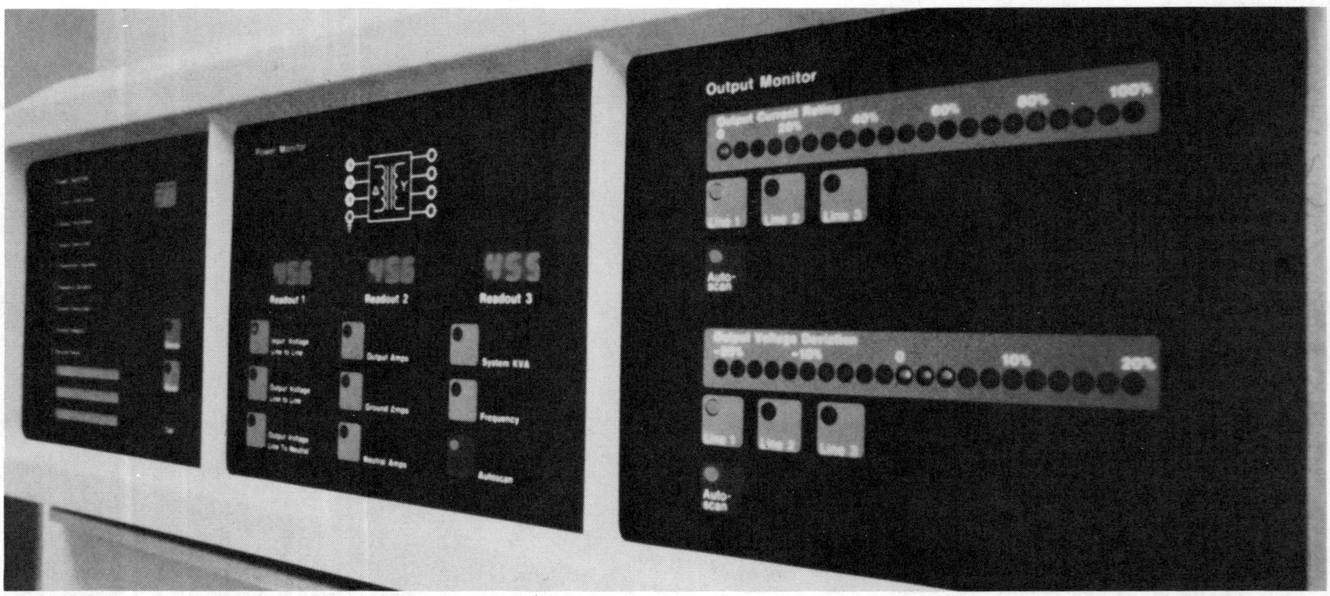

LED AND DIGITAL PANEL give continuous readouts of input and output voltages, loads, and status for each phase. Readouts can be selected, or automatic scanning will continuously switch through entire set every few seconds. Alarm panel monitors unit and environment and alarms at any potentially dangerous condition. User can choose parameters to be monitored as desired.

been plagued with a variety of other power problems—all of them rather mysterious and coming from unidentified sources—irregularities that resulted in computer errors, rerun difficulties, and other annoyances. Considering the probable further decline of power quality in the Detroit area in future years, it was evident that the problems could only increase.

The minimum solution

If additional power capacity had been the only problem, an easy and inexpensive solution was at hand. The addition of a second 225-kVA constant-voltage transformer in the computer room, at a cost of about $12,000 plus installation, would have provided enough power to accommodate the estimated 325-385 kVA needed by 1985. Hamlin quickly rejected this approach, which he says provided adequate power but solved only about 20% of the problems of power variations and outages. While constant-voltage transformers are somewhat effective in eliminating power fluctuations, they are still vulnerable to blackouts, severe or extended brownouts, extreme voltage sags or surges, and transients. In addition, the throwover times for the 4800-V primary transfer switch and 208-V computer room transfer switch would still be too long to prevent the computer from "crashing."

The maximum solution

If cost were not a consideration, an uninterruptible power supply with a diesel generator backup might be considered the ultimate system. Such a system could handle almost all power problems, but it also carried a price tag of about one million dollars. While a UPS system would guarantee power continuity and provide excellent power quality, the installation takes up a great deal of space and is expensive to operate and maintain.

A UPS system, with a 15-min battery backup, would permit transfer from one utility source to the other without affecting the computer operation. In addition, a UPS would provide clean power to the computers despite transients, sags, and similar disturbances on the building power system. However, it would, in the event of loss of both incoming services (a highly unlikely occurrence) require an orderly shutdown of the computer during the 15 minutes the batteries could supply power. Such a system would protect the computer against about 99.5% of all power problems.

Addition of diesel generator to back up the two utility services to the UPS would add to the cost but would provide power to the computer even if both utility sources failed. Since the generator could come on line in less than 10 sec, the additional cost could be partially compensated by reducing the battery capacity to 5 min or even less, just enough to assure power from the generator. This would increase protection to cover about 99.7% of all power problems. The diesel generator, of course, would require periodic maintenance.

With the UPS system, provision would be required for power monitoring and distribution. Providing for future expansion is difficult and expensive, and flexibility is limited.

Motor-generator set

Another idea considered was using a motor-driven generator (MG) set, the motor receiving power from the line and driving the generator (actually, alternator), which would provide power to the computer. The inertia of this system, especially if equipped with a flywheel, would provide power to the computer for several cycles on loss of input power to the motor. This power would tend to decrease rapidly, how-

Fig. 2. Final expanded installation

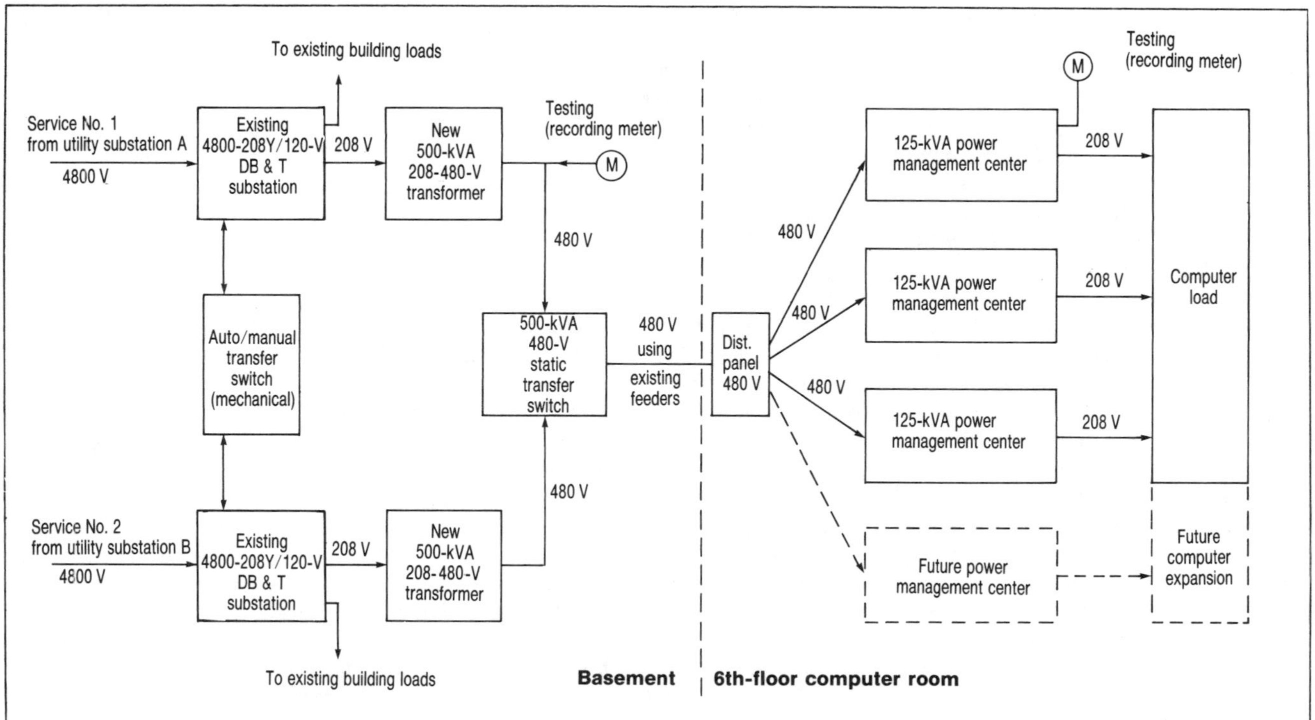

POWER MANAGEMENT CENTER with covers removed shows transformers, inductors and capacitors that synthesize a 12-step output voltage wave of clean 60-Hz power. Wide input voltage variations, noise, and transients have little or no effect on output voltage, which makes this equipment desirable for computer power systems.

ever, and it is questionable whether power to the computer would be acceptable long enough to override the transfer time of the low-voltage mechanical transfer switch. This difficulty could be overcome by using a static (nonmechanical semiconductor type) transfer switch, which could transfer the motor to the alternate power source in about ¼ cycle (4 milliseconds). This would assure continuous power to the computer.

In addition, the MG set would completely isolate the computer from disturbances on the building power system, provided that the input voltage to the motor did not drop too low for long periods and cause motor overheating, as in a severe brownout.

A combination of MG set and static switch, with provision for computer power monitoring and distribution, would cost about $200,000 installed and would protect against about 95% of the power problems. However, maintenance of the MG set would require a computer shutdown.

The optimum solution

A static switch, capable of switching in 4 milliseconds (.004 sec) provided part of the final answer. However, even a 4-millisecond interruption in power is still intolerable to the computer, and a means of bridging the gap had to be found.

A new power center and ac power synthesizer design, using only ferromagnetics (transformers and inductors) and capacitors, and employing no semiconductor devices, was found to be capable of providing continuous "clean" power through an outage of up to 1 cycle (16.7 milliseconds). It bridged the transfer switching time with a generous safety factor. The power management center (PMC) and its ac synthesizer use utility power as merely an energy source to generate an independent output of 3-phase, 60-Hz ac power, free of noise and other disturbances. In addition, each unit provides for individual load-circuit switching and overcurrent protection, complete power monitoring, and flexible cable power distribution to the computer. Input voltage variations of as much as ±40% from nominal are tolerated. The oscilloscope tracings (see photos) compare fluctuating input power with the clean output of a PMC unit. This PMC also offers excellent protection against "spikes," such as those caused by switching or lightning, and even against loss of one incoming power phase. Because the synthesizers consist of only dry-type transformers, inductors, and capacitors, they are essentially maintenance-free.

The system finally selected by DB & T is shown in Fig. 2. The three main features of the new system are:

- Two 500-kVA, 208-V to 480-V step-up transformers (one for each power input source)
- A 500-kVA static transfer switch capable of transferring power from source to source in ¼ cycle (0.004 sec).
- Three 125-kVA power management centers to eliminate power fluctuations and bridge the 4-millisecond gap caused by throwover of the static transfer switch. Provision has been made to install a fourth unit when the power demand exceeds 375 kVA.

The installed cost of the system was about $270,000—or one quarter the cost of a conventional UPS system. It is estimated that this installation will solve 99.5% of all computer power problems. Note that a 480-V line carries all the power to each of the three power management centers. The PMC units convert this input into three separate 208-V circuits, which are then distributed at three distribution modules, by means of flexible cables, to the computer equipment.

When the system was completed, engineers from the bank and the PMC manufacturer installed sensitive recording meters at the 480-V output of one 500-kVA transformer and at the 208-V output of one PMC. They then threw the circuit breaker on one of the two 4800-V primary power inputs to the building. The static switch automatically went to the alternate power source. In two trials, the voltage variations were negligible; in a third trial, the voltage drop was less than 3%—too small a difference to upset any computer. No transient or other disturbances appeared on the power to the computers, although the interruption clearly showed at the 480-V transformer output. The recording meters were line-disturbance analyzers capable of recording all voltage and frequency vari-

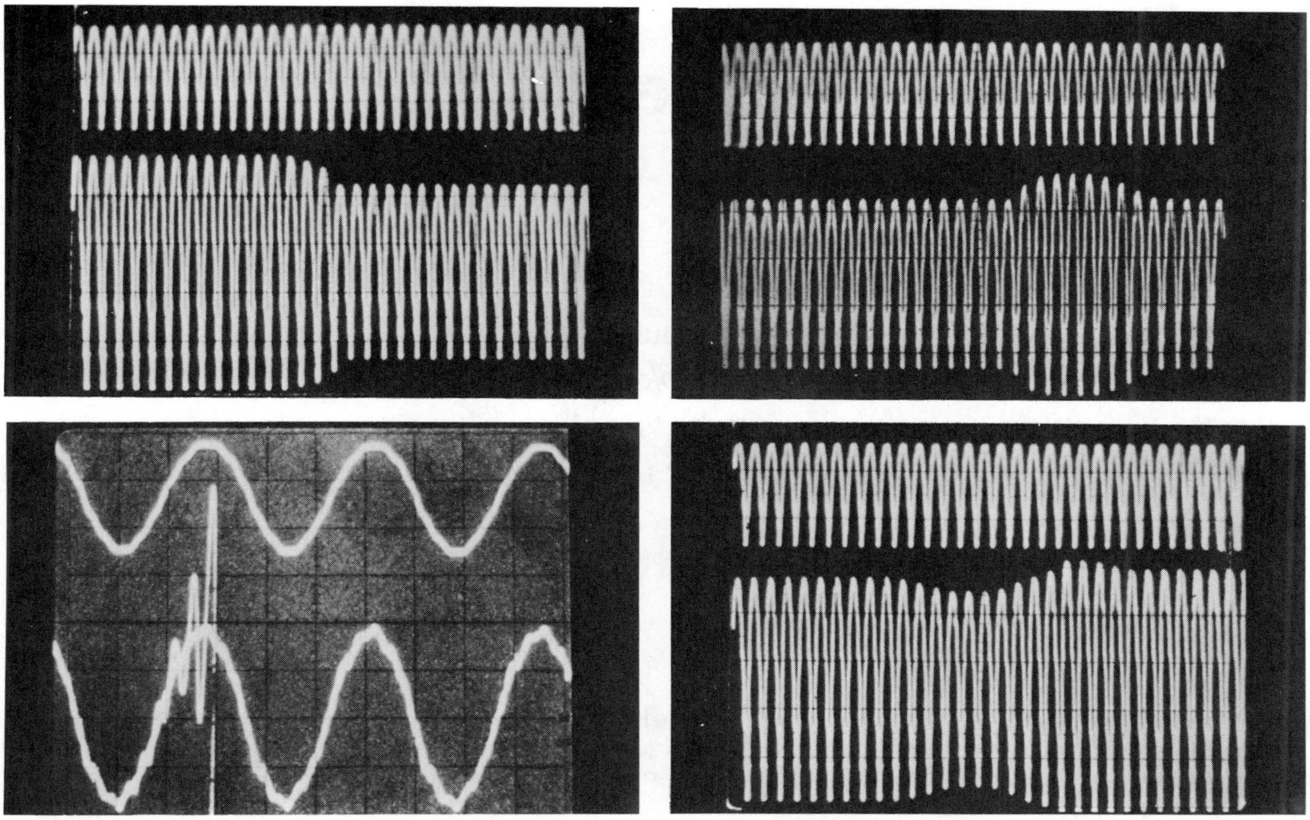

OSCILLOSCOPE TRACINGS show varying voltage inputs to power management center (lower trace in each photo) and negligible effect on output voltage (upper trace). Upper left: drop in steady-state voltage, which continues for extended period of time, as in a brownout. Lower left: severe transient, or spike, such as caused by lightning or switching surges. Upper right: brief voltage increase, or overvoltage, as could occur on shutting down a large load or on utility variation. Lower right: voltage dip, or sag, from utility variation or starting a large motor. In all cases, the PMC output voltage remains steady and undisturbed.

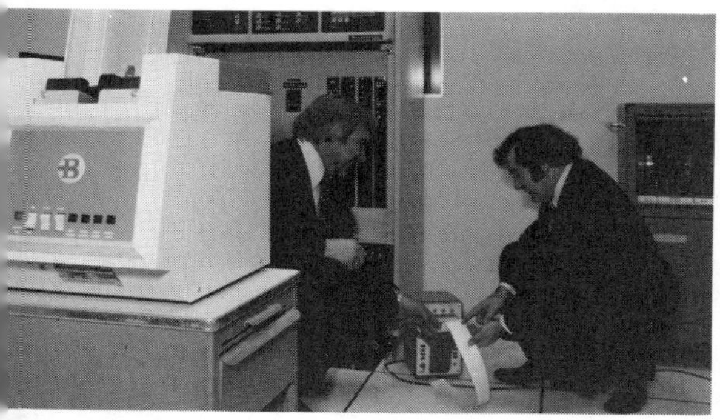

LINE DISTURBANCE ANALYZER is connected to the power management center (against wall) to read its output during computer power system test. Big smiles on the faces of Dick Hamlin, Detroit Bank and Trust Co. Properties Officer, and Paul Gowdy, DB & T Co. Data Processing Manager, are because the analyzer printout shows no problems or disturbance to the computer power although the 4800-V input power to the bank has been switched from one utility source to the other several times.

ations and spikes as brief as 800 microseconds, monitoring all three phases, and giving a hard-copy printout.

Not every computer room has access to two separate power sources. However, if a UPS system is under consideration, the possibility of bringing in power from a second substation definitely should be considered as an alternative, in view of the total costs involved.

The DB & T computer power arrangement will meet their requirements through 1985, even though the quality and reliability of utility power is likely to deteriorate. If the computer room is again expanded, or even moved to a new location, the power management centers—with their flexible cable power distribution—will be easily adaptable to new assignments. In addition, since the PMC units are portable, they can, for tax purposes, be depreciated rapidly as equipment, rather than slowly as part of the building construction. Also, they are eligible for investment tax credits.

Since the installation was completed by the contractor, Industrial Electric Co. of Detroit, there have been absolutely no problems with the computer from "dirty" power. In addition, just recently the utility supply lost an incoming phase. The static transfer switch sensed the single-phase condition and transferred to the alternate utility source, while the PMC filled in the transfer gap without any effect on the computer. With the old installation, the computer would probably have crashed and certainly would have lost its operating data. DB & T is extremely pleased with its comparatively low-cost but fully effective power conditioning system.

Modular centers for computer power

"Energy management" must aim at efficiency, flexibility and economy in power distribution—*as well as power* consumption. *Here's a look at a new concept for supplying electrical circuits to data-processing equipment, in a way similar to modular wiring in hung ceilings.*

NEW SYSTEMS have been developed recently for distributing, monitoring, and isolating electric power in facilities housing computers and associated equipment (peripherals). Heart of the systems is a computer *power distribution unit* (PDU), or simply power unit.

The first PDUs were manufactured about 1974; however, since that date, their popularity has increased dramatically as more engineers, architects, and data-processing executives have become familiar with the many benefits provided.

The alternative to a power unit is conventional built-in building circuiting, or "hard wiring," as it is usually called. When computer rooms are hard-wired, a feeder is run into the room from a utility connection or an uninterruptible power source (UPS). The feeder terminates at one or more panelboards bolted to the computer room wall. The panelboard usually contains a main breaker and branch circuit breakers. Any NE Code-recognized wiring method is used to run from the wall-mounted panel to each computer and its peripheral equipment.

The PDU, on the other hand, is supplied with a junction box that is mounted out of sight under the computer-room raised floor. The main feeder from the utility-fed building distribution system or from a UPS is wired into this junction box. A flexible cable supplied with the PDU carries the power from the underfloor junction box into the PDU cabinet and terminates at a main breaker. The cabinet is attractively designed and looks like a piece of computer equipment. A super-isolation, electrostatically shielded transformer is supplied to properly condition the power for computer applications. Branch circuit breakers provide protection for circuits to individual computers or peripherals. Power finally feeds to each computer-room device through flexible cables that run under the computer-room floor. While conditioning and distributing electricity, the PDU is also monitoring and reporting the status of electrical power to keep the data-processing manager advised of any problems that may develop.

Computer power design

Computer rooms have very special electric power requirements. A high-quality, computer-grade equipment grounding arrangement is absolutely imperative with today's high-speed data-processing equipment. The new generation of computer hardware, with its micro-chip technology, simply will not stand poor power. Any electrician can ground computer equipment and, in fact, the code *requires* a ground. However, the code does not say the ground must be "effective," or computer-grade, or broadband, or single-point.

There has been a tendency of some electricians to install hard-wired systems with grounds run to the closest water pipe and/or to nearby building steel. This practice satisfies the code but may introduce loops into the grounding system along with accompanying electrical noise and data confusion. The problem is, simply, that you can satisfy the code without satisfying your computers' special grounding requirements. Even in those cases where a hard-wired system is designed by a highly qualified electrical engineer, the material must be cut and fabricated by workmen on the jobsite. In large systems requiring miles of wire and hundreds of feet of steel conduit, you may not get exactly what the designer put on paper. One faulty connection can upset the most carefully engineered plan.

Power distribution units are pre-engineered by people who devote all their time to computer-room power problems. They are assembled in the factory where exacting quality control, close supervision, and testing is possible. With power units, you are guaranteed a single-point, isolated grounding system where all of your computers "see" the same grounding potential over a broad frequency range.

Because computers are sensitive to all sorts of electrical interference including RFI, common mode noise, and transients, a PDU is designed to minimize these problems through proper grounding, isolation, and controlled manufacturing. A typical hard-wired system is designed only to distribute electrical power.

Most codes require an emergency shut-down system in computer rooms. Typically, a large red pushbutton will be installed beside each exit door in the room. This button sends a signal to the heating, ventilating, and air-conditioning (HVAC) system and to all computers and peripherals to shut down in the event of an emergency. For instance, in the event of a fire, this system allows the last person through an exit door to shut down the entire computer room by pushing one button. It is a good idea, but it is expensive to engineer and construct with a hard-wired system. Some codes require low voltage (24 volts) while others do not. Some HVAC systems must be fed a 120-volt signal to shut down. On such an application there appears to be very little that is standard.

All power units come with an emergency shut-down system pre-engineered and included as standard equipment. Power units provide a no-worry method of satisfying this important code requirement. They also can be interfaced with other building systems, such as smoke detectors and water sprinklers, for monitoring or control functions.

PDU flexibility

Most computer rooms are constantly being rearranged as data needs change. Rearrangement may result from the addition of faster processors or the addition of more memory units or simply the desire to rearrange equipment to provide a more efficient work flow. Each time a change is made, the flexibility of a PDU is evident. Since power is distributed under the raised floor through specially designed flexible cables, the computer and its peripheral equipment can be moved to any new location at will.

With a hard-wired system there is no flexibility. Any move must be done by an electrical installer, who brings new wire and new conduit on the job, where it is cut to fit and secured in place—an inflexible extension of an inflexible system. The trouble and expense of bringing in an electrician for each move keeps many DP managers from experimenting with different equipment arrangements in search of the most efficient layout. For example, it may be desirable to widen an aisle

TYPICAL PDU is a floor-standing enclosure that looks very similar to computer equipment and contains an isolating transformer to feed circuit-breaker panelboards that supply circuits to computers and peripherals—as well as providing monitoring, status-indication and other intelligence output on the conditions of power supply to the data processing system.

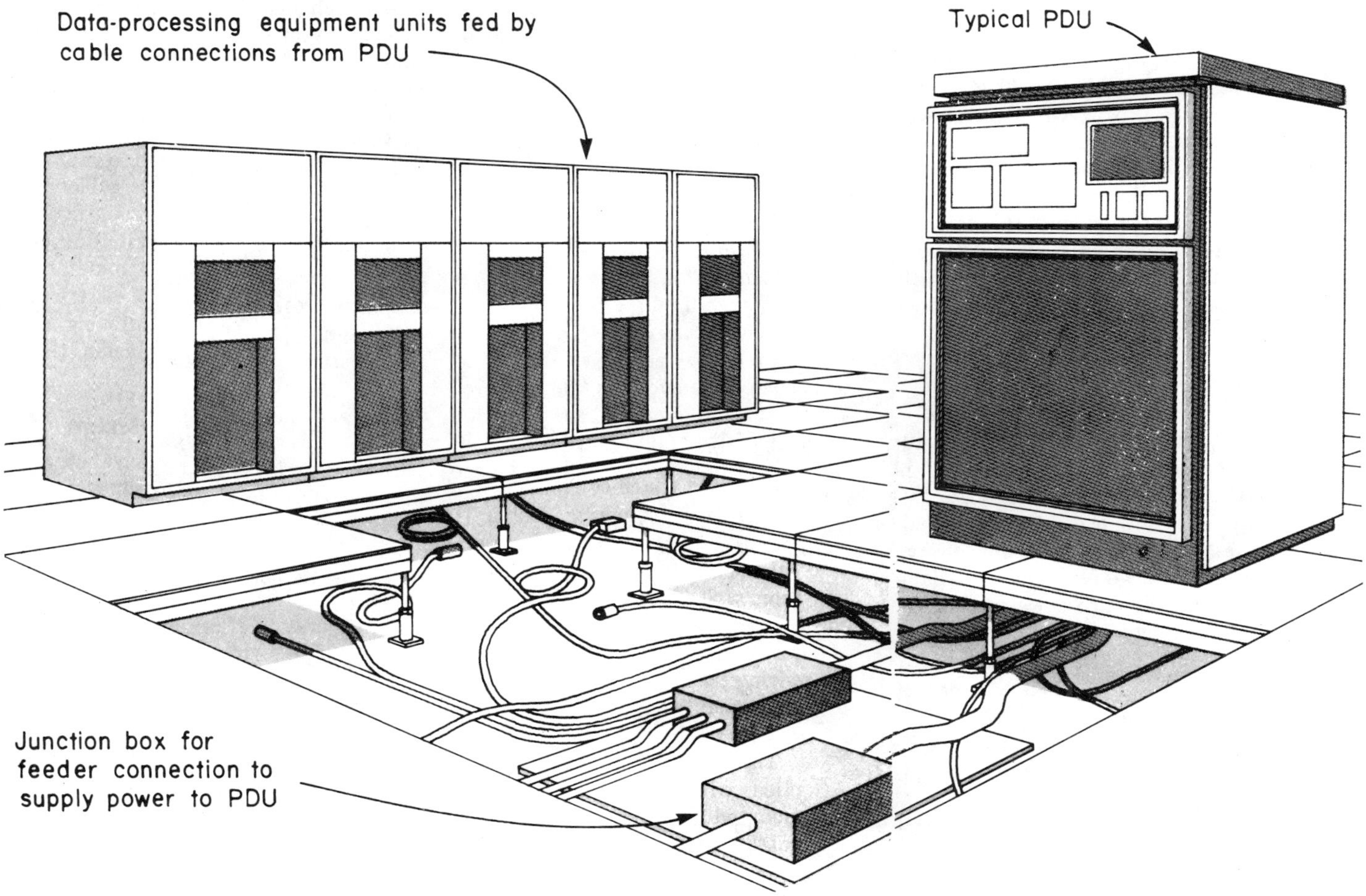

between two rows of peripherals to allow easier personnel access. If power units are used, this move becomes very simple. The DP manager simply unplugs his equipment, moves it to the new location, and reinserts the plug in his power unit's flexible cable. If this same DP manager has a hard-wired system, it may be necessary to have the engineering department or building superintendent first make a change on the electrical plans or schematic drawings. Then the job may have to be sent out for two or more bids, depending on company policy.

In the case of a new computer facility, power unit flexibility becomes a great asset. If the facility is to be hard-wired, it will be necessary to plan the arrangement of the computers and peripherals very carefully. Since the system will not be flexible, it will not be easy to move things around after the facility is up and running. Someone will have to spend a lot of time and "crystal ball" the future to preplan the most efficient layout. It is hard to do this with a large facility and be 100% accurate because of the infinite number of possible equipment arrangements.

If power units are specified for this same new facility, it will be possible for the DP manager to move his equipment around almost at will after he has had a chance to see how his employees interact with the new system. And power units do not become part of the building structure, so the user can take them along when he moves to a new location.

Monitoring status

Everyone understands that computer rooms are sensitive to all sorts of electrical problems. Included in this group are surges, brownouts, sags, outages, and fast transients. Some electrical problems have been known to damage software or burn out circuit cards.

Unfortunately, in too many places the quality of electrical power seems to be deteriorating. We are probably going to have more difficulties in the future than we have had in the past. To cope with such deterioration at a time when DP equipment is operating at ever-increasing speeds, it is necessary to have good management data on the status of computer-room electrical power. There is certainly no way to take contingency action or track down electrical problems if the DP manager does not know what problems exist.

Power units solve this problem nicely. Different models are available that have the type of monitoring and annunciation capabilities to fit in with any computer facility. New models are now available with microprocessor-controlled monitoring systems as standard equipment, providing an enormous amount of management information never available before at a reasonable price. As an example, the DP manager can simultaneously have displayed on a cathode ray tube (CRT) the following data:

(1) Input voltage from outside the computer room—typically 480 volts—between each pair of hot lines (three displays)
(2) Output voltage from the power unit to the computers, line-to-line—nominally 208 volts—(three displays)
(3) Output voltage from the power unit to some peripherals, line-to-neutral—nominally 120 volts—(three displays)
(4) Output current in each hot line (three displays)
(5) Output current in the ground
(6) Output current in the neutral
(7) Apparent power (kva)
(8) Instantaneous power (kw)
(9) Power factor
(10) Energy consumption (kwhr).

Additionally, a *system fault monitor* warns of:

(11) Phase loss
(12) Phase reversal
(13) Ground fault
(14) High voltage
(15) Low voltage
(16) Frequency error
(17) High transformer temperature.

Also, a *building alarm monitor* can watch sensitive building systems critical to the computer environment, such as:

(18) Sprinkler systems
(19) Halon systems
(20) Smoke detection equipment
(21) Security system
(22) Air conditioning
(23) Water chillers
(24) Motor-generators
(25) UPS equipment.

Finally, a *power line disturbance (transient) monitor* records fast transients displaying:

(26) Transient magnitude
(27) Transient duration
(28) Time-of-day of disturbance
(29) Date of disturbance.

A permanent record of transient activity (items 26-29) and system faults (items 1-17) is made by a paper-tape printer to watch the computer power 24 hours a day and give the DP personnel the type of meaningful record necessary for proper power management. None of this information is available with a hard-wire system.

PDU system costs

The bottom line cannot be overlooked. Some power-unit customers say it costs 5% less for material and installation to use power units rather than hard wiring. Some say 25% less. That depends on the system and the model chosen. A system factory-assembled on a production line simply costs less than one built on the job piece by piece.

There are also some hidden savings in specifying power units that should not be overlooked:

(1) *Planning cost.* If a hard-wired system is chosen, some DP manager has to spend a lot of time carefully laying out each piece or equipment, since the hard-wired system is not flexible and changes in equipment position cannot be made easily. With power units, less planning is necessary because the flexible cables allow rearrangement if it becomes desirable.

(2) *Electrical design costs.* A hard-wired system requires many hours of work by an electrical engineer on a drawing board to design the layout. Power units are pre-engineered at the factory; consequently, the user saves the entire cost of this phase of electrical design.

(3) *Tax credit.* Power units are purchased like any other piece of computer peripheral equipment. They qualify for the 10% investment tax credit. Hard-wired systems become part of the building structure and do not qualify for the tax credit.

(4) *Rapid depreciation.* Power units can be depreciated for tax purposes like other pieces of computer equipment. Usually the Internal Revenue Service will allow a 5- to 8-year write-off, depending on the industry in which the user is involved. Since hard-wired systems are part of the building structure, the usual period for write-off is between 22 and 30 years, if the user owns his own building. If the hard-wired user is leasing his premises, the hard-wired system must be treated as a leased-hold improvement and depreciated over the life of the lease. Obviously, you get your benefits from Uncle Sam much quicker by using power

FLEXIBLE DISTRIBUTION of power to and from a PDU is readily achieved by cables and other NE Code-approved wiring methods run in the raised-floor space of the computer room.

units; and, in these days of high interest rates, time is money—a lot of money.

(5) *Savings from related components.* Since a PDU has its own transformer, higher-voltage feeders (480-volt rather than 208) can be run into the computer room. The smaller 480-volt feeder costs substantially less than a 208-volt feeder. This feature becomes particularly important if the computer facility has a UPS. The bypass switchgear on the UPS is much cheaper at 480 volts than at 208 volts. Also if the UPS output is at 480 volts, it will be possible to bypass directly to the utility without using an extra transformer as would be the case with a 208-volt-output UPS for a hard-wired system. Some UPS manufacturers charge an additional 3% to produce 208-volt output units that can feed a hard-wired system. In fact, some UPS users say they saved enough on a 480-volt feeder system to pay for their power units.

PDUs offer something for everyone. They are pre-engineered for the electrician, are flexible with good monitoring capabilities for the DP manager, are available at low cost for the purchasing agent, and provide good tax policy for the company treasurer.

Uninterruptible power system serves computer center

Combination of lead-acid storage batteries, charger, and inverters assures stability and dependability of computer supply.

By A. B. PFLIEGER,
Exide Power Systems Div.,
ESB Inc., Philadelphia, Pa.

AN UNINTERRUPTIBLE ac power supply system incorporating a bank of 120 lead-acid standby battery cells was used at the new computer center of the Firestone Tire and Rubber Co. in Akron, Ohio.

Operational since the latter part of 1968, an IBM System/360 Model 50 computer is teamed up with eight other computers in the center to form a fully integrated computer communications system. Memory and logic circuits of the System/360 are directly connected via telephone lines to other computers and data terminals in 110 Firestone plants, warehouses and district offices throughout the country. A $2-billion business relies on the 100% efficiency of this system and its ability to ask questions, issue orders, write notes, check files, and perform many other fantastic tasks 24 hours a day.

The high-speed, super-sophisticated, complex electronic machinery has proved its ability. But a precaution is necessary: the solid state "third-generation" hardware of such systems is extremely vulnerable to the slightest interruptions of electrical supply.

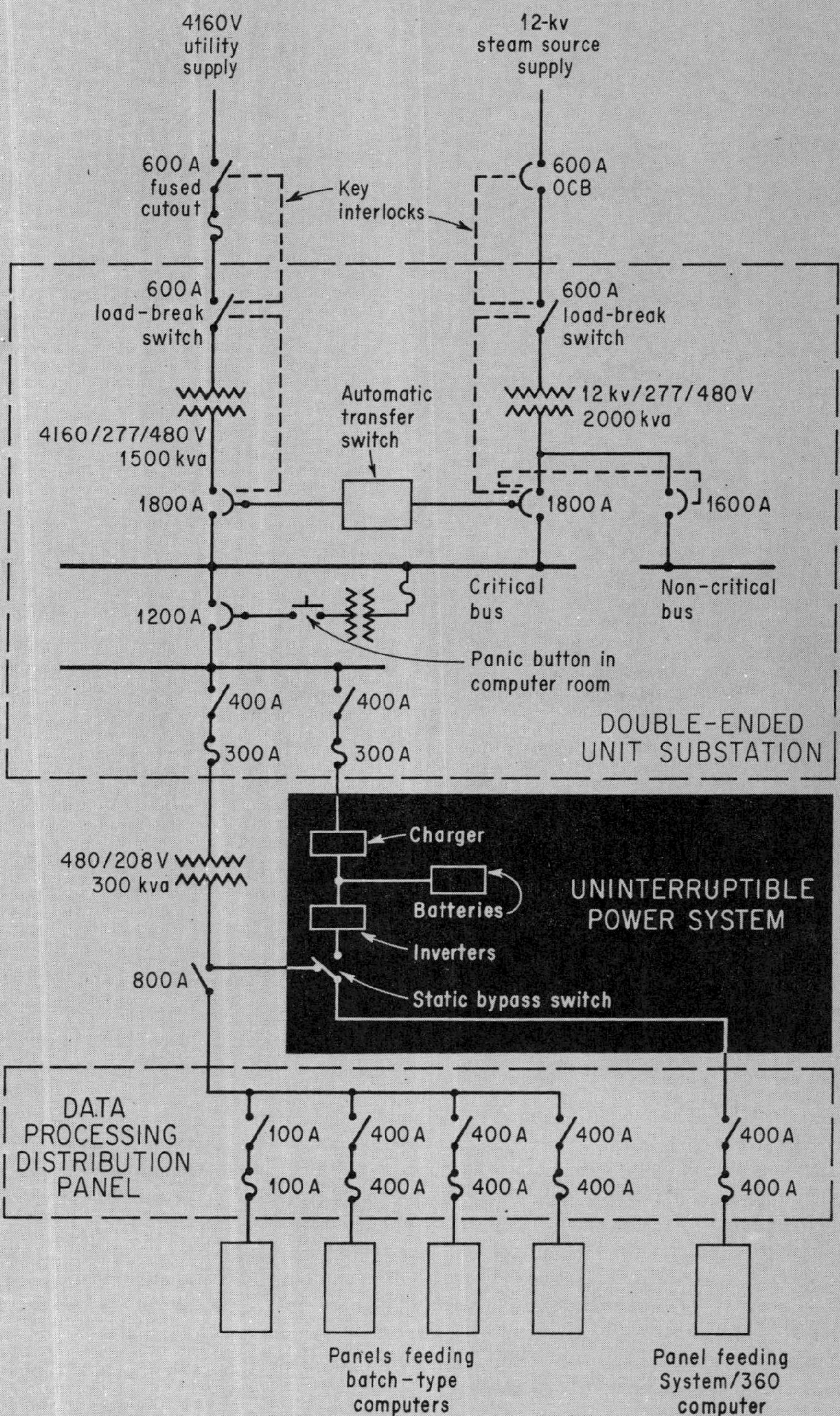

SINGLE-LINE DIAGRAM shows major elements of distribution system and connection of uninterruptible power components.

STATIC INVERTER system rated 90 kva changes battery power to ac, providing regulated output voltage and frequency.

Company engineers had provided two separate supply systems—Firestone's own electric generating plant, equipped with four steam-driven turbine-generator sets, and a standby alternate power supply for critical circuits of the computer center from the Ohio Edison Co. Both of these supplies feed into a double-ended unit substation in the basement of the computer center. If the power supplied by the steam plant deteriorates in quality or is interrupted, a circuit-breaker switching system transfers the critical circuits to the standby system.

However, although this transfer interrupts the current flow for only a fraction of a second, the 6- to 8-cycle loss is unacceptable to the System/360 operations. A power outage of only ½ cycle (8 milliseconds) is sufficient to cause a loss of data being processed or stored amounting to a "blackout" of 512,000 positions of magnetic core memory. While most of this data is program information which could be quickly replaced, nobody knows how much data would be affected in the transmission lines connected to the computer.

That's where the uninterruptible power system (UPS) comes in. Power is fed to the System/360 computer through the UPS. If and when a power transfer occurs from the steam-powered source to the standby utility supply, the UPS supplies "no-break" power to the computer, the electronic circuits of which don't even notice the changeover. The same applies when power is switched back from the standby line to the prime source.

The battery cells provide the emergency power needed during the tiny interruption. They also can supply enough power to operate the computer for 10 to 15 minutes (depending on how many system components are operating) if *both* electric feeds to the substation fail. If such an unlikely situation developed, this short period of battery-powered operation would be used for an orderly shutdown of the computer's "real-time" operations. (Real-time processing involves the transmission of data to and from the computer by means of telephone lines from remote locations, while "batch-type" computers process one job after another entirely within the confines of the installation.)

Chief components of the uninterruptible power system, in addition to the battery cells, are two battery chargers rated 648 amp-hrs at an 8-hr discharge rate and a 3-phase static inverter system rated 90 kva. A built-in static switching device is used by the system when it is necessary to remove the UPS from the line for any reason, providing a maximum interruption of 0- to-0.25 cycle, well within the computer's operating limit.

The UPS system is more than a source of emergency power, however. Connected in the system at all times, the combination charger/battery/inverter eliminates voltage variations and transient conditions that can cause computer downtime. The System/360 requires a 3-phase, 208-volt supply that may not deviate from the stated voltage by more than +10% or −8%. Power that varies from these specifications can cause erratic behavior of components, affecting the reliability of data in process. Core data may be lost from memory; information may be dropped during transfer; physical damage to hardware could result; or the program may recycle.

The UPS system "conditions" the

BATTERY CHARGER converts ac input to regulated dc which maintains battery at full charge and provides inverter input power. A portion of the 120-cell battery bank is at left.

power supplied to the computer by eliminating voltage spikes and interruptions and by constantly monitoring the frequency to ensure maintenance of 60 cycles. Voltage spikes are corrected in the battery charger and by absorption into the battery cells, which act as electrical "sponges." These corrections occur at the speed of electricity, unnoticed by the computer.

Frequency is normally more reliable than voltage in electrical supply systems, but frequency deviations can and do occur. The UPS static inverter cabinets contain stable transistorized oscillators capable of maintaining the inverter output frequency to within ±0.50 Hz. (Extremely critical loads can be protected with a quartz crystal oscillator controlling the inverter's frequency output within a tolerance range of ±0.01%.)

A synchronizing circuit matches the utility frequency and phase to the inverter's output, with each supply monitoring the other. An over-and-under frequency relay is used to protect the computer against "excursions" of the substation's frequency output. This relay disconnects the inverter from the synchronizing circuit if the power feed experiences a frequency deviation beyond the design parameter of the computer load.

Power fed to the UPS is converted from ac to dc for float-charging of the battery cells. This is the process by which a standby battery is prevented from losing its charge. (Any battery will gradually lose its reserve of chemical energy, even with no load connected to it, unless it is kept on float charge.) The charger system constantly monitors the battery's voltage (nominally 2.0 volts per cell) and corrects any decrease by reversing the chemical processes that cause the loss. The inverter is directly supplied with dc from the chargers, where it is converted to ac and fed to the computer.

The accompanying single-line diagram shows the major distribution elements of the power system. Critical circuits, in addition to the computer circuits shown, include one of two 250-hp chillers and such auxiliary services as air-handling units, exhaust fans, water pumps, etc., that relate to the computer installation. Note that only the System/360 computer is served through the UPS. The remaining eight batch-type computers at the installation are not involved with communication-line (real-time) transmission; hence their requirements are not quite as critical as those of the System/360.

Protection of the real-time circuits is thus the prime reason for use of the UPS. However, it also provides protection for the batch operations performed by the System/360 when it is not engaged in real-time processing. A power interruption could cause the loss of one or even two hours' processing time, at a cost of several hundreds of dollars per hour, while operations that were in process are repeated. Such unscheduled time losses can cause backups in the schedule, resulting in late invoicing, shipping delays, and other headaches. To date, no such losses have occurred.

The Austin Co. of Cleveland, Ohio, designed the installation in conjunction with Firestone's manager of electrical engineering, E. L. Smith, and senior electrical engineer, Thomas Roden. The system was installed by electrical contractor W. J. Campbell, Inc., of Akron, Ohio.

Gas turbine power back-up

Compact, lightweight power sources proved to be the answer to complex standby power installation problems in a New York office building.

Gas-turbine generator system

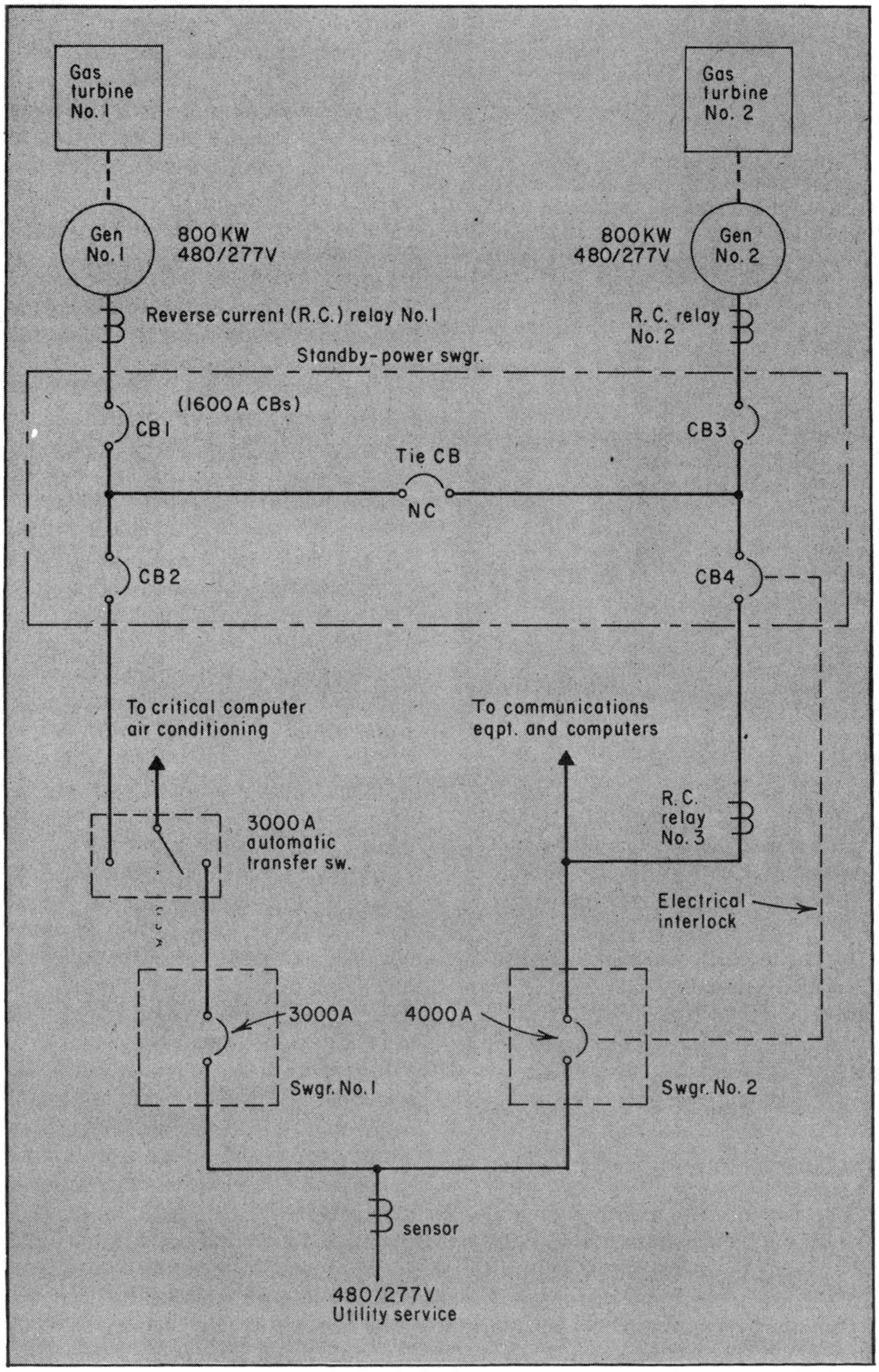

Two 800-kw generators, each driven by a gas-turbine engine, provide reliable auxiliary power at the RCA Global Communications, Inc. (RCA Glob-Com) headquarters building in New York City. The turbine-driven generators supply 480/277-volt standby power to extensive communications equipment and associated computers installed on several floors in the 42-story structure.

During initial planning, Steve Latargia, manager of facilities engineering for RCA Glob-Com, had planned to use diesel-engine generators to power the system; however, lack of space, inadequate structural support, and a number of other problems in the existing building indicated that installation of diesel equipment would be difficult and costly. After investigating the characteristics of gas-turbine generators, it was decided that they would result in the simplest and most economical installation.

The standby power system was installed to provide backup power for administration and engineering offices, and air conditioning for all computer rooms, certain computer systems, and other equipment related to the firm's worldwide communications network. Some computers are powered through a UPS system backed by a 375-kva diesel-engine generator set. Power via a UPS system for other computers, however, could not be justified because of their importance to overall operations. A standby power system was the answer.

Design and installation

RCA Glob-Com facilities encompass over 250,000 sq ft on the first seven floors of the 42-story skyscraper. Their leasing arrangement permits almost any alteration necessary, but also requires installation and maintenance of any service facilities required.

During early design stages, a 1000-kw generator was thought to be sufficient to handle the standby load, and the basement near the main switchgear seemed to be the best location for the generator. However, space for a diesel generator of this size was not available. Several smaller units were contemplated, but the cost and complexities of installation ruled them out. Also, most of the first floor is leased by major banks and tearing up the terrazzo floor, together with installation problems and interruptions to bank operations were highly undesirable.

800-KW GAS-TURBINE GENERATOR, one of two installed, is being inspected by Robert Kaible, supervisor of plant utilities, and Steve Latargia, manager of facilities engineering, both of RCA Global Communications, Inc., New York City. Enclosure panels are removed for inspection.

Several other locations for the diesel-generator were considered. The roof was available, but costs for hoisting, structural reinforcing, and installation of circuit runs and piping were excessive. Office areas would not be appropriate for structural reasons and because of unwanted noise and vibration created by the unit.

On the contrary, gas-turbine generator sets are ideal for the installation conditions encountered. For a given kva rating, they are compact, lightweight and easy to install. Latargia decided to install two 800-kw units to provide for future growth as well as to obtain additional reliability. If one unit is out of service for any reason, then the other unit can handle most of the standby load.

The two turbine generators were installed in a second-floor electric room directly over supporting beams. The existing structure was found to be sufficient to support the units, which weigh about 12,000 lb each. Also, installation of the turbines was simplified because they could be hoisted through window space on the second-floor exterior. A 10-ton hoist was used and placement of the units required less than a day.

The units were bolted directly to the floor; concrete foundations were not required since gas turbines produce little vibration. This was important because the units were located directly over the bank. The turbines provide a relatively quiet installation with the predominant sound in the high-frequency range. However, this noise is satisfactorily attenuated by equipment enclosures. Because the units are air cooled, no cooling-water plumbing was required, providing an additional cost savings. Considering reductions in installation and structural costs, it was estimated that total installed costs of the power sources resulted in a 50% savings.

System components and operation

Each gas-turbine generator set consists of a 1200-hp turbine engine, a reduction drive, and a generator mounted on a steel sub-base. Kerosene fuel was selected because it has a long storage life. A 6000-gal tank is buried in the street outside the building and a 270-gal day tank is installed in a closet adjacent to the electric room. Although fuel consumption of a gas turbine is higher than that required by reciprocating engines, it was considered acceptable because the system is for auxiliary power only and is expected to operate intermittently.

Each unit is furnished with a battery bank to supply power to the turbine starting motor and to turbine control circuits. Rated 400 amp/hr, 24 volts, the batteries are rack-mounted adjacent to each unit.

In the electric room, a control lineup consists of a turbine/generator control section for each unit and a synchronizing section for generator synchronization during parallel operation.

Five 1600-amp drawout power-circuit breakers provide for automatic operation of the standby power system and afford overcurrent protection. The CBs are grouped in a switchgear assembly at one end of the electric room.

As shown in the accompanying diagram, the local utility supplies 480/277-volt power to two separate

12,000-LB TURBINE-GENERATOR SET is hoisted through window area on second floor by 10-ton crane. Second unit rests on flatbed truck in front of building. Both units were put in place in less than a day.

STANDBY POWER SWITCHGEAR (above) contains five 1600-amp power-circuit breakers to provide for automatic operation of standby system and to afford overcurrent protection. Two CBs are connected in series with each generator output and one serves as a tie breaker.

CONTROL PANEL, (above, left) being checked by A. Nagia, engineer for RCA Global Communications, Inc., consists of generator synchronizing section at left and two sections at right each providing control for a turbine-generator set.

COMMUNICATIONS EQUIPMENT (left), computers, and air conditioning for computer rooms are major loads supplied by standby power system.

main switchboards. One feeder circuit, protected by a 3000-amp CB in switchgear No. 1, supplies power to air-conditioning equipment for critical computers through a 3000-amp automatic transfer switch. In switchgear No. 2, one feeder circuit, protected by a 4000-amp circuit breaker, supplies power to communications equipment, associated computers, and other equipment. A tap from the standby power source circuitry is made on the load side of the 4000-amp CB.

If the utility power source fails, a signal from the voltage sensor causes the 4000-amp circuit breaker to open. This action is delayed for 10 seconds to be certain that the utility outage is not momentary. When the 4000-amp CB opens, a signal is sent to the starting circuit of the turbine engines initiating their start-up. Turbine-generator No. 2 starts first. When generator No. 2 comes up to speed with correct voltage and frequency outputs, CB-3 and CB-4 close, supplying generator power to the load. Meanwhile, turbine-generator No. 1 has been started, and when its generator output is correct, a synchronization circuit parallels both generators as CB-1 and CB-2 automatically close. The 3000-amp transfer switch is furnished with a "voltage-seeking" feature. A voltage monitor will automatically position the switch to where voltage is available. Thus, after CB-1 and CB-2 close, the switch automatically connects the generator source to the load.

Reverse current relays No. 1 and No. 2 sense any imbalance of generator current due to nonsynchronization or other causes, and will open the tie CB if imbalance current is excessive. Reverse current relay No. 3 detects any current that might flow from the utility service in the event that the electric interlock fails and the 4000-amp CB and CB-4 are closed at the same time. This relay sounds an alarm and opens CB-4 via a separate circuit.

The electrical contractor on the project was Arc Electrical Construction Co., New York City.

UPS assures

Total reliability for computers

Uninterruptible power system combines gas-turbine generators, rectifier-inverter and battery banks, and well-planned switching and control to provide maximum dependability for a large industrial computer center.

COMPUTER operations at the new Combustion Engineering Inc. data center in Windsor, Conn., are backed by an uninterruptible power supply (UPS) designed to provide an extremely high degree of service continuity. The system also isolates computer equipment from the ac supply, eliminating voltage and frequency variations that could impair proper computer operation.

The 50,000-sq-ft facility offers engineering, scientific and business EDP services to its divisions in Windsor and in a number of locations throughout the country. Because of the critical nature of the work performed at the data center and because of the costly effects of interruptions in computer operations, company officials called for a high magnitude of reliability in the design of the mechanical and electrical systems.

Design of the power distribution system had to dovetail with the UPS system. Major elements include the normal incoming power supply, an alternate supply consisting of two 800-kw gas-turbine generators, main switchgear with a nonessential-load bus connected through a tie breaker to an essential-load bus, three main busways providing for alternate routes of supply to computers, and the UPS system, which is sized for a computer load of approximately 750 kva with 250 kva redundancy.

When computers are on normal operation, the local utility delivers power at 22.5 kv to a 2000-kva transformer installed adjacent to the building. The transformer steps down the primary voltage to 480/277 volts. Secondary power, carried by seven 750MCM conductors per phase and neutral, extends through the outside wall into the main switchgear. Protected by a 4000-amp main circuit breaker, the switchgear bus is divided into two sections, one designated for essential loads and the other for nonessential loads. A 3000-amp tie breaker, interlocked with auxiliary generator power breakers, connects the two sections.

From the essential power bus, 480-volt power is carried by a 2000-amp, 3-conductor busway to the UPS room. This bus supplies the UPS system, a utility bypass circuit and a 208/120-volt busway through a 300-kva transformer, and 400-cycle motor-generator sets.

The UPS system supplies power at 208/120 volts, completely free of power-line transients, to a 3000-amp bus. Branch circuits to computers and related equipment are tapped from this bus through plug-in fused switches.

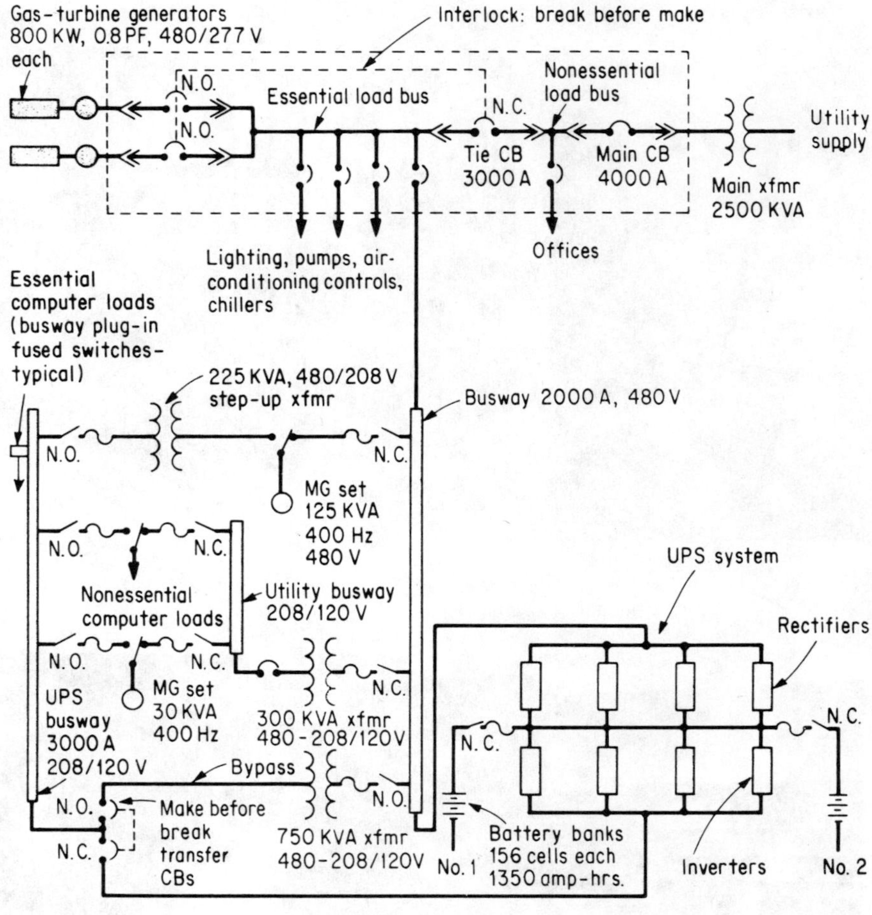

ONE-LINE DIAGRAM

System components

Four 250-kw solid-state rectifiers convert incoming ac power to dc to provide charging voltage for two battery banks. These banks, rated 600 kw for 5 minutes, supply voltage for sys-

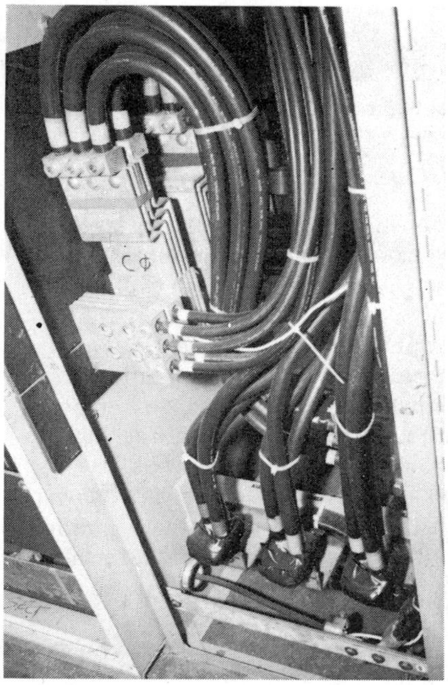

COMPUTERS and data-processing equipment (above, left) installed in 50,000-sq-ft data center are protected from transients, which could cause inaccurate operation, by UPS system. Twin 800-kw gas-turbine generators provide power in the event of power failure.

2000-KVA TRANSFORMER (above, right) pad-mounted in fenced-in enclosure steps incoming primary power to 480/277 volts for distribution. Gas-turbine generators are in background.

GENERATOR TERMINATION (left) is made by three 500MCM THW conductors per phase connected to line side of CB (bottom) and six 4/0 conductors connected to neutral bus (center). Two such installations exist, one for each generator. Larger conductors at top are six 500MCM per phase connecting load terminals of circuit breakers to main buses in the switchgear.

400-CYCLE MG SETS (below) are mounted on floor under UPS essential-power bus, which supplies computer loads. Conductors from bus plug-in switches are installed in large flexible metal raceway, permitting busway disconnects to be plugged into adjacent busways if required.

tem operation during transfer from utility to generator supply, assuring uninterrupted power to computers. Four 250-kva, solid-state inverters convert the battery dc power to ac for computer use. Other important components include twin power circuit breakers interlocked to provide make-before-break action, permitting the utility supply to bypass the UPS system and feed computer load directly if required.

Design criteria for the UPS system specified duplicates of critical components, circuits and modules. In instances where it was not practical to provide duplicate items, highly re-

MAIN SWITCHGEAR, rated 4000 amps at 480/277 volts, is installed in an electrical equipment room. Generator control and line circuit breakers are in foreground.

CONTROL CONSOLE (center) permits one operator to monitor and control UPS system. At far left are twin 2000-amp circuit breakers interlocked in computer bypass circuit. Farther down this equipment line-up are system rectifiers and solid-state inverters.

LEAD-CALCIUM BATTERY SETUP consists of two banks, each made up of 156 cells. Banks are rated 350 volts, 1350 amp-hr, and are capable of carrying the computer load for 5 min. Disconnect switches on each wall are rated 800 amps.

liable long-proven components were used. Because the computer load is limited to 750 kva, only three 250-kw rectifiers and inverters are required to carry the load. One rectifier and inverter are redundant and are available to supply power in the event that one of the on-line units should fail. During operation, all units share the load equally; thus there is no interruption of power should a rectifier or inverter be dropped out.

A second order of redundancy is provided in the inverter stages. Each inverter consists of twelve stages, each stage supplying a portion of the total load. The system permits continued operation of the inverter even with loss of one stage. Thus, failure of a thyristor, diode or logic signal will instantly be isolated by its own fuse, permitting the remaining stages to operate with only a very slight difference in overall inverter effectiveness.

When inverters are operated in such a redundant (parallel) system, it is of prime importance that the frequency of all inverters be identical. This is achieved through a master frequency generator, which sets the frequency of the entire system. Because this master frequency reference is the signal to which all the inverters synchronize, it must be extremely reliable. The master reference consists of three oscillators (each on individual plug-in cards) operating in a redundant mode, such that loss of one will not interrupt system operation. Three oscillators are required to give a two-out-of-three logic to determine if any oscillator is in error. The master oscillator operates from the critical bus so that it too benefits from uninterruptible power. It is normally free-running except during the make-before-break transfer, when it is synchronized to the line.

Each of the two battery banks consists of 156 lead-calcium cells, an output of 350 volts dc. Lead-calcium type was selected because these batteries can be operated continuously and do not require an "equalizing" charge for reliable operation. Each bank is protected by an 800-amp fused disconnect switch.

A control console containing metering, alarm indicators, mimic bus and controls is installed in the UPS room. Its primary function is to provide a single location to monitor the complete UPS system. It also houses the master frequency generator.

System operation

Should the normal utility power fail, the tie circuit breaker automatically opens. While the gas-turbine generators are in the process of starting and coming up to voltage, the battery bank maintains power to the computers. When the generators are up to proper voltage, their line circuit breakers close, and output voltage is synchronized to the UPS system supply and to critical mechanical equipment. Only one generator is required to power the UPS system; however, to obtain highest reliability, a second unit is on-line to assure continuous operation should one fail.

During emergency-mode operation, loads on the nonessential bus are dropped, while loads on the essential bus continue to function. Essential loads include four 235-ton chiller units that provide chilled water for the computer room air conditioners. There are 15 conditioning units in the room to maintain specified temperature and humidity levels. Each unit is furnished with a filter box, cooling and dehumidification coils, a blower, reheat coils, and a steam humidifier. Control is provided by thermostats and humidistats located in the air supply to each conditioner.

In the UPS room, a variety of circuit arrangements permit alternate selection of power supply routes, providing for further dependability of operation. As shown on the modified one-line diagram, three busways are installed parallel to each other. This arrangement permits removal of bus plug-in switches from one busway and connection to another. The 3000-amp UPS busway and the utility busway are installed in close proximity, and circuits are carried in large flexible metal raceway long enough to be trained to either busway as desired. Double-throw switches are provided so that motor-generator sets can be supplied from either of two sources, as shown. The m-g sets provide 400-cycle power required by some of the data-processing equipment. Additional details are provided by accompanying photos.

Electrical contractor on the project was Baldwin-Stewart Electrical Co. of Hartford, Conn. Consulting engineer was C.E. Maguire, Inc. of Waltham, Mass.

Constant-power system for computers

By J. V. ORTIZ, Senior Electrical Engineer
Syska & Hennessy, Inc., Engineers, New York City

EMERGENCY POWER for bank electrical loads is furnished by seven diesel electric sets capable of serving a combined load of 5400 kw. Two units serve as standby for the two constant-power motor-generator sets, while the balance would supply other critical building loads in the event of a major outage.

Flywheel kinetic energy, backed by diesel standby power, insures uninterrupted data processing supply.

ELECTRIC power failures, ranging from long blackouts to transient deficiencies lasting milliseconds, will not affect essential operations at the New York office of the First National City Bank of New York, thanks to a 5400-kw comprehensive emergency power system and two 500-kva constant-power motor-generator sets installed by the bank.

Transient effects are eliminated by the M-G sets, which are equipped with inertial flywheels to smooth the power flow. Seven back-up diesel electric sets will automatically supply emergency power in less than ten seconds after a power failure for the bank's computers, air conditioning, ventilation, lighting and elevators.

Uninterrupted power is crucial. More than 6000 employees work at this location, 23 major computer systems function around the clock and millions of transactions involving checks, stocks, payrolls and loans are processed daily.

Unlike other electro-mechanical devices, computers are extremely sensitive to power failures and environmental conditions. In most cases, computer power is specified as a 3-phase alternating current of 60 hertz frequency, 120/208 volts, using 3-phase conductors and a grounded neutral.

According to manufacturers' data, computers malfunction when the voltage exceeds plus or minus 10% of normal and/or when the frequency varies more than one-half cycle above or below normal. They also malfunction when the shape of the phase-voltage sine wave is distorted by peaks or dips lasting more than $\frac{1}{8}$ cycle or the equivalent of 2 milliseconds. This is understandable in view of the widespread use of silicon-controlled rectifiers and other solid-state electronic devices in computer circuits. Problems are also caused by harmonics contamination exceeding 5% or by variations in phase vector displacement exceeding four electrical degrees above or below the normal 120-degree displacement. Should the voltage vary from 10% to 20% of normal for periods of time ranging from one millisecond to 100 milliseconds ($\frac{1}{16}$ cycle to 6 cycles), computers will make errors without shutting down. This seems like a very short time, but a modern computer could, for example, make between 16,000 and 266,000 imperfect additions in this time interval.

Finally, environmental conditions must be kept constant. Computers can malfunction when the room temperature and humidity changes rapidly from the ideal conditions of 75F and 50%, or when the 60-to-85F operational temperature range and/or the operational relative-humidity range of 40-to-60% is not maintained.

Considering the essential and precise power requirements, it is obvious that power supplied by the utility or emergency diesel generators is not suitable for an error-free computer operation. Although utility power comes from transformer networks with the highest degree of firmness and reliability, records show that power failures of long, short, and transient duration occur at a minimum rate of 52 times per year or an average of once a week. The great majority of the outages are in the form of voltage peaks or dips of a transitory nature lasting less than a second, or about 60 cycles. All of them affect computer operations.

In fairness to the utility, it should be stated that these power deficiencies are largely unavoidable. Routine switching operation of generators, transformers, transmission lines, etc.; maintenance and repairs of facilities; new construction; and unscheduled outages of equipment are daily occurrences in any large utility system. These functions require, for the protection of men and equipment, the opening and closing of circuits which cause voltage and frequency transients.

The availability of diesel emergency power during a utility outage is by itself of no practical value to the computers. By the time the diesel generator controls sense a utility power failure and start cranking

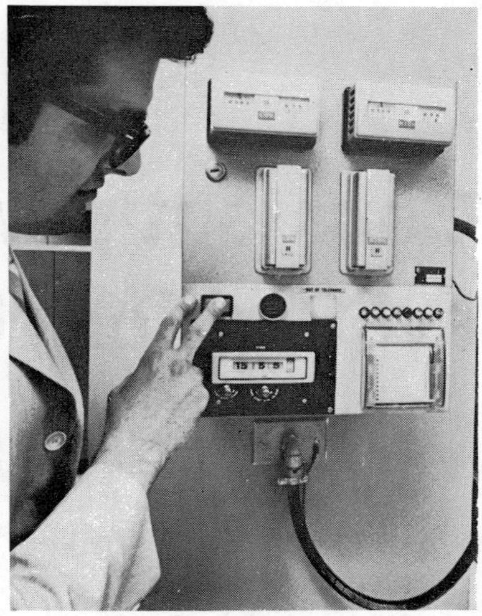

INERTIAL-TYPE constant-power units consist of wound-rotor driving motor, flywheel, eddy-current coupling, and generator. Two such sets provide a total of 1000 kva for computer supply. Monitor installed in computer area (photo at right) constantly checks voltage, frequency, temperature and humidity. Should any function deviate beyond tolerances, a light flashes, a horn sounds and a recorder charts the date, time and amount of deviation.

up, all computer operations would already be impaired and the system would be in a shutdown course. Even if diesel generators were kept running on a continuous standby basis, the act of automatically transferring the power supply from the failed utility lines to emergency diesel generation is sufficient to cause error-producing transients and a shutdown of computers. Moreover, the quality of the diesel generator power as related to critical voltage and frequency variations under changing loads is likely to fall short of computer requirements.

This is where the inertial-type power buffering system comes in. It has two major components: (1) a power input section, consisting of a wound-rotor driving motor connected to a flywheel and the input side of an eddy current coupling; and (2) a power output section consisting of the generator connected to and driven from the load side of the coupling.

The driving motor is energized normally by the utility lines. Its speed is automatically controlled to maintain the speed of the input section at a nominal 1200 rpm. Through the magnetic field of the coupling, power is transmitted from the input to the output side of the coupling and its connected generator. The power transmitted is continuously modulated to maintain the generator voltage and frequency at a constant level under changing loads.

When the utility power fails, the eddy current coupling permits instantaneous use of the kinetic energy stored in the flywheel. The inertial force of the input section moving at 1200 rpm is released by the coupling to maintain the generator at the nominal speed of 900 rpm for as long as 15 seconds. During the 15-second period and under full load, the flywheel is likely to slow down from 1200 to 1000 rpm.

As most of the power failures are transitory and lasting less than a second, the inertial system will automatically correct the majority of utility power deficiencies without the aid of the diesel generators.

If the utility outage lasts more than two seconds, the diesel generator is automatically started, brought to full voltage and frequency, and switched by means of an automatic transfer switch to drive the input motor. This two-second delay is necessary to avoid false starts on utility power transients. Engines get up to full voltage and frequency in an average of five seconds, while the transfer from utility failure to emergency power is accomplished in an average of 8 seconds. The diesel engines are prelubricated and equipped with jacket water heaters to ensure rapid starting. Under diesel power, the motor will rapidly recover its initial speed of 1200 rpm and replenish the stored energy.

When the utility power is regained, the system will automatically transfer from diesel generation to utility power. The diesel generators will be shut down and the system will be ready for the next emergency.

The system is equipped with bypass circuits to transfer the computer load from the motor-generator set to the utility lines and vice versa, should the need occur.

This constant-power system was chosen over other available systems because, for this application, it was found to offer technical and economic advantages. It was particularly important to avoid the use of equipment or circuits which might become self-contaminating.

The use of a well-designed output generator assured a computer power supply with good voltage regulation, low harmonics content, accurate wave form, fixed vector displacement, and adequate reserve capacity for momentary overloads. A unit used in the computer room to detect and record undesirable power and environmental deficiencies showed that during the first week of operation the system shielded the computers from the effects of several transients.

The installing electrical contractor was Arc Electric, New York City.

POWER CONDITIONING FOR AUTOMATED MANUFACTURING

Joseph R. Knisley

UPS and on-site generation provide instant, reliable power in the event of a utility outage.

CONTINUOUS and clean power for computers, distributed process controls, HVAC systems, and security systems at the R. J. Reynolds Tobacco Co.'s (RJR) cigarette packaging plant in Tobaccoville, NC, is provided by an array of uninterruptible power systems (UPS).

At this 2,000,000-sq-ft facility, the largest and most modern plant of its kind in the world, the firm aims for lowest cost and highest quality product through automation. The controls include 100 programmable logic controllers located within some of the many motor control centers within the facility.

Each of the various brands of cigarettes made at Tobaccoville uses a unique blend of tobacco to create particular smoking characteristics. Many different types of tobacco combined in precise quantities are required for each product. While being mixed, the tobacco is constantly subjected to moisture analysis by sensors that report back to the distributed control system in each area, thereby assuring that uniform blending of tobaccos and flavoring ingredients is being achieved.

From the final bulkers, the tobacco is pneumatically conveyed to 72 machine complexes (laid out as 72 separate processing lines in the "making and packing" area) where cigarettes are formed, packaged, and palletized.

Delivering unreliable power to the computer equipment supervising the manufacturing process could result in scrambled digital data, overstressed semiconductor components, or loss of software program sequence; thereby leading to off-normal equipment performance and reduced quality.

Clean, reliable power needs

In focusing on methods for protecting the sensitive electronic equipment from power aberrations, the RJR engineering department, which creates master specs and engineering guidelines on projects, had to weigh economic considerations in the selection of this equipment. They recognized that line-voltage regulators, isolation transformers, dedicated lines, power distribution centers, and the like would not bring them to a 100% level of assurance. Thus, to shield all critical loads from even a brief loss of power and from any type of noise, the overall protection afforded by an uninterruptible power system (UPS) was specified. This choice was based on the company's electrical engineering department's previous experience at existing tobacco processing plants, in which similar types of automated equipment had been installed.

A life-cycle cost analysis was prepared in selecting the UPS units, so that site preparation, maintenance requirements, and equipment operating efficiency also were considered in the evaluation procedure.

The continuous 3-phase UPS system rated 150kVA serves the primary processing computer room and is typical of the 300kVA total capacity that is now serving the administrative computer room at the Tobaccoville facility. The output voltage is 3 phase, 480VAC and the utility input voltage is 3 phase, 480VAC.

Fig. 1 shows a block diagram of the 150kVA UPS single-module system. During normal operations, the unconditioned AC input feeds the rectifier that converts AC to DC, float charges the battery bank, and supplies DC to the static inverter. Anytime that normal power is lost, the battery delivers DC voltage to the inverter.

In the event a rectifier/charger or an inverter section malfunctions, the control circuitry initiates a rapid transfer of the load to a bypass line without interruption. This is called a "make before break" transfer or "reverse transfer" and is a key design element or feature in the UPS unit. Three events occur just about simultaneously in the rapid transfer. As the voltage on the inverter output is decreasing, a static transfer switch, using silicon control rectifiers (SCRs), operates within 4 ms (nearly $\frac{1}{4}$ cycle) to begin feeding the bypass line power to the load. At the same time, the control circuitry initiates opening of the normally closed output switch and closing the normally open bypass switch with which it is ganged. This switching action is motorized.

Once the bypass line power is flowing through the fully closed bypass switch, the static transfer switch is removed from the power flow because it is not rated for continuous operation. By sending a control signal to the SCR gates, each gate circuit initiates the power turn-off through its SCR. The UPS is always maintained in synchronism with the AC input line to permit the transfer.

Because the bypass line feeds "raw" (unfiltered and unregulated) power to the computer, it is a temporary emergency or maintenance

The Tobaccoville facility, which makes extensive use of computers in virtually every step of the manufacturing process, attains a high degree of power reliability. It is served by the RJR 80MW utility plant, in foreground, that cogenerates with the local power company. Circle shows one of the transmission towers supporting the utility's 103.5kV incoming line, which traverses the site.

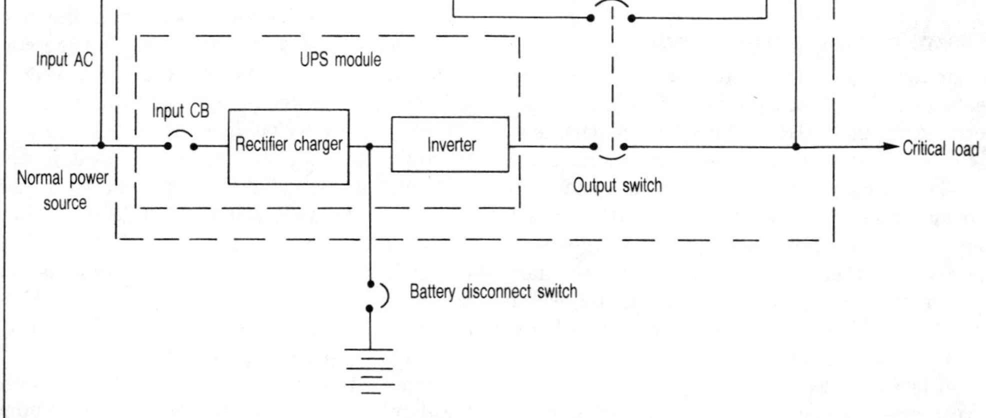

Fig. 1 The 150kVA UPS, single-module system permits a transfer from normal AC input to a bypass line by simultaneously turning on the static transfer switch, closing the bypass switch, and opening the output switch. Located above the static switch, a manual disconnect switch provides a way of bypassing both the UPS module and the static switch for maintenance or troubleshooting on the entire system.

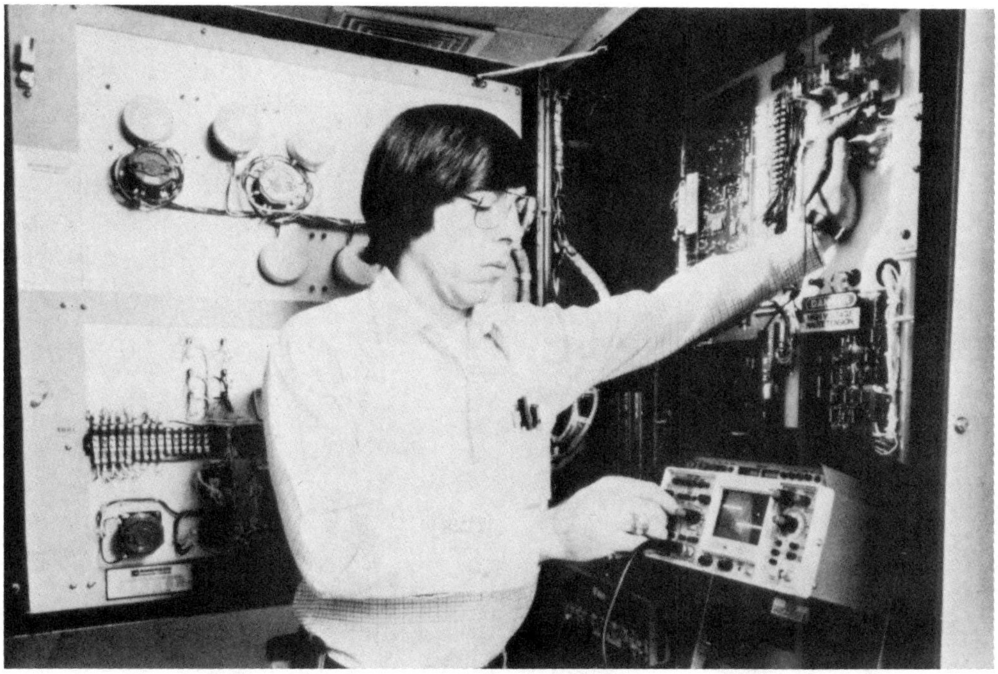

An RJR electrical technician checks waveforms relating to one of the 150kVA on-line UPS logic boards using an oscilloscope. With an on-line system, the load is continuously served from the UPS, which buffers all disturbances on the incoming line. Each inverter section is made up of two 3-phase bridge inverters, and output of the inverter pair on each phase are added to create a step wave or staircase-like approximation of a sine wave that requires relatively little filtering.

condition only. Whenever a UPS unit malfunction occurs and/or the load is switched through the bypass line, alarms are sent to a supervisory workstation in the plant's control room to permit rapid correction of this off-normal condition.

Monitoring and protection

Close attention was given by the RJR engineering staff to the technical details associated with the selection and integration of the UPS into the plant's electrical system. By its inherent design, the step wave inverter is not protected against short circuits on its output, and it must rely on the control circuitry to cut back output power quickly to a level that will not damage inverter components. At the same time, the UPS has limited overload capability. If it is exceeded in magnitude and/or duration, the unit will "current limit." The action of current limiting reduces the output voltage, which in turn transfers the load to the bypass line. Thus, the bypass line, static switch bypass circuit breaker, and power-source switchgear were specified and rated to carry and/or interrupt this overload.

The static switch also has its own overload limit. If this limit is exceeded before the bypass circuit breaker closes, the static switch fuses operate to protect the static switch.

Diagnostic circuitry for troubleshooting and system alignment aids is provided on RJR's UPS unit. Operating conditions are visually read from control panel meters and a lineup of LEDs. A mimic bus of the entire UPS single-module system on the control/monitoring panel shows the layout of elements in the UPS module, including the CBs and their status.

Off-normal LED alarms include overloads, overtemperatures, control power failure, blown fuses, output over/undervoltage, battery discharge, DC overvoltage, input power failure, control power failure, and disabled static switches. These alarms also are communicated to a station in the plant's control room.

Modularity feature

Since the RJR plant uses other UPS units from the same manufacturer and since the plant's maintenance program functions in conjunction with a company-wide spare parts inventory and ordering system, substitution and switching of parts in the event of failure can be carried out easily. All of the UPS units of this design with ratings from 50 to 750kVA use the same control card, which is an advantage if a semiconductor in the card fails.

Because the UPS equipment is modular in design, a UPS can be doubled in capacity on-site by linking a duplicate rectifier/inverter section with the original system without need for reconstruction.

Construction and installation

The 150kVA UPS unit is located in a well-ventilated, air-conditioned room, and adequate horizontal clearances as well as vertical clearances are maintained around the cabinet. Four air-circulation fans are mounted in the top cover of the cabinet section holding the rectifier and inverter components. Forced air circulation is used to make sure that semiconductor components are never subjected to excessive elevated temperatures. The unit is rated for operation in ambient temperatures up to 40°C.

The lead-calcium battery bank is located in

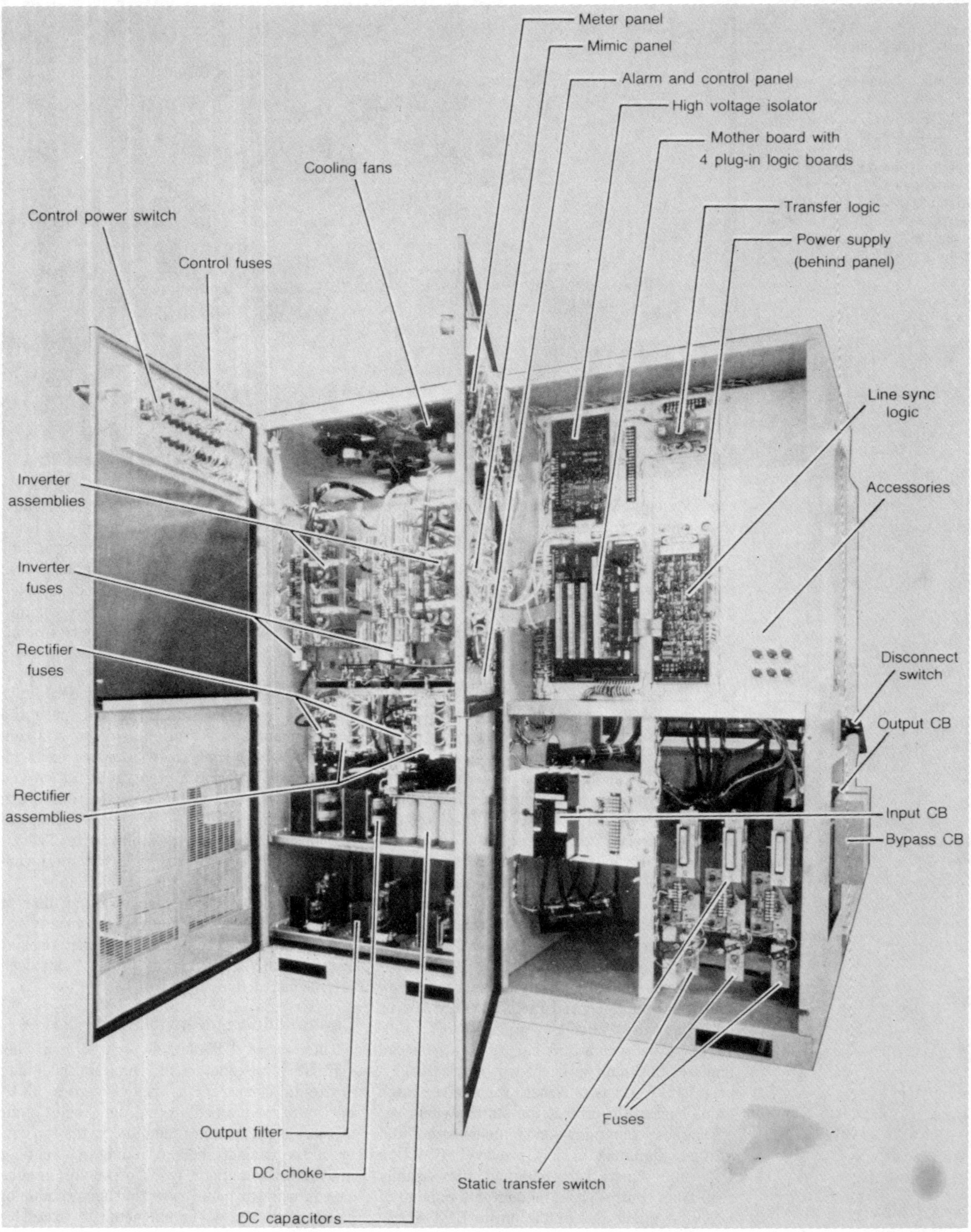

UPS module consists of the rectifier/charger and inverter section which are mounted in the left cabinet. A module can be combined with other modules to increase kVA capacity. The system control section is located in the right cabinet. Important components are identified to show the accessibility of the various elements. The equipment is housed in a NEMA 1 metal enclosure suitable for floor mounting, and the structural frame has provision for hoisting, jacking, and fork-lift handling. Associated switchgear is rated for the application.

its own ventilated room near the UPS cabinet. Consisting of 184 cells, connected in series to produce a nominal discharge of 405VDC, the battery array provides rated kVA for approximately 20 to 30 min.

All batteries are rack-mounted and their internal connections made when the batteries were installed. Field work consisted of connecting the cables from the battery racks to the UPS module and interconnection between batteries. To assure full output and rated battery life, all field connections were made finger-tight and then torqued to specifications.

As part of RJR maintenance procedures, the batteries are visually inspected on a quarterly basis. In addition, the connections between cells are checked by infrared scanners for overheating and for proper torque during scheduled maintenance procedures. At the same time, resistance measurements are made at terminals to uncover possible hidden corrosion.

Power distribution system

In addition to the application of UPS equipment, a cogeneration scheme and a highly reliable plant power distribution system are also used to gain a high level of reliability.

The plant's electrical distribution system (15kV resistance grounded) has switchgear primary and emergency busses and is laid out to provide power under almost any contingency. 13.2kV service is brought from RJR's on-site power plant where two 40MW generators cogenerate with the local utility.

From the switchgear, a series of dual feeders are routed to the manufacturing building. Within the building, distribution is a radial, primary selective system, and each substation within the manufacturing building is fed by a primary and an emergency feeder. The substations have two electrically operated switches (with a common fuse) that will automatically transfer to the emergency line in case of primary line failure.

In addition, each substation's 480V secondary power distribution system is protected by high-resistance grounding that prevents overvoltages and permits the addition of selective ground-fault detection devices. In the event that a feeder suffers a ground fault, no shutdown occurs on the first fault since ground-fault detectors only alarm the condition. The maintenance force has an opportunity to locate and correct the fault with a minimum of service discontinuity.

Clean power sources

The plant's preventive maintenance program also includes procedures for recording power-line disturbance occurring on the power system to pinpoint sources of power-line aberrations created within the plant and solve problems well in advance of breakdown. Power-line disturbance monitors and event recorders are widely used for troubleshooting each of 144 power distribution panels (PDP) located on the making and packing floor. The PDPs are used to house the CBs for all branch circuits powering the nearby making and packaging equipment. In addition, each panel contains two 480/120V step-down transformers, one of which is an isolated transformer for serving computer equipment. An electrostatic shield between the primary and secondary windings of the isolation transformer prevents the passage of high-frequency transients (common-mode noise) of short durations. The disturbance monitors and other sensitive instruments are plugged into a receptacle from the isolation unit. The other step-down transformer is a standard type for convenience-power receptacle outlets.

Rust International Corp., a Birmingham, AL, engineering firm, assisted the RJR project team in the development and construction of the plant. Morrison-Knudson Co., Boise, ID, was construction manager on the project.

Power distribution panel contains the CBs for all branch circuits powering the nearby packaging machinery as well as CBs for a step-down transformer serving a convenience duplex receptacle (lower right). The panel also contains an isolation transformer that provides a source of "clean power" for test instruments such as an event recorder. A convenience duplex receptacle served by the isolation transformer is located inside the panel.

UPS and m-g sets combine to protect major computer center

Transient-free, regulated power for critical computer and emergency loads is insured by an uninterruptible power supply and diesel-generator plant.

By CHARLES H. KRIEGER, EE
San Francisco, Calif.

THE IMPORTANCE of computer operations to the Fireman's Fund Insurance Co., San Francisco, is evidenced by the increase in floor space devoted to them—from 2200 sq ft 20 years ago to an entire acre today, plus an additional 250,000 sq ft for computer support facilities. This growth was largely responsible for the construction of a new, recently completed corporate headquarters building in Lucas Valley, Marin County.

Electrically, the original computer system in the old building was served by a 225-kva substation, which eventually was tripled in capacity. The computer center in the new home office is supplied by a 1500-kva, 12 kv-480/277-volt isolating substation supplemented by a 2650-kva emergency diesel-generator station and a 750-kva uninterruptible power supply (UPS) that eventually will be expanded to 1000 kva as the computer load increases. Three additional utility-supplied substations furnish power to normal building lighting, heating, cooling, cooking, and mechanical equipment.

As shown by the accompanying one-line diagram, the computer load receives clean, transient-free power from four 415-Hz motor-generator sets. The three 900-amp batteries associated with the UPS modules serve as standby power during the 10-sec interval between failure of the utility supply and availability of the engine-generator output, since the maximum outage time the computers can tolerate without loss of data is 30 cycles.

The detailed and costly concentration of design on a dependable, consistent power supply was necessary because of the supersensitivity of computer solid-state and electronic circuitry to electrical supply malfunctions. Computer errors can occur if voltage varies by more than ±10%, if frequency varies by more than ±½ cycle, or if voltage sine waves are distorted by peaks and dips that last more than a few milliseconds. Additionally, trouble can occur if harmonic contamination develops, if phase displacement angles vary appreciably from their normal 120 deg, if ambient room temperature and humidity conditions vary measurably from the ideal 75F and 50% RH, and if atmospheric conditions inside computer cabinets are not reasonably close to the outside ambients.

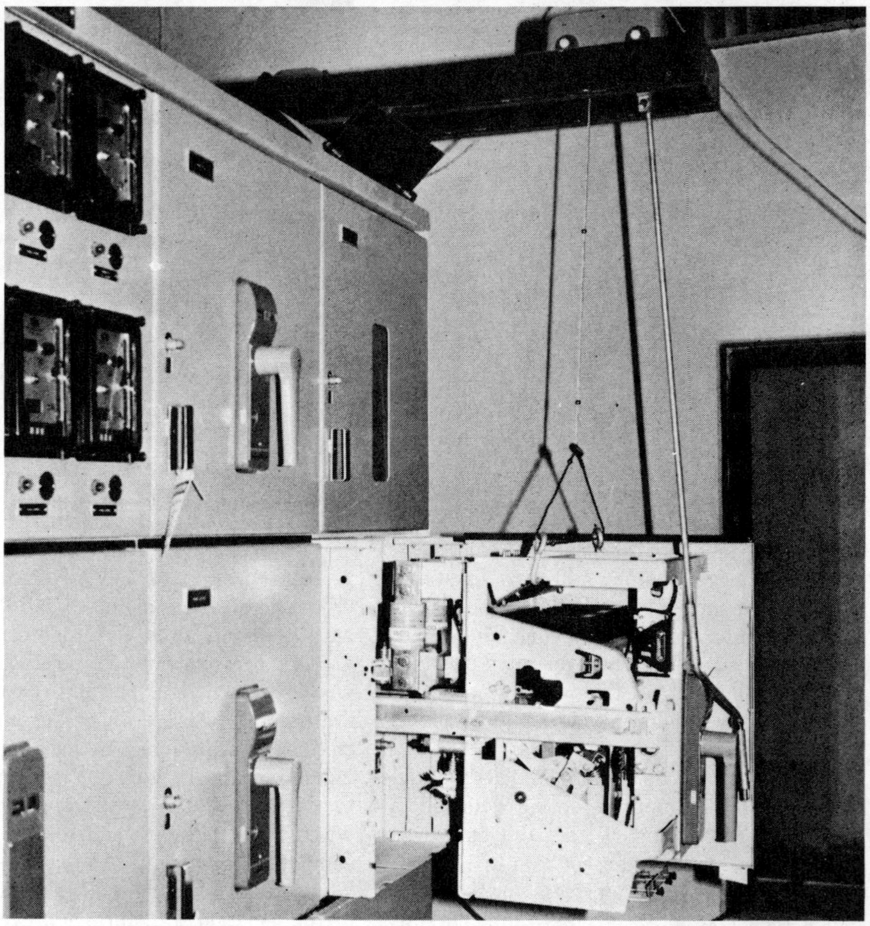

ISOLATING SUBSTATION serving UPS system receives primary power from 12-kv closed loop. Bus structure on secondary side of 12-kv/480-volt transformer can also obtain power from emergency generators. Power is carried by 2500-amp busway to UPS switching center. Draw-out feeder CBs are combined with current-limiting fuses for added high-interrupting-capacity protection.

Minor momentary power interruptions and line-characteristic variations are practically unavoidable on utility systems; therefore, normal utility service is not sufficiently precise to insure constant, error-free computer performance.

Critical loads served by the UPS include 512-kva of 60-Hz computer demand plus 164 kva supplied by the 415-Hz motor-generator sets. The three 250-kva UPS units are sufficient to handle the total present load of 676 kva. While a fourth unit eventually will be required and would

TWIN GENERATORS (above, left) equipped with compressed-air starters and electrically synchronized provide emergency power in event of utility outages. Each 1325-kva diesel-driven unit has sufficient capacity to meet all present plus anticipated future UPS needs, so that no critical-load shortages should result from short-term shutdowns for inspection, maintenance or repairs.

THREE 900-AMP, 405-VOLT BATTERIES (above, right) serve as standby power for computer loads in the event of a utility power failure until engine-driven generators can take over.

TRANSIENT-FREE REGULATED POWER for sensitive computer applications (left) is obtained by feeding either rectified ac or battery-supplied dc into solid-state inverter sections of UPS modules such as this 250-kva unit.

POWER DISTRIBUTION CENTERS (right), with wide, hinged doors on three sides, facilitate access to UPS, normal lighting, receptacle, emergency and miscellaneous power panels isolated from each other by full-height interior barriers. Dry-type, 480–208/120-volt transformers are mounted above panels. Hung ceiling and raised floor (added later) permit free wiring access to these specially designed assemblies.

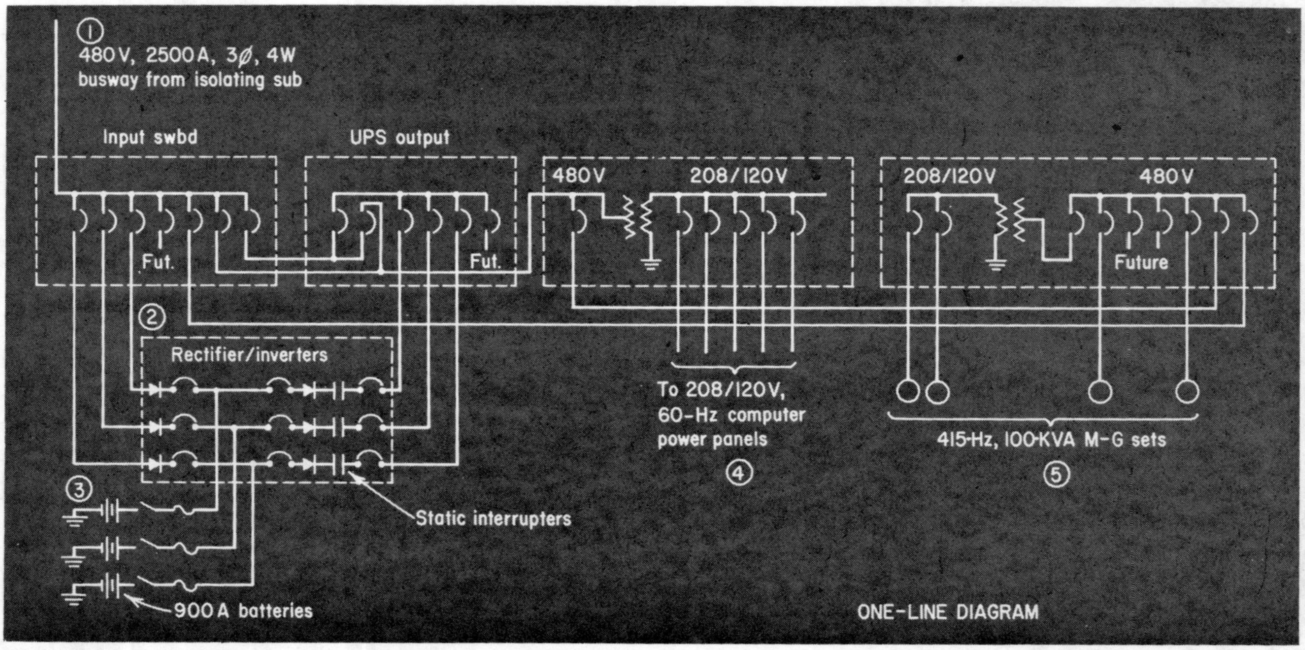

UPS SWITCHING CENTER obtains power from isolating substation (1) where input normally is from a utility-source loop or, in emergencies, from twin diesel-driven generators. After rectification, power is fed to dc/ac inverters (2) that can also receive power from separate 900-amp batteries (3). With this arrangement, stabilized transient-free current is available for computer peripheral equipment (4) and 415-Hz motor-generator sets (5) feeding computer circuitry.

be desirable as a spare and for rotation in service during maintenance or repair periods, an alternate arrangement was proposed and accepted that saved the initial cost of a fourth unit. Switching arrangements were provided (see diagram) whereby both the m-g load and the peripheral equipment load can be served directly from utility power if it becomes necessary to take one or more of the UPS units out of service. The rotational inertia of the m-g sets maintain power-supply stability during the brief 3-cycle periods when circuit breakers are opened and closed to accomplish the transfer. The 12-kva overload carried by the two 250-kva UPS units in service during such periods creates no problem.

The standby batteries that sustain loads during brief utility service interruptions consist of three 900-amp, 405-volt banks, one for each UPS unit. Three banks were installed instead of a single 2700-amp assembly to avoid any possible hazards from the higher concentration of power and to simplify maintenance.

Emergency generators

The emergency power installation consists of two electrically synchronized, 1200-rpm, 1325-kva, diesel-driven generators. The engines are started by compressed air and synchronized automatically to furnish 3-phase, 4-wire, 480/277-volt power.

If during an emergency only one diesel-generator starts (or if one generator fails when both are running), all loads are dropped except the UPS system, which will continue to operate until all computer work in progress is concluded, after which an orderly shutdown of the UPS system can be conducted if necessary.

When both diesel-generators are restored to service, UPS loads are connected first, after which other loads may be added in predetermined sequence.

If one 250-kva UPS module is down and the 415-Hz m-g sets are being driven by utility power, it is possible to start one diesel-engine generator manually and connect it to both the UPS units and m-g sets.

Power distribution

To minimize line losses in conductors interconnecting the computers, intercabling distances were kept as short as possible by locating related electrical and electronic components either above or below each other on different floors, with conductors routed through floor sleeves. Large, free-standing power centers were designed to facilitate this approach. Located beside main structural columns of the building, each of these centers contains an 800-amp, 208-volt UPS distribution panel, a 225-amp UPS branch circuit panel, and a 480/277-volt emergency panel feeding fan-coil units and emergency lighting, and a 208/120-volt, 4-wire panel controlling normal house power receptacle circuits and miscellaneous loads.

Incorporated in these centrally located distribution assemblies is a dry-type, 480-208/120-volt transformer fed through a shunt-trip breaker in the emergency panel that trips automatically when the diesel generators are required to pick up emergency loads.

A unique feature of these power centers is that they extend downward through the raised floors of the computer areas to the structural slab and upward into the ceiling plenum, providing unobstructed access from both above and below. Prepunched knockouts in all accessible vertical surfaces facilitate connection of conductors.

Electrical plans for this computer and communication building were developed by Charles H. Krieger and Associates, Inc., San Francisco. Fishbach and Moore, San Francisco, was the electrical contractor. The UPS consultant was Karl Immer, New York, N.Y.

DEPENDABLE POWER SERVES DOW JONES OFFICES

Joseph R. Knisley

Multiple power sources keep this computerized facility constantly operational.

DESIGN parameters for the new editorial offices of the Wall Street Journal and other Dow Jones financial services in the World Financial Center, New York, included the requirements that certain lighting, HVAC, and computer communications equipment be operational regardless of any problems that might occur with the serving utility, even an extensive power outage.

Since the facility was to contain a communications center that constantly receives business information from around the world and daily transmits newspaper copy via satellite to regional printing plants throughout the United States for daily printing of the Journal, a system for contingency operations had to be designed into the power distribution system.

To achieve the necessary requirements for ultradependable power and for contingency operations, five discrete power distribution systems were installed within the nine floors occupied by Dow Jones. These systems are shown in Fig 1.

1. The *base building power system* provides power to the air-conditioning units and to several panels on each floor. This base building power system is served from the utility network that delivers power throughout the Financial Center complex for such central equipment as elevators, the general building air-conditioning equipment, and ten-

Dow Jones & Co., Inc., occupies 400,000 sq ft in the 41-story building in the foreground. Three standby generators are located within the dome roof (arrow) of the adjacent 9-story gatehouse building. Four major office towers make up the 6-million sq ft office space development of Olympia & York Equity Corp.

ant power. Power is received at 13.2kV and stepped down at several points to utilization voltages of 480/277V and 208/120V. This power source is subject to any and all power abberations experienced by the utility, including total failure.

Power panels are identified as HV (480/277V) or LV (208/120V), plus the floor number and closet location A or B. The HV and LV panels on each floor serve lighting, convenience receptacles, and other nonvital loads.

2. The *dedicated Dow Jones power system* consists of two main 480V cable risers deriving their source from the same point as the base building power system. One serves as the normal utility input to the 12th-floor automatic transfer switch for the supplementary refrigeration equipment room; the second is the normal utility input to the 15th-floor automatic transfer switch for the uninterruptible power supply (UPS).

3. The *base building emergency system* serves only exit signs and minimal egress lighting on the Dow Jones floors. These are served from two panels (HV and LV) in the 12th-floor electrical closet A. Other elements served by this system that are of concern to the Dow Jones operation are the stair lights, selective elevator control, and building alarm systems.

4. The *dedicated Dow Jones standby diesel generator system* is routed through and serves virtually all Dow Jones office spaces and thus permits certain data-processing equipment, lighting, and air-conditioning equipment to be operational even during a prolonged loss of utility power. This system utilizes either the base building power system or the Dow Jones dedicated power system as its normal power input.

In the event that all or part of these sources fail, the respective transfer switches signal the three 1200kW generators to start. The load is transferred by each affected transfer switch when the generator power becomes available. Panels connected to this system are encoded as follows.

SDP designates the *standby distribution panel* and is preceded by the floor number. This panel serves as the distribution point for all other standby distribution panels on the floor.

SHV is the *standby 480/277V lighting panel* taken off the SDP and generally serves the 277V lighting critical to office operations.

SLV, the *standby 208/120V panel,* is taken off the SDP and generally serves furniture-mounted indirect lighting powered via the raised floor system and such miscellaneous equipment as copiers that are critical to operations but are not subject to damage due to power fluctuations or momentary outages.

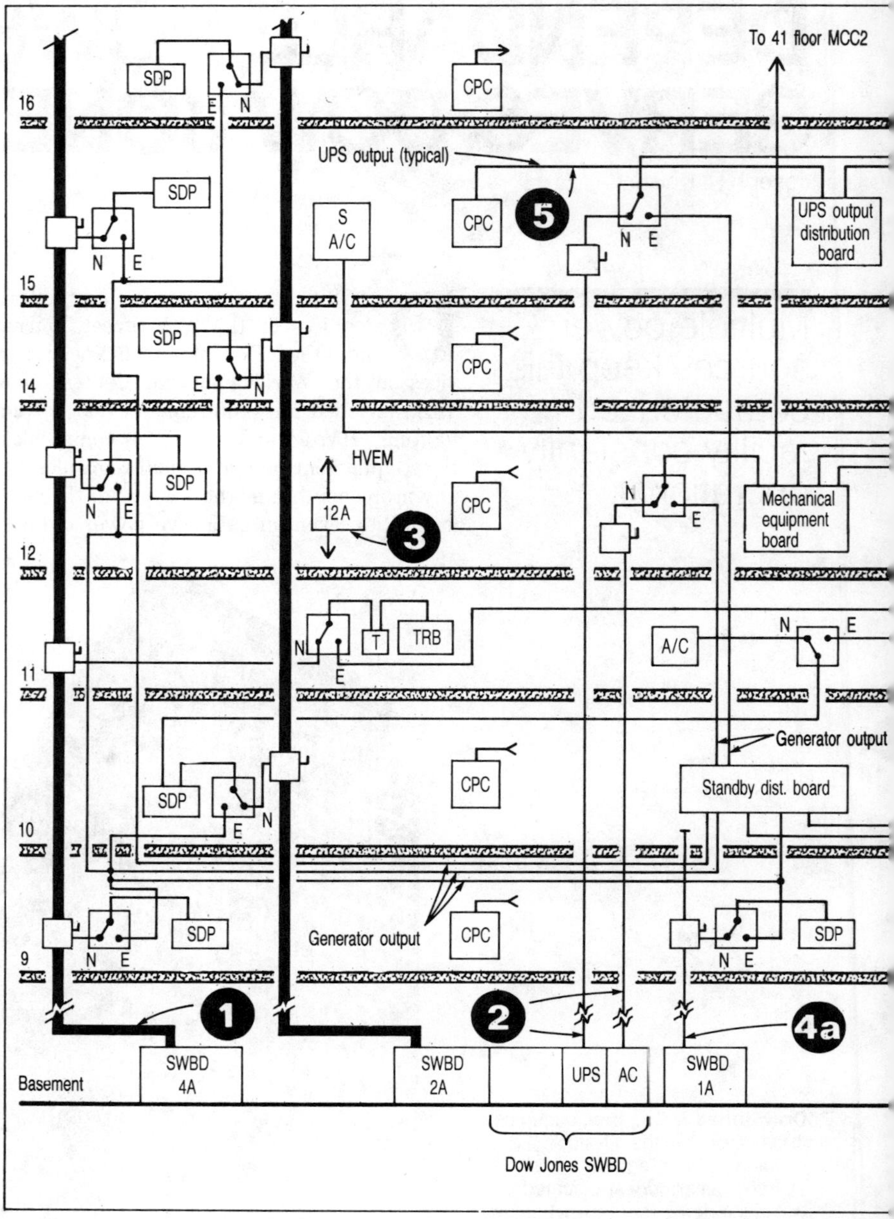

Fig. 1. One-line diagram shows the five discrete power distribution systems for ultrareliable power and contingency operations on the nine floors occupied by Dow Jones & Co. They are (1) base building power system, (2) dedicated Dow Jones power system, (3) base building emergency system, (4) dedicated standby diesel generator system, and (5) uninterruptible power supply (UPS) system.

SAC is the *standby air conditioning panel,* also taken off the SDP. It serves floor-mounted (and some ceiling-mounted) supplementary air-conditioning units arranged to provide adequate environmental conditions for critical equipment or spaces. These include the packaged units on each floor and the units in the computer room and data/communications equipment rooms.

Packaged air-conditioning units located in closets or alcoves adjacent to the office areas permit a tolerable comfort level to be maintained for personnel in these occupied spaces if utility power is lost for an extended period. The consulting engineers researched the equipment in this field and specified a unit that operates through chilled water with a

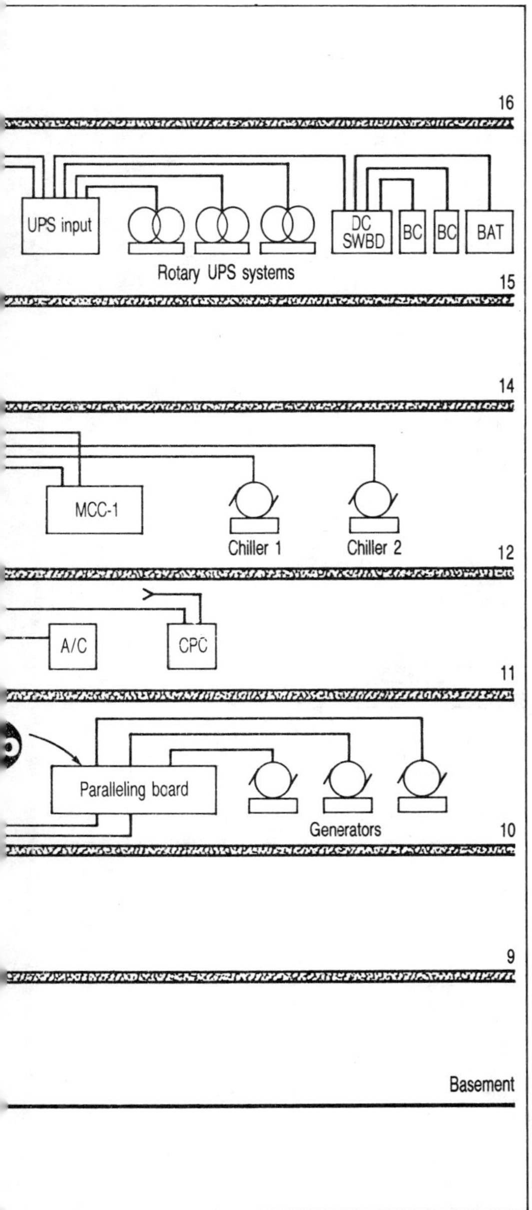

sufficiently low noise level for the office environment.

The standby air-conditioning power system also provides the emergency input to the transfer switches for the supplementary refrigeration plant and the UPS.

5. The *uninterruptible power supply (UPS) system* receives its input from a transfer switch (as previously mentioned) and, using a motor-generator system to internally regenerate the AC source, delivers clean, regulated power to the computer equipment regardless of outages and disturbances on the utility power supply. The system consists of three paralleled 360kVA motor-generator units located on the 15th floor. The UPS units in turn are supported by a battery system that can provide power for 15 min under full load to the rotating equipment in the

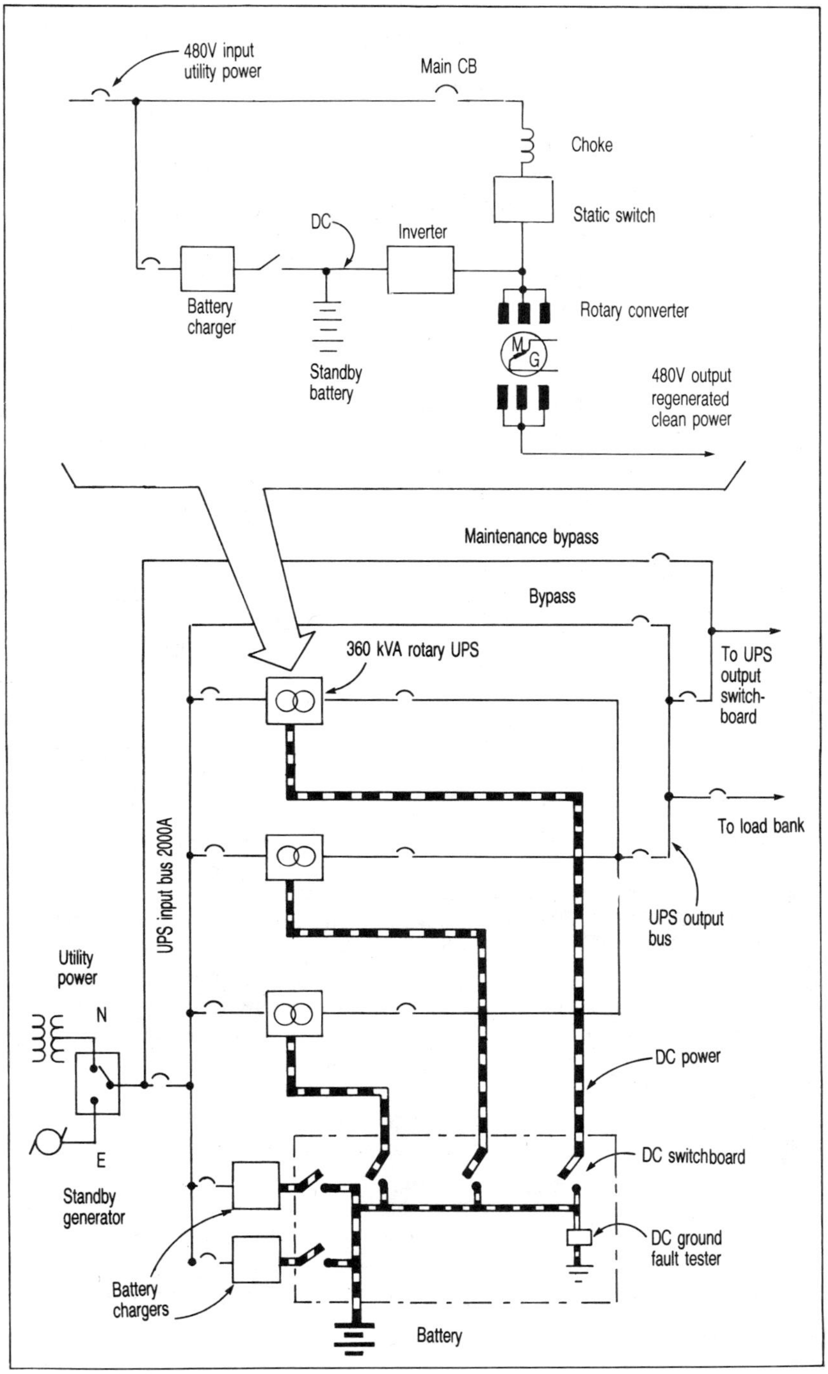

Fig. 2. Schematic shows trio of rotary UPS units and the extensive battery bank that provides power for an extended period if the utility power is lost.

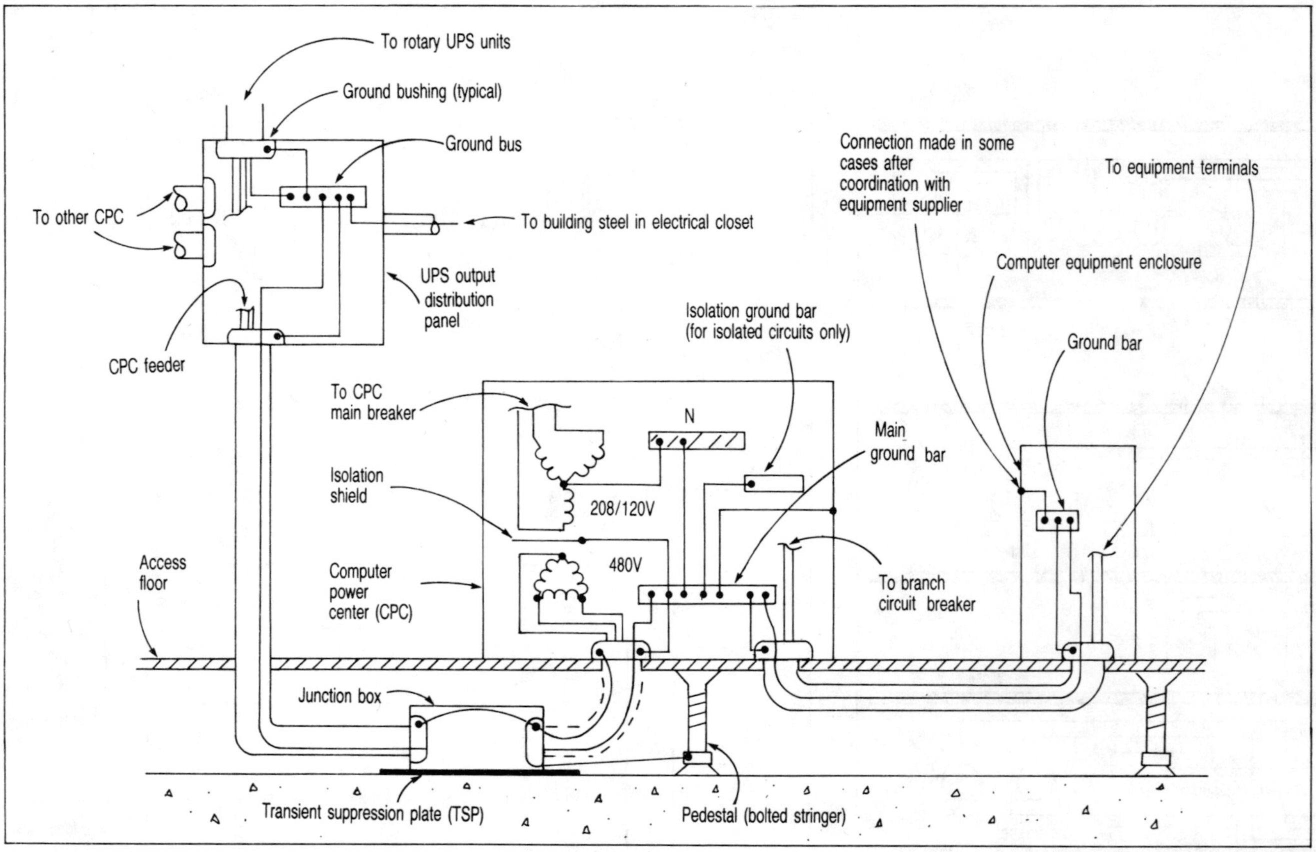

Fig. 3. Computer grounding and power wiring from the UPS to a CPC serving a typical computer room is shown in detail. Located close to the loads it serves, the CPC shielded isolating transformer has a single-point ground, in conformance with NEC Sec. 250-26 on the grounding of separately derived AC power systems. The transformer secondary ground point is also the point where the interconnected parts of the computer's grounding system are connected to other ground conductors extending beyond and outside of the computer room. A single-point ground prevents unwanted ground loops by providing only one conducting path between it and all external equipment.

event of a utility failure.

Each of the three cabinet-enclosed 360kVA rotating units consists of a motor driving an AC generator, or alternator. It provides the pure sine wave required for proper operation of the extensive computer systems, being completely isolated electrically from the transients, harmonics, spikes, and dips that are characteristic of the utility's service.

Power-line voltage changes of ±20% can be experienced, and the voltage to the load will still be maintained at ±1% of nominal. The rotational momentum of the elements on the common shaft permits the set to ride through short-duration power-line dropouts.

As shown in Fig. 2, which is a single-line diagram of the UPS array, electrical power is fed via a static switch and choke to the rotary converter and then out to the load. Alternate slots of the converter stator are wound with input and output windings. When the converter is energized, a rotating magnetic field is created, causing the DC-excited rotor to run at synchronous speed. By controlling the electrical characteristics of the rotor, the output of the secondary stator winding is controlled.

The rotary UPS is a relatively low-impedance power source capable of clearing subcircuit protective devices under fault conditions much more readily than other systems. For example, it can supply up to seven times full-load current under fault conditions. Thus, a short circuit within a computer or peripheral can be cleared by its branch-circuit breaker without the need to take the UPS system off line and without affecting output to the remaining computer load.

Because of the two sets of windings on its stator, the converter has a large frame size. However, it is smaller and lighter overall than comparable systems, because its flywheel action is provided by the motor's ventilator rather than by a separate flywheel.

Since the system has numerous transfer and bypass switches and circuit breakers, a single rotary unit can be removed conveniently for maintenance or repair. As shown in Fig. 2, the utility input follows two current paths through the system: 95% of the current is through the static switch and choke; the remaining 5% is through the diode rectifier, standby battery, and inverter. With only minimum current through the rectifier/inverter, the unit is kept in an active state without subjecting the thyristor circuits to unnecessary heat stress. Either path is capable of supplying full power to the stator input windings should the other fail. If utility power fails, the stored energy in the rotating section supplies power until the off-line battery is connected to the inverter.

When the rotating section of a unit has to be serviced, a bypass around the rotating section is utilized. Because the bypass breaker is an electromechanical device and not a static switch, complete isolation between input power and output power is maintained when it is in the open position. An external maintenance bypass is also provided so the entire system can be made available for maintenance and inspection. Since the three rotating sets operate in parallel, any one of the units can be shut down for maintenance and inspection.

Computer power design

Output of the UPS system is directed to a switchboard containing 11 feeder circuit breakers, each of which serves an individual computer power center (CPC) that delivers power to localized loads. The CPC is a free-standing cabinet that looks like a piece of computer equipment.

A shielded isolation transformer in the CPC steps down the 480V supply to the 208/120V equipment utilization voltage, and circuit breakers provide protection for branch circuits to individual computers or peripherals.

A high-quality, computer-grade equipment grounding arrangement is used in conjunction with this modular power-distribution scheme at all of the CPCs used throughout the 15th-floor computer room, where high-speed communications circuits operate. All of the computers "see" the same grounding potential over a broad frequency range, and the 18-in.-high floor stringers are part of a signal reference grid.

The larger free-standing computer equipment is hard-wired directly from the CPC, as shown in Fig. 3. The general office areas use an 8-in.-high access floor system, and each CPC has a junction box mounted out of sight within the raised-floor cavity. A feeder from the UPS output switchboard is wired into this junction box. The plan view in Fig. 4 shows the junction box from above. Computer-grade grounding is not used here because the data circuits operate at lower speeds.

The 120V duplex outlets of all CPC-originated circuits are identified by the orange color of the receptacle body and are mounted in a service outlet box that also contains data and telephone circuits (Fig. 5). All equipment connected to this system is served by clean power and is backed up by the battery system of the UPS units and standby diesel generators.

Power/data distribution

Because of extensive communications needs, the Dow Jones office space is largely occupied by electronic workstations from which information is rapidly transferred and stored. To deliver the required services—

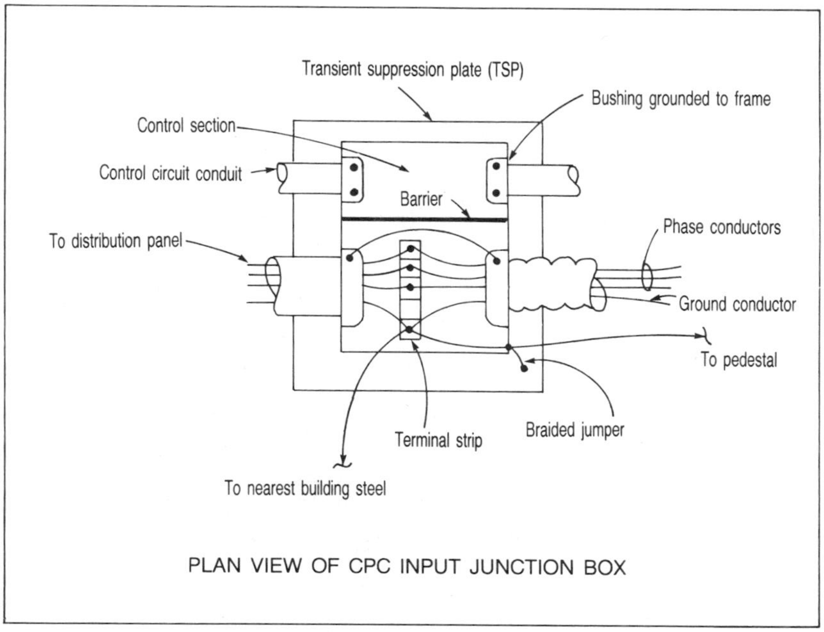

Fig. 4. Detail of CPC input junction box shows the copper braid run from the box enclosure to the transient supression plate, which is installed under each junction box in the underfloor spaces, including those JBs serving the supplementary air-conditioning units. The steel plate reduces high-frequency noise currents that might otherwise be coupled into the underfloor cables from subfloor reinforcing steel.

Generator paralleling board, which handles control and paralleling of the three standby generator sets, is checked by Frank Lorenz, chief electrical engineer and associate at Edwards and Zuck. Scheduled exercising tests on the generators are run with output delivered to an air-cooled load bank.

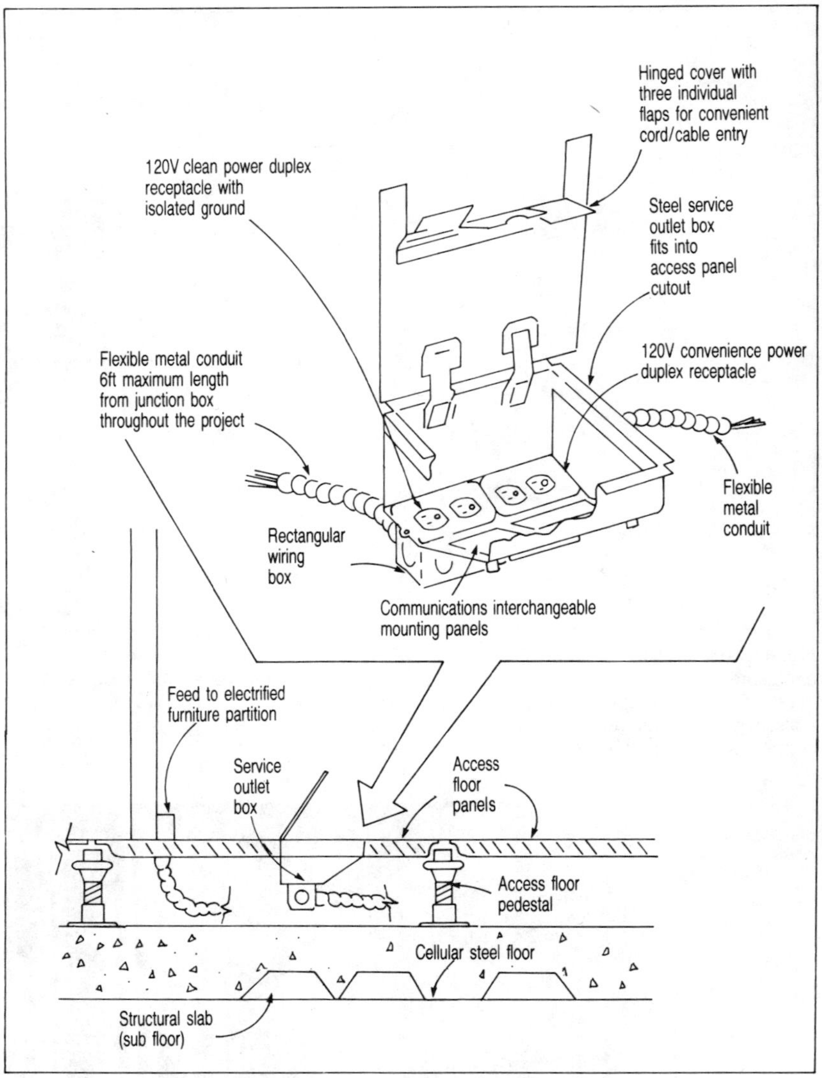

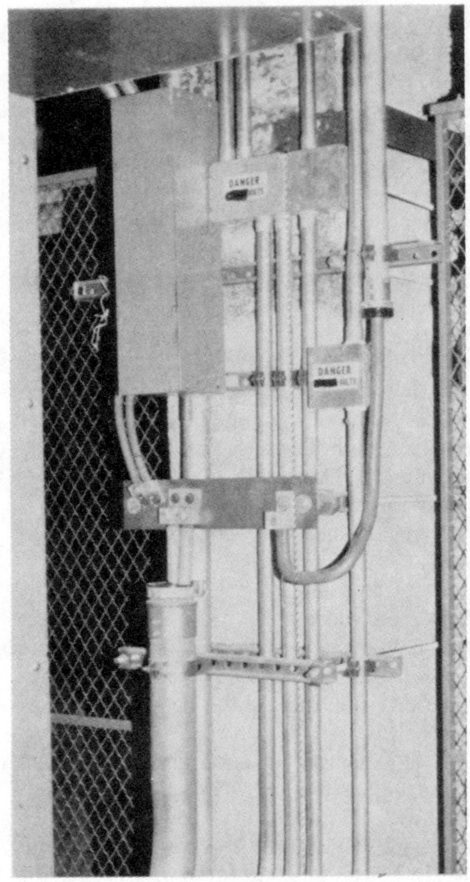

Grounding detail shows the grounding conductors carried from the generator sets and other equipment to a common ground point in the dedicated Dow Jones switchgear 11 floors below. The 500MCM conductors in the 4-in. conduit at left extend downward and connect equipment on 12th and 15th floors with the main service switchboard in the basement.

Fig. 5. Underfloor delivery of power, data, and phone circuits is accomplished by the access floor service module, which provides effective wire-management capabilities for a workstation. The module assures capacity expansion and has interchangeable mounting panels on which coaxial connectors, telephone sockets, 25-pin connectors, priority network cabling systems, and any other type of device can be mounted. Both a cellular steel floor raceway system and a raised floor system are used in combination to achieve effective and highly versatile wire and cable distribution. Each panel is supported by four pedestals, which are held in place with adhesive. The panels are butted together and leveled; carpet squares complete the installation and hide panel joints.

Standby diesel generator, rated 1200kW, is one of three in a sound-attenuating enclosure in the penthouse space above the ninth-floor extension. Air for combustion and general ventilation is achieved through motorized louvers. Special isolation mounts and a floating floor slab supporting the gen sets prevent transmission of vibration to other areas.

power, telephone, and data communications lines—a cellular steel-deck system, an access raised-floor system, and a conduit and box grid system mounted on the slab are used to achieve the desired degree of wire management. Additionally, some cabling is carried in cable trays within the ceiling plenum.

The cellular deck is used to carry normal power branch circuits and telephone circuits. The grid system carries computer and 120V generator-backed lighting power conductors in conduit to boxes located throughout the floor.

All other conductors are run beneath the access flooring, consisting of modular floor panels supported by adjustable steel pedestals that form an underfloor cavity.

The pedestal assembly consists of a head plate, a post, and a base plate. Both the cast-metal head unit and the post or shaft are threaded to permit setting a precise height adjustment by running the head up or down on the shaft.

Power, data, and telephone services are delivered to a station in the general office area via the access floor service outlet box mounted on the underside of a panel (Fig 5). The outlet box is about $7\frac{1}{2}$ in. deep, and thus a $\frac{1}{2}$-in. clearance is maintained between the bottom of the box and the structural slab so that communications cables can be run anywhere without impediment. The service outlet box is of modular construction and contains various mounting plates. Data/ communications receptacles can be installed within the box.

Two duplex receptacles are mounted in wiring device boxes fastened to the under-

Rotary UPS enclosure is one of three units in a room that also contains two free-standing air-conditioning machines, all operating in a temperature- controlled environment. Designed for continuous duty, the motor and generator are mounted on a common shaft. The calculated mean time before failure (MTBF) of the single stator design exceeds 500,000 hrs.

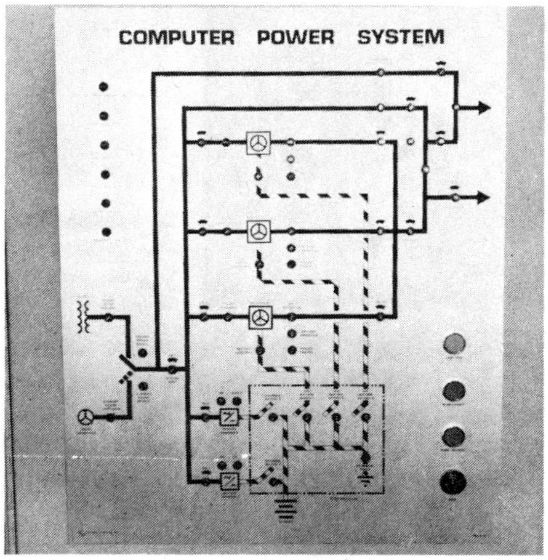

Status/warning board contains a single-line diagram of the complete paralleled system with appropriate panel lights at major components to indicate normal or off-normal conditions. The status of circuit breakers and isolation or bypass transfer switches can be easily seen by operating personnel, and any necessary contingency plans can be quickly implemented. A lead-calcium battery bank providing a nominal discharge level of 500VDC is located in a separate room, where a ventilation system is used to clear the space of hydrogen emitted from the cells.

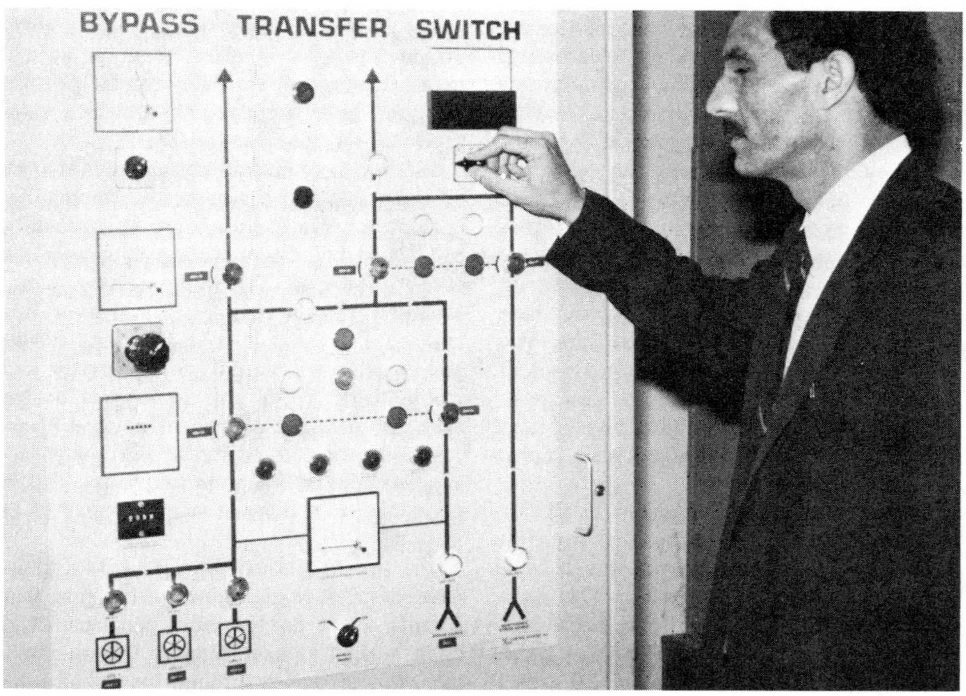

Bypass transfer switch control board, which shows status of the power system within the UPS bank, is reviewed by project-engineer Bob Almovodar. The board facilitates isolation of a UPS system component for scheduled maintenance or for setting up any contingency operations that might be necessary. A water-cooled load bank, located in a nearby closet on the same floor, can receive the output of the three parallelled UPS units during an exercising test procedure.

Ground-fault test panel located at top of UPS output distribution switchboard permits making a test on any one of the 11 feeder circuits, each of which serves a computer power center.

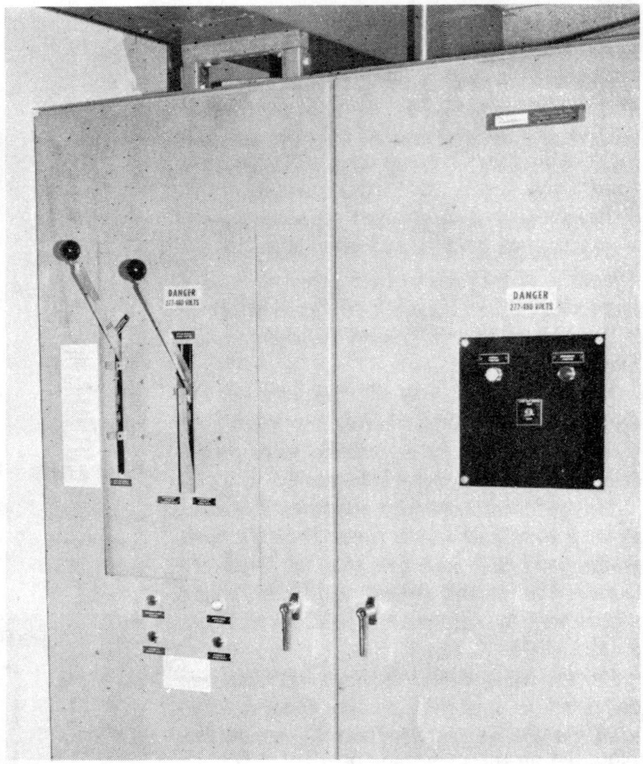

Automatic transfer switch in the UPS room provides connection to the standby generators if normal power is lost. An integral, isolated bypass switch within the enclosure permits maintenance work to be done on the switch, which is rated at 1600A.

side of the service outlet box and are accessible from within the box. One receptacle provides 120V power loads; the other provides the "clean" power system for computer-based loads. Therefore, each receptacle is served from a separate branch circuit.

Throughout the clean-power system, an isolated grounding conductor is run with the power conductors. Computers are inherently sensitive to electromagnetic interference (EMI) and noise generated by equipment connected to the normal power system. Normal grounding conductors can provide the route by which these disturbances can reach the computer equipment. The dedicated power system at Dow Jones uses what is called a "quiet" ground—an insulated, green-colored grounding conductor carried throughout the "computer-grade" branch circuits and back to a ground bus in the CPC enclosure. This bus, in turn, is tied into the service grounding electrode of the transformer secondary, to which the grounded circuit conductor (neutral) is also connected, since it is a separately derived system (Fig. 3).

A publication called *Guidelines on Electrical Power for ADP Installations* from the National Bureau of Standards was used in designing the grounding system. This document is commonly referred to as *FIPS (Federal Information Processing Standards Publication)* PUB 94.

Contingency operations

A well-thought-out plan for contingency operations was established during the design stage to provide a practical combination of redundant capacity, alternate sources, and bypass of susceptible components. This scheme minimizes any possible interruption of electric service to important equipment. A contingency condition is one in which a given system operates in other than its normal mode. Contingency capability can be achieved by three basic methods, all of which were used at Dow Jones.

The first is redundant equipment, whereby all components of a system operate simultaneously but are oversized so that, should a component fail, the remaining equipment can support the load. The utility service to the Financial Center complex operates in this manner. For example, a single utility feeder and transformer can fail simultaneously, and the network is still able to support design load. All electrical equipment is sized to accommodate a 25% additional load above design calculations for up to two hours. This is in addition to a planned spare capacity of at least 20%.

The second method provides for an alternate source during a failure condition so that a replacement for the failed component can be installed. The most common way to implement this method is through the use of auto-

Typical CPC, a free-standing enclosure that looks very similar to computer equipment, contains an isolating transformer to feed circuit-breaker panelboards that supply circuits to computers and peripherals, as well as providing monitoring, status indication, and other information on the power-supply conditions.

matic transfer switches. (See Fig. 1 for their locations.)

The third method is a manual bypass or substitution of a failed component. This procedure provides for an alternate source but only after manual inspection and review determines the appropriateness of the transfer. The equipment used for this method includes manual transfer switches, maintenance bypass breakers, and isolation bypass switches. The plan includes, for example, having an extra CPC in storage should the transformer or some other component in a CPC fail. Thus, either a part or a complete computer power center can be substituted.

A status/visual-alarm panel in the UPS room monitors computer functions (high temperature, module failure, etc.). A number of sensor outputs is also sent to a monitoring board at a central control station. Most of the alarms are simply to provide early warning of an off-normal condition so that staff personnel can investigate and analyze the situation. They are not intended to cause instant reaction.

The electrical design was carried out by Edwards and Zuck, P.C., consulting engineers, with Bob Almodovar as project engineer. Installing electrical contractor was Arc Electric Co., Inc., with Hank Fromm as general foreman. All firms are in New York City.

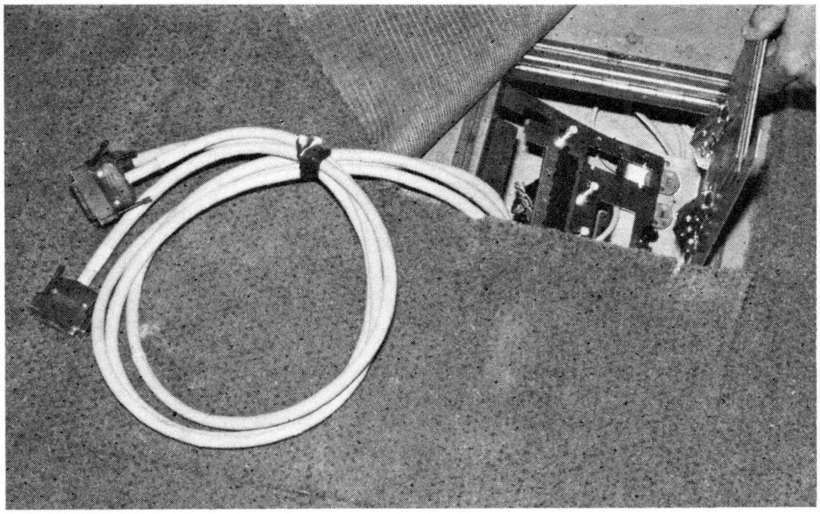

Access floor box provides convenient wire-management features for feed-in of branch circuits serving duplex receptacles, along with data and communications cable. The receptacle shown, with an isolated ground, serves the computer loads. The large, dark panel has cutouts for insertion of small plates fitted with coaxial, phone, or other type of termination devices. Box is lowered into a matching frame secured to the 24-in.-sq metal panel of the raised-floor system.

Power feed to a vertical panel of the electrified furniture partition that forms a workstation is achieved by bringing branch-circuit conductors up through the modular-floor panel from the access space below.

UPS systems provide control-circuit reliability

The need for constant availability of steam for pipe-tracing, process heating, and space conditioning required protection against utility outages.

UNINTERRUPTIBLE power supplies installed recently at the Welland Plant of Cyanamid Canada Inc. in Niagara Falls, NY demonstrate how such equipment can be applied as a reliable voltage source for ordinary electric control circuits.

At this plant, the reliability of steam supply is vital. Steam is used for heat-tracing pipelines as well as for process and space heating. The freeze-protection function is particularly important in one of the facilities within the plant dedicated to the production of ammonia. At no time is this more critical than when the electric power supply to the plant is interrupted during the winter months. However, it is precisely during ice storms that the utility lines are most vulnerable. A recent freezeup in the ammonia plant during such an outage caused management to undertake a serious review of ongoing procedures with the goal of preventing a recurrence. The unusual nature of the problem involved made the application of UPS systems a must.

By ALFRED BERUTTI

The problem

Auxiliary fans and pumps of three boilers used for generating steam operate on 25 Hz; auxiliaries for a fourth boiler operate on 60 Hz. Theoretically, if either the 25-Hz or the 60-Hz utility supply were lost, at least one boiler would be available to generate steam. Even if both the lines were interrupted

Fig. 1. 120VAC sensing and control circuits

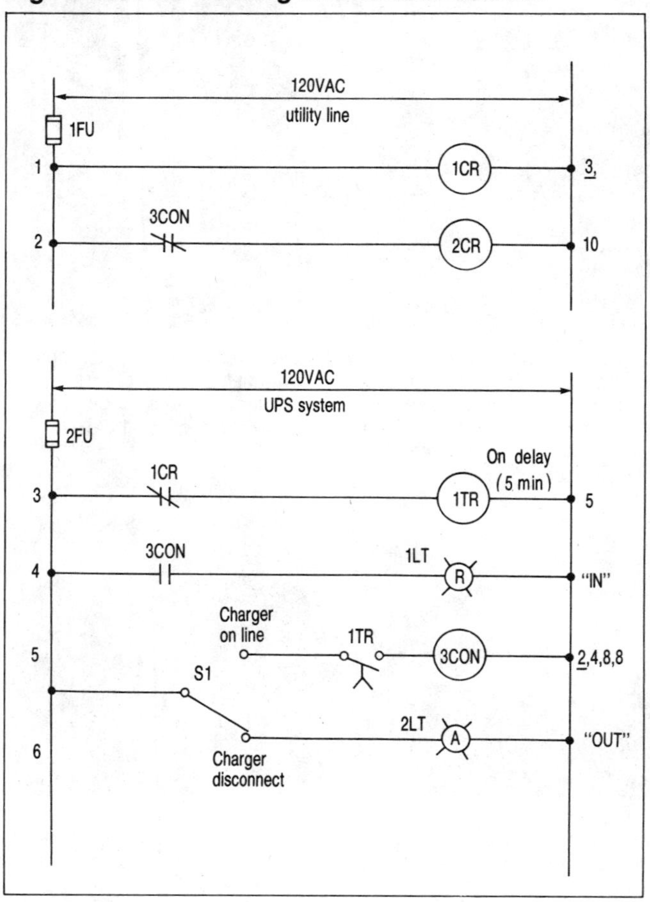

Fig. 2. 60VDC control and DC bus

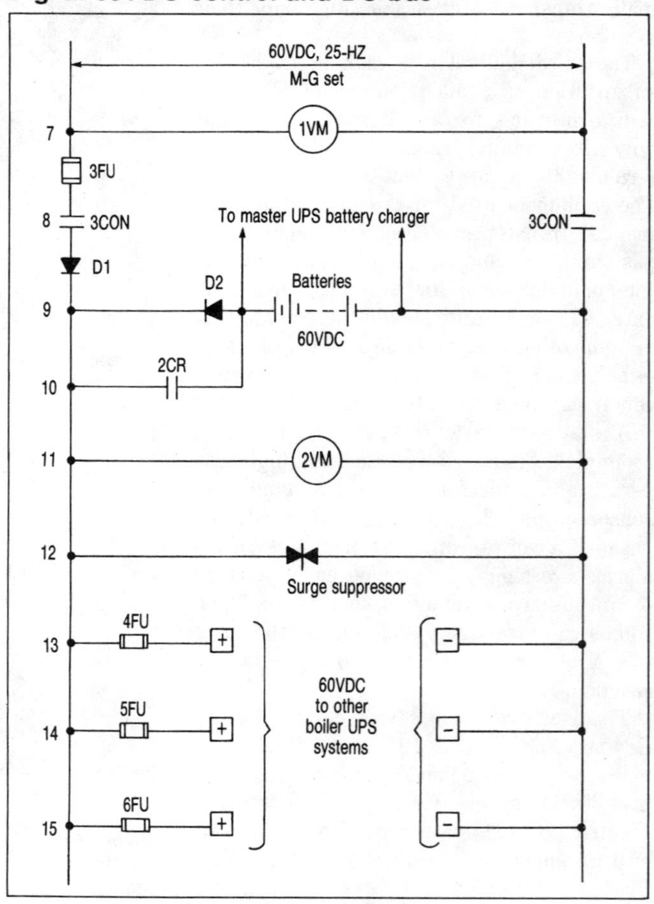

the auxiliaries would still be able to function, since critical units are driven by steam turbines as well as electric motors.

There was, however, a weakness in the system. Over the years, all electrical controls, including each boiler's flame safeguard and monitoring systems, were converted to 60 Hz because of the unavailability of 25-Hz relays, timers, and other parts. As a result, loss of the 60-Hz supply would force all the boilers to shut down because large steam boilers cannot be operated without their safety controls.

A UPS solution

One of the plant's electrical project engineers, Dave Bailey, was given the responsibility of finding and implementing a solution for the boiler-control reliability problem.

The modification that would provide the most fail-safe source of control power called for the installation of individual UPS systems for each of the four sets of controls. An analysis showed that installing four 1800VA units and their auxiliaries would be far too costly, since catastrophic shutdowns were rare. Further study showed that one of the major factors involved in the cost of a UPS system was that of the batteries and the battery-charging equipment. Bailey decided to explore alternatives that would incorporate the best features of individual UPS systems with lower cost and equal reliability.

An obvious alternative was a single UPS system to back up all four sets of boiler controls. But such a unit had the disadvantage of "putting all the eggs in one basket." Although the UPS is a generally reliable piece of equipment, failure of the UPS at the same time as an outage on the power lines would leave the plant exposed as before.

The alternative finally adopted was more complex but less expensive. Four 1800VA UPS units were purchased. Three of the units were to have no batteries and charger; the fourth was designated as a master unit and came with batteries and charger. All four systems were connected to a common DC bus and thus DC will be available for the UPS inverter sections if 60-Hz AC power outages occur. To minimize the cost, the battery bank was rated to provide approximately 12 min of power at 60VDC.

Such a short period of battery backup obviously would be inadequate to ride through a prolonged 60-Hz power outage. However, a 25-Hz motor-generator set, located in an old 25-Hz substation control building, to provide 60VDC power to monitoring and control circuits was available. The generator, which was no longer being used, was relocated to the steam plant and its output connected to the DC bus.

How the system works

Figs. 1 and 2 are schematic ladder diagrams of the controls involved in the boiler-control power system finally installed. The four UPS systems are energized with normal voltage available from the utility. The AC input

A

THE CYANAMID PLANT boilers, before the introduction of the UPS systems, had been able to ride through long periods of power outage. Mechanical equipment such as the 175-hp motor-driven boiler feedwater pump (photo A) operating on a 25-Hz line separate from the other utility feed permitted the boiler to run even when the 60-Hz line was out of service. Even if both lines were down, steam-turbine pumps (photo B) allowed pumps, fans, etc. to provide essential boiler needs. The weakness of the system was that boiler flame safeguard and other controls (photo C) depended upon a vulnerable utility line operating at 60 Hz. The system was developed to provide control power during outage affecting this source of power.

B

C

power is rectified to DC and then inverted to AC in each system. The coils of control relays 1CR and 2CR are energized. A contact of 2CR (line 10) is closed and power is fed to the other three inverters via the common DC bus. At the same time, the batteries located in the master UPS unit are floated.

Upon loss of the AC line, control relays 1CR and 2CR are deenergized. A contact of 1CR (line 3) closes, and the coil of timing relay 1TR is energized. The UPS systems continue to provide 120VAC via the inverters now fed by the battery bank located in the master UPS unit.

Meanwhile, steam-plant operators are alerted to the loss of the AC line and make whatever transitions are necessary to mechanical equipment to keep the boilers in operation. The 25-Hz M-G set is started and brought up to the proper voltage. The operator then flips switch S1 to the "charger on line" position; contactor 3CON is energized after the delayed closing of the 1TR contact; and DC power to all the inverters is now supplied by the generator through contacts 3CON (line 8).

Power restoration energizes 1CR; its contact opens and deenergizes 1TR. 3CON is deenergized, the M-G set is disconnected from the equipment, and DC power is once again fed from the rectifier/charger. The operator then manually shuts down the M-G set and switches S1 to the "charger disconnect" position to prepare for the next outage.

PLANT ELECTRICIAN Bob Embleton services the batteries installed as part of the master UPS system just prior to testing the operation of the units. The 25-Hz, 575V, 5kW motor-generator set that supplies 60VDC to the inverters during prolonged power outages is to the left of the master UPS unit.

FINAL CHECKOUT of one of the three slave UPS systems is conducted by (l. to r.) electrical foreman Don Wall and electricians Bob Embleton and Paul Washnuk. To reduce the cost of the installation, a unique system was devised that required only one set of batteries and a battery charger to be mounted as part of master UPS unit.

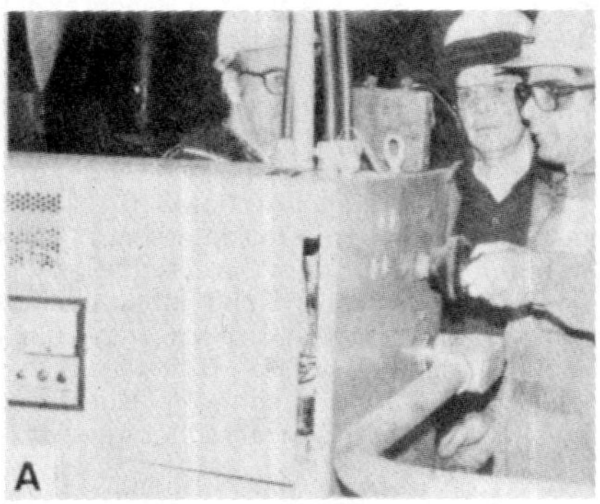

M-G SET is brought on line and the voltage adjusted on a V-O-M (photo A). Once its output voltage matches the requirement, boiler-engineer Don Arsenault flicks a switch on the master UPS changeover panel (photos B and C), and the M-G set assumes the role of supplying the 60VDC needed by the UPS inverters. Electrical project engineer Dave Bailey monitors the operation of the flame safeguard and safety shutdown interlocks (photo D) during the startup operation. The installed system proved to be reliable under all simulated outage conditions.

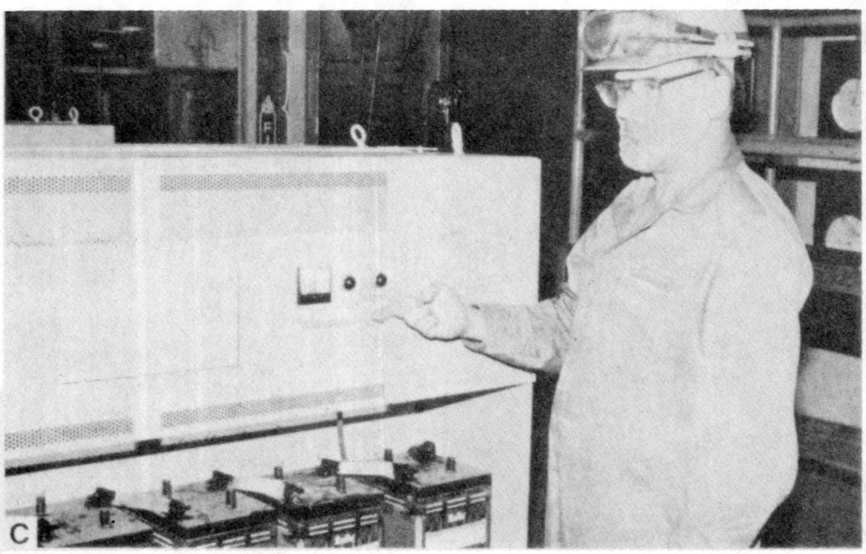

Expansion of a bank computer facility

Over a period of more than 20 years, computers serving a bank grew from a small-capacity vacuum-tube system to a modern microchip-based system of vastly greater capacity. The electrical distribution had to be modified to keep pace with that growth and with the increased importance of system reliability.

THE ONE word that characterizes the data-processing industry over the past 20 years is "change." A prime example of this is the Marine Midland Bank computer installation in Syracuse, NY, one of three similar facilities that serve the bank throughout the state.

The initial design and construction of this dedicated data-processing facility was done in the early 1960s, and the building was occupied in 1963. The computer systems installed were of the vacuum-tube type, which was the state of the art at that time, although much new equipment was becoming transistorized. Most of the building's air-conditioning system capacity was required to provide the proper environmental conditions for the computer because of the large amount of heat given off by the vacuum-tube equipment.

The power supply to the building came from a utility-owned 500kVA pad-mounted transformer, with a simple 208Y/120V radial distribution system within the building to all loads, including the computer and associated mechanical equipment. The need for continuous operation was taken care of by a 175kW diesel-engine-generator set. The engine-generator was started manually, and the part of the electrical load that was regarded as essential was transferred manually to the generator when it was up to speed.

Experience had shown that most power outages in the area where the computer center was located were the result of lightning strikes, since the utility supply lines were overhead. Therefore, when deteriorating weather was seen or forecast, an orderly shutdown of the computer was made at an appropriate point in the computer run, the computer and environmental loads were transferred to the generator, and the computer was brought back on line. The diesel-generator carried the entire computer and environmental load until the weather conditions improved. Then the process was reversed, manually transferring the computer back to the utility power and shutting down the generator. Although a total of about six hours of computer production was lost in the power-up and power-down procedure, this mode of operation proved to be generally satisfactory at the time.

Marine Midland Bank's use of data processing grew rapidly, along with that of most financial institutions. In the mid-1960s, the obsolete vacuum-tube equipment was retired and replaced with solid-state equipment. Considerable new and more-modern computer equipment was added, along with the necessary additional air-conditioning equipment. To provide back-up power for this new installation, a second diesel-engine-generator, rated 225kW, was added. The existing power-distribution system remained unchanged, and the new generator supplied its loads through a similar, but entirely new and separate manual transfer-switch arrangement.

This lasted only a few years, until

BERNARD P. CARR is the chief electrical engineer for Galson and Galson, PC, Consulting Engineers, and an associate in the firm. He received his BSEE degree from Syracuse University in 1949.

Fig. 1. Distribution system before latest revision

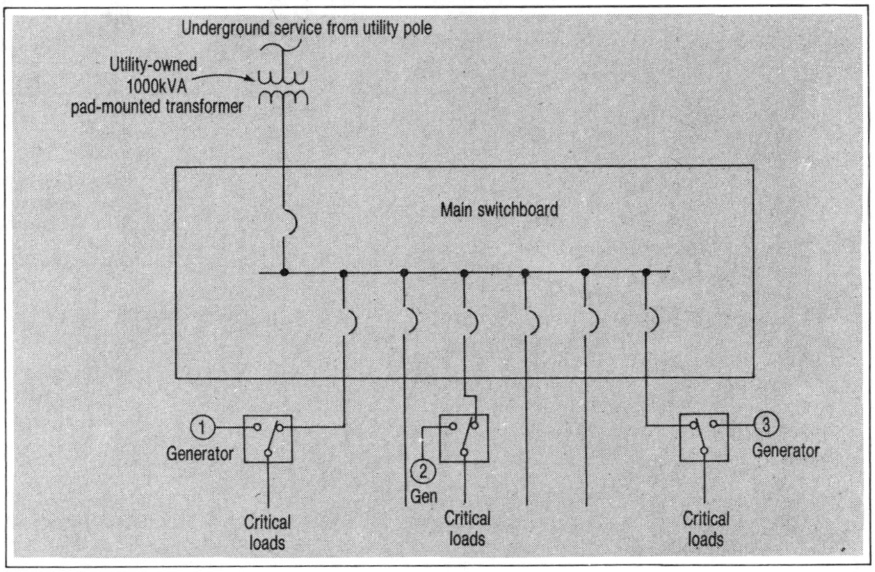

By BERNARD P. CARR, P.E.

the late 1960s. By then Marine Midland had expanded its customer services and its own use of data processing, and in addition offered some data-processing capability to other institutions in the area on a shared-time basis. It was again necessary to expand the computer facility. Furthermore, computer equipment had changed rapidly, and new capabilities and new methods made a complete revision of the bank installation necessary. Not only was the computer facility expanded, but the entire physical plant was enlarged, adding to both the front and rear of the building to approximately double the area. The changes were completed in 1970.

The power supply to the building was doubled, to 1000kVA. To minimize disruption of the ongoing data-processing and other bank functions, the existing power-distribution system was maintained intact at the time of the expansion. New equipment was simply added to it as additional radial distribution. As with all the distribution-system changes over the years, work was planned and executed in segments that could be completed on a weekend shutdown to avoid loss of bank operating time.

A third diesel-engine-generator, rated 535kW, was added during this expansion. It supplied its own separate loads through a manual transfer-switch arrangement similar to that of the other two generators. As in the past, the generators were manually brought on line and transferred to the computers when the weather was deteriorating. The one-line diagram of the building power system at this time is shown in Fig. 1.

Computers and data-processing equipment continued to become more sophisticated, faster, and of higher capacity. At the same time, the Marine Midland Bank and other financial institutions were transferring more and more of their transactions to electronic facilities. As the dependence upon data processing increased, so too, did the vulnerability of service to depositors and other customers of the bank. In the late 1970s, Marine Midland decided to develop the data-processing facility in Syracuse as a backup for other such facilities in the Marine Midland system. The intent was that in the event of a major disaster at another facility, the Syracuse installation would have adequate electrical and environmental systems to be capable of performing all the necessary depositor and customer functions for that facility, as well as its own. The bank believed that it could bring in additional data-processing equipment to do this within 30 days.

This required the entire electrical distribution system at Syracuse to be updated, increasing capacity and automating the emergency-power system. The criteria in upgrading the electrical distribution were increased capacity and flexibility, with improved reliability, while making maximum use of existing equipment. Constraints were imposed on the project by the nature of the facility. Work could not interfere with normal around-the-clock operations, so all required outages had to be scheduled between Saturday evening and 7:00 am Monday, during which time the facility is shut down. The system had to be ready for full operation on Monday morning when normal activity was resumed.

After initial engineering studies, basic decisions were reached to:

- purchase power at 13.2kV.
- purchase a new lineup of 13.2kV switchgear, with a fused main service disconnect, primary metering, and two fused feeder switches.
- purchase from the power company the existing utility-owned 1000kVA transformer, in place, and purchase from normal sources a second transformer of similar rating.
- purchase a new secondary-voltage switchboard to be served from the new 1000kVA transformer, with a tie to the existing switchboard.

Fig. 2. Present distribution system

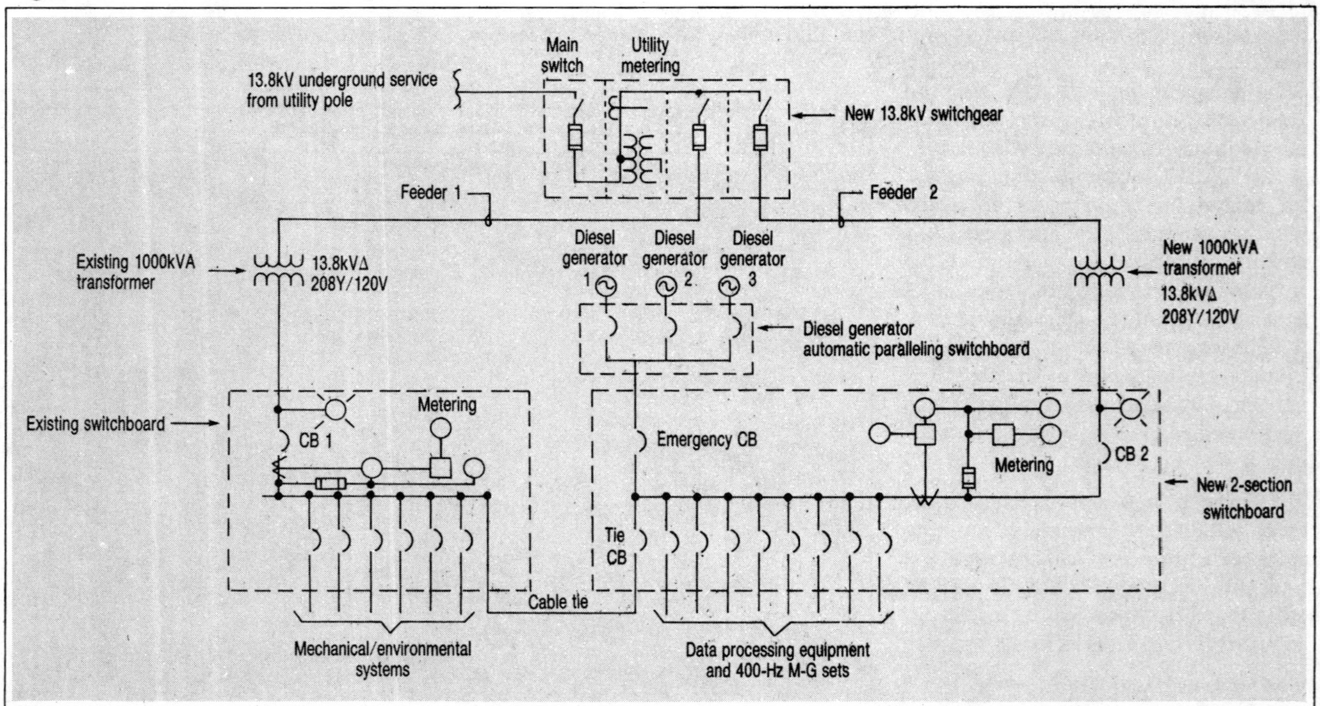

AUTOMATIC GENERATOR PARALLELING switchboard is capable of bringing all three dissimilar diesel-generators up to speed, paralleling them, and putting them on line through the new distribution switchboard. Generators, one shown at left, have modified governors and voltage regulators and cross-current compensation to make this possible.

CONTROL CONSOLE, located in the computer room, monitors computer power system and environmental equipment, with alarms to indicate any problems. It is possible to start and parallel all three generators and bring them on line from this console if desired. Service from normal power sources may also be restored from this unit, if systems have been on generator power.

- modify the existing diesel-generator sets for automatic operation and paralleling, connecting them to the new switchboard.

These extensive changes and the need to avoid outages except on weekends, required preparation of a detailed, step-by-step schedule of work by the consulting engineers, Galson & Galson, PC, of Syracuse, NY, in conjunction with the electrical contractor, Bec Electric, Inc., also of Syracuse. The integration of an existing switchboard from one manufacturer, a new switchboard from a second manufacturer, a diesel-generator paralleling panel from a third manufacturer, and a control console fabricated and assembled locally took careful planning and required a complete and detailed interconnection diagram from the consulting engineer.

The one-line diagram for the new system is shown in Fig. 2. The new secondary switchboard included a tie breaker and automatic-transfer scheme for putting the entire load on one transformer if the other transformer is lost. Either transformer can carry the full load of the building. An emergency breaker on the new switchboard can transfer the entire system to the paralleled diesel generators. Because of limitations of the available space, the new switchboard had to be built in two sections.

Since the three diesel-generators had been purchased separately at different times and had been operated independently, it was necessary to modify the governors and voltage regulators and to install cross-current compensation to permit parallel operation. A detailed and comprehensive functional and operational description of the system was prepared and became part of the procurement specifications for the new secondary switchboard and for the diesel-generator paralleling panel. The manufacturer of the switchboard was made responsible for the modification of the main breaker in the existing switchboard and for providing the undervoltage and lockout relays and any other equipment and wiring necessary to make the automatic-transfer scheme operational. Similarly, the manufacturer of the generator-paralleling panel was made responsible for all necessary modifications to the three generator sets to permit parallel operation.

The normal operation of the system is with both transformers in service, each feeding its own secondary switchboard. The tie breaker and emergency breaker are open and the diesel generators shut down. The data-processing equipment loads, including the M-G sets for the 400-Hz computer equipment, are all supplied from the new switchboard. The environmental equipment is supplied from the old switchboard. Upon loss of one of the sources, the associated main breaker trips and the tie breaker closes, putting the entire load on the remaining transformer, which has sufficient capacity. If both sources are lost, the diesel-generators start and are automatically paralleled; then, in a timed sequence, main breaker No. 2 trips, the emergency breaker closes, main breaker No. 1 trips, and the tie breaker closes. This puts the entire system on the emergency generators. When normal power returns and is stabilized, restoration to normal power is done manually.

A control console is located in the main computer room. It has a graphic display, with indicating lights and alarms, and monitors the status of the electrical system and major parts of the environmental system at all times. In addition, it is possible from this console to initiate the starting and paralleling of the diesel-generators, to effect a 2-stage transfer to diesel-generator power (first the computers, then, if desired, the environmental systems), and to restore loads to normal power as required.

The only modification that will provide any large degree of increased reliability over the present system is an uninterruptible power supply. In the past, the cost of such equipment and its installation has not been justified by the increase in reliability. However, as more and more bank operations depend on the computers, this situation is changing. A UPS system is being considered and will probably be the next major step in the evolution of this installation.

Modern UPS powers utility control center

Uninterruptible power supply uses logic circuitry to assure continuous, transient-free power for computerized control of electrical distribution network.

COMPANY engineers specified a modern, high-quality UPS system incorporating redundant components to obtain the high degree of system reliability required by the Philadelphia Electric Co.'s extensive distribution network. Transmission lines rated from 60 to 500 kv and serving a 2500-sq-mile territory are controlled from the firm's computer-supervised center in the heart of Philadelphia. The UPS supplies the continuous, closely regulated power required for proper operation of the computers and solid-state/electronic controls.

The control center consists of three identical computers, associated pe-

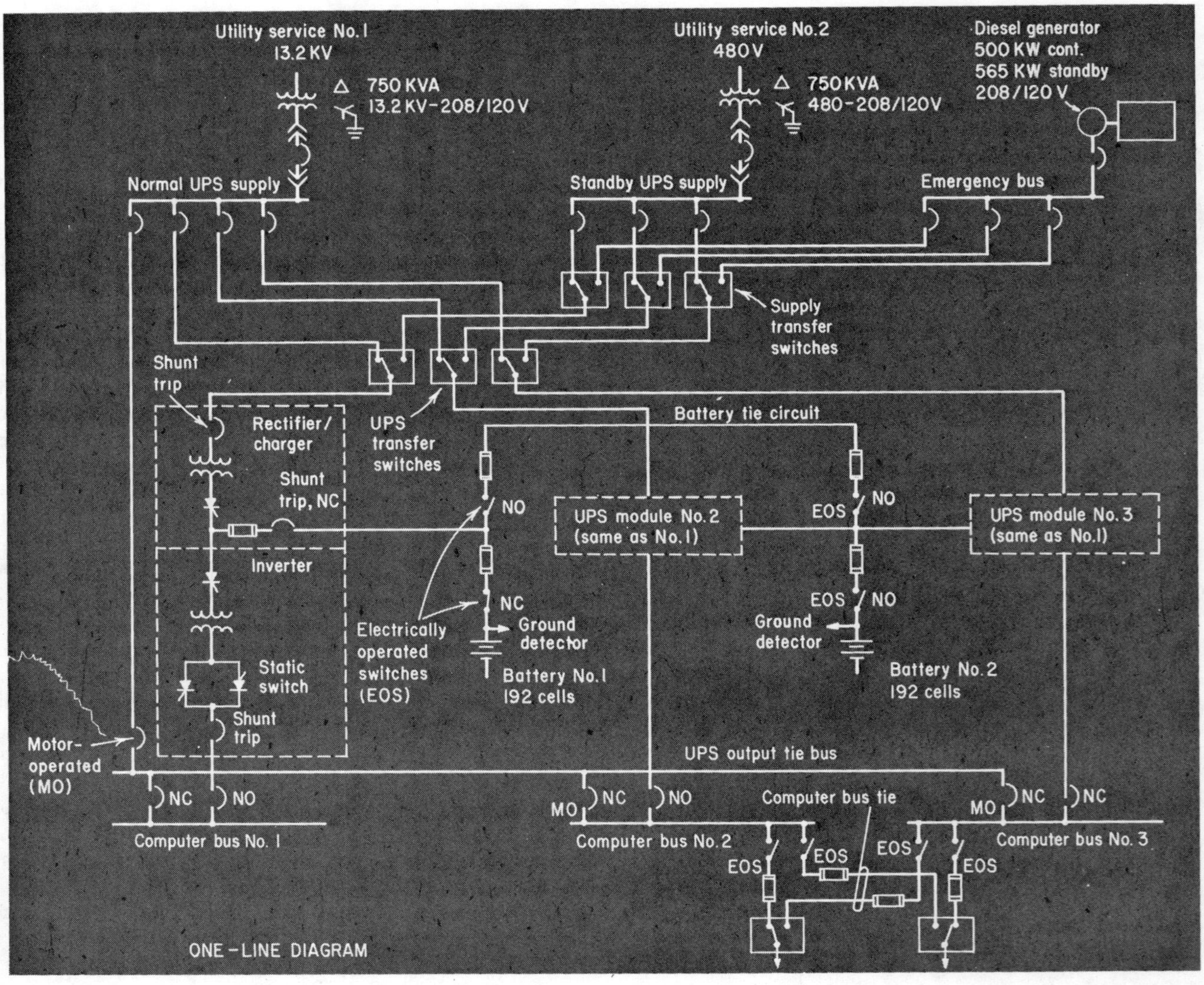

ripheral equipment, and six system operating consoles furnished with a variety of control panels. Field data on generation, line voltages, circuit-breaker status, etc., is supplied to the computers and monitoring instrumentation from over 1600 points. Data is displayed on banks of cathode-ray-tube monitors at each control console. Computers continuously integrate all information, and dispatchers call for displays of schematic diagrams and graphs or data readouts as required to provide for the most efficient and dependable system operation.

Power to the UPS can be obtained from either of two separate incoming utility lines or from a 500-kw diesel-engine-driven generator set. As shown in the one-line diagram, service No. 1 is the normal supply, and service No. 2 is used if service No. 1 should fail. If both utility lines are out, the engine-generator set becomes the source of power. Power is supplied at 208/120 volts to a total computer load of 315 kva. The generator is sized to handle the air-conditioning load required for computer operations under emergency "must-run" conditions.

UPS components

Three rectifier/inverters, each rated 160 kva, supply power to the computer load. The kva ratings selected permit two of the three to carry the entire load when one is out of service. Two lead-calcium battery banks rated at 320 kw for 30 minutes supply voltage for system operation during transfer from utility to generator supply and can, if ever required, supply power to the full load for 30 minutes.

Six electro-mechanical automatic transfer switches are installed ahead of the UPS units. Three provide for power transfer between the utility normal and standby supplies; the other three transfer power between the utility standby supply and the engine-generator source.

At the output of each rectifier/inverter is a static (thyristor) switch rated to handle the full short-circuit current available. In the event of a downstream fault, sensing circuits remove gating voltage from the switch, providing the extremely fast opening required to assure that the fault does not affect operation of the computers. The switch will open in ¼ cycle or less.

SCHEMATIC DIAGRAMS (above) displayed on monitors at dispatcher's control station are created by computers from information sent from remote terminal units on utility distribution system. UPS remote console is at right of monitors.

MASTER CONTROL CONSOLE (below, left) has mimic diagram of UPS system on front panels. Recording wattmeters for each UPS module are at bottom of panel. Logic circuitry contained in the panel automatically directs operation of the UPS system.

RECTIFIER/CHARGER (below, right), one of three, supplies charging voltage to battery bank at left. The two 192-cell lead-calcium battery banks can carry total computer load for 30 minutes or one computer and critical associated equipment for two hours.

THREE STATIC INVERTERS, (bottom photo) each rated 160 kva, convert dc battery power or dc output from rectifier/chargers to ac for computer use. Although all three units are normally in use, only two units are required to carry the 315-kva computer load.

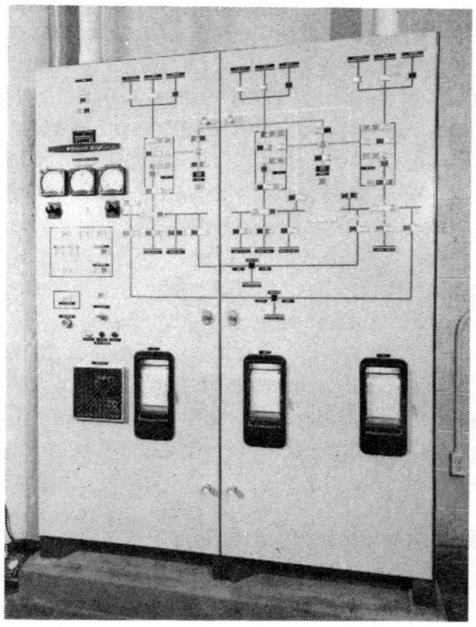

Circuit breakers at the input and output of each rectifier/inverter provide additional overcurrent and fault protection, but their primary purpose is to permit isolation of an inverter if signaling devices indicate a malfunction. These breakers are furnished with shunt-trip coils for remote operation. A fused, shunt-trip CB, installed at the battery supply point to each rectifier/inverter, permits it to be isolated from the battery bank if required. These breakers are not provided with trip elements; the series fuses provide fast-acting overcurrent and fault protection. Electrically operated fused switches were selected for similar isolation control on the battery supply circuits and to permit paralleling or interchanging battery bank supplies to the rectifier/inverters.

Located near the UPS system is the master control console, which is furnished with a mimic diagram of the UPS, a system status annunciator, recording wattmeters, logic circuitry for automatic operation, and manual controls. In the central system operating station is a UPS remote-control console to permit direct operation by the dispatchers if required.

Solid-state logic circuitry controls the UPS system, permitting operation in normal, standby, emergency, or bypass modes. All circuit breakers and switches operate automatically as conditions dictate.

Normal operation

From utility service No. 1, 208/120-volt power is carried to the normal UPS switchgear. Protected by a circuit breaker, each of three main feeders extends from the switchgear through a transfer switch to a rectifier/inverter. All three rectifier/inverters (UPS units Nos. 1, 2 and 3) are supplied in parallel. Normal output is to three separate computer buses paralleled through motor-operated circuit breakers and a UPS output tie bus as shown in the one-line diagram.

The two battery banks are on float charge, the charging voltage being supplied by the UPS rectifier. Normally, battery bank No. 1 is connected to UPS No. 1; battery bank No. 2 is connected to UPS Nos. 2 and 3. A tie circuit permits one battery to be taken out of service for maintenance or allows paralleling or interchanging of battery banks between UPS units. Battery voltage is a nominal 432 volts dc, with both legs above ground. Ground detectors will indicate and sound an alarm if one leg becomes grounded.

The UPS tie bus serves a number of functions. It is supplied from the normal UPS supply switchgear through a circuit that permits any or all of the UPS units to be bypassed. This bus is connected to three motor-operated circuit breakers, each supplying a computer bus. The supply bypass circuit is furnished with a normally open, motor-operated breaker in series with a circuit breaker at the switchgear.

A number of different circuit arrangements can be obtained through this bus. For example, if UPS No. 1 fails, the logic circuitry will check that units Nos. 2 and 3 are operating properly. If so, the output breaker at UPS No. 1 will be opened automatically, and UPS units Nos. 2 and 3 will supply power to computer bus No. 1 through their respective computer buses, the motor-operated CBs, and the tie bus. The motor-operated circuit breaker in the bypass supply line will remain open if this sequence is chosen by the logic circuitry. In another option, any one of the UPS units can be bypassed and its computer bus supplied directly from utility service No. 1 through the bypass circuit. Other arrangements can be selected by the logic system, which is being fed information from all components to assure the best possible configuration of circuits and use of components.

Emergency operation

If one of the UPS units fail, the other two can carry the entire computer load while the defective unit is being repaired. In the unlikely event that two are down at the same time, one UPS can carry a "must-run" load of selected computer equipment. All decisions are made by the logic circuitry.

The logic system directs all operations in event of utility power failure. If utility service No. 1 fails, the UPS transfer switches transfer to the standby power utility No. 2 service. If this service is lost, the diesel-generator set will automatically start and will provide power through the supply transfer switches. While the generator is coming up to speed, the battery banks supply the UPS systems, permitting uninterrupted operation of the computers.

If both utility power sources fail and the computer operates in the "must-run" or restricted mode, the logic system will initiate many actions to assure continued operation.

First, the three UPS units immediately switch to battery power. The computers continue to run properly on transient-free power, without interruption. If required, batteries are capable of supplying power to the entire load for 30 minutes.

Next, the engine-generator starts automatically. When voltage is up to the proper level, transfer switches connect the UPS system to the generator.

Meanwhile, the logic circuitry, which is kept advised of all actions being taken via signals from all equipment, prepares the computer equipment for operation in the "must-run" mode. The load supplied includes the minimum amount of equipment needed to keep the control center functional—one computer, associated peripherals, control consoles, and minimum air conditioning. All of the above actions occur within the first few seconds of the outage, except for the powering of the air conditioning, which is energized about 30 seconds after the start of the outage.

The point of no return is at two minutes after the start of the outage. If power returns on either utility line before the 2-minute deadline, the logic circuitry will return the systems to normal operation. All loads that were intended to operate from the engine-generator set will be transferred back to their normal power source.

If the outage continues for more than two minutes, the logic is committed to convert the computers and UPS systems to the "must-run" condition. The computer software enters a decommitment routine, and the three computer output buses are paralleled through the UPS output tie bus.

At outage-plus-three minutes, the logic circuitry signals two of the three computers to begin sequencing off. This action allows two computers to power-down properly but maintains fans for cooling. Fifteen seconds later, the logic initiates a load-shedding program, opening shunt-trip CBs to remove loads not associated with the "must-run" mode. The logic circuitry also has prepared for system operation via one of the three UPS units.

At 3¾ minutes after the start of the outage, one UPS will be ready for energizing from the generator set. The logic circuitry selects which of the three UPS units will be used.

At four minutes, the "must-run" system is in operation, the generator supplying power via the selected UPS. The status of all components is reported on both the master and remote control consoles. In the event that the engine-generator set is lost, the battery banks can supply power to the "must-run" load for up to two hours. When utility power is restored, a manual procedure is used to return normal UPS service to all computers.

The UPS system was designed by Philadelphia Electric Co. engineers in cooperation with those of the UPS manufacturer. Electrical installation was by Morris Newmark & Bros., Inc., electrical engineers and contractors of Philadelphia.

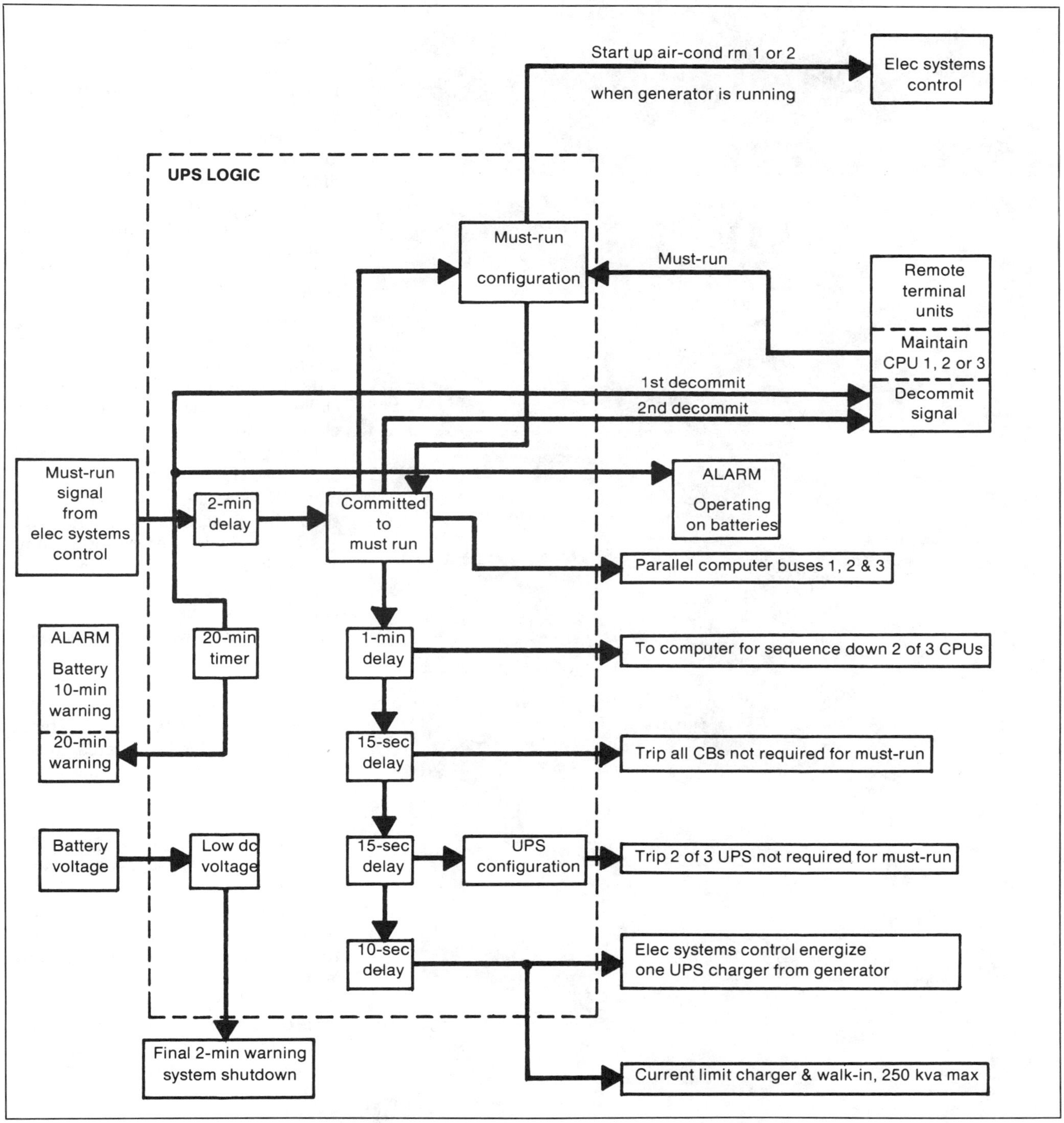

LOGIC FLOW DIAGRAM for "must-run" operation illustrates sequence of events taken into account by logic circuitry in the event that both sources of utility power are lost.

Reliability is key to distribution design for airline reservations

Well-thought-out design details for service and distribution networks permit all-electric reservations center to cope with power supply disturbances.

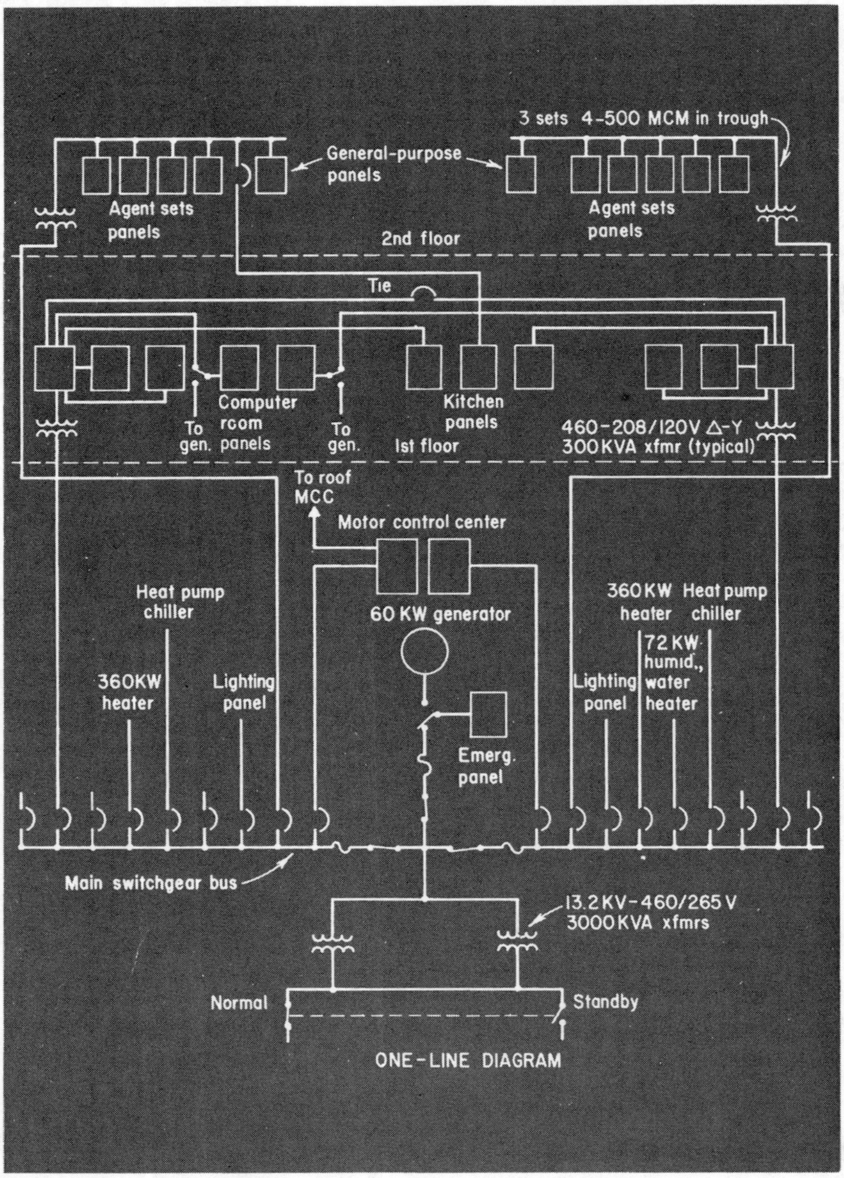

THE DECISION to go all-electric at the Eastern Airlines telephone reservation center in Woodbridge, N.J., made reliability of power the primary consideration in the service and distribution system design. Equipped to handle a capacity of 3000 calls per hour, the 94,000-sq-ft concrete structure provides 800 reservation stations plus electronic and telephone equipment rooms on the second floor, with dining, training, and office areas on the first floor.

At the outset, the client indicated the need for quality design to assure safe, uninterrupted operation of the building, which serves eight northeastern states and Ontario, Canada.

The choice of all-electric design was arrived at by first considering four different energy systems, keeping in mind the high degree of environmental comfort control desired for both employees and electronic gear:

1. Gas heating with absorption cooling
2. Gas heating with electric cooling
3. Internal source heat pump with gas auxiliary heat
4. Heat-pump units providing both chilled and hot water, with auxiliary electric heat.

Heat pumps with auxiliary electric heat were chosen for reasons of economy—the system was ⅓ less costly than the most expensive alternate, gas heating with absorption cooling. An internal heat reclamation system used with the heat pump provides for both heating and air conditioning. Two auxiliary 360-kw boilers boost water temperature when required.

With a total connected load of 3555 kw in the design for lighting, HVAC, reservation equipment, and kitchen and accessory loads, isolated generation was considered but rejected for three basic reasons:

1. A tight construction schedule—the building was to be operational less than 11 months after groundbreaking.
2. The computer unit would require very rigid frequency control and call for extensive and sophisticated regulating gear; owning and operating costs for an on-site system would be greater than purchased power.
3. The owners wanted to keep maintenance and operation personnel requirements to a minimum. (This is contracted to an outside firm.)

Concentrating on the importance of reliability of utility-provided

TWO SWITCHGEAR SECTIONS, each with capacity for nine feeder CBs, are installed on either side of large 3-bay center portion containing sequence-shedding equipment, CTs, and a 3-pole, 4000-amp, solid-neutral, bolted pressure switch for isolating each breaker section. Panel lights give graphic indication of equipment status. Five spare breakers or breaker positions are for expansion.

MOTOR CONTROL CENTER serving HVAC equipment is divided into two sections, one CB from each of the two switchgear sections feeding each half. For further reliability, the 5, 10 and 50-hp pump motors in the HVAC each have a standby motor and pump paralleled within the piping network, which is displayed on panel at left.

power, the consulting engineers incorporated the following features in the electric design:

1. Dual primary lines, with throwover transfer switch
2. Automatic load-shedding procedure in emergencies
3. Secondary spare service conduits
4. Secondary split busway to dual switchboard
5. Provision for installation of an emergency 60-kw generator.

Design details

The service consists of two sets of underground primary feeders to a below-grade transformer vault next to the building. One set of cables is the preferred (or normal) power feeder, the other a standby line (or alternate) available in case the preferred feeder is out of service. A throwover switch within a utility-supplied enclosure transfers the building load to the standby feeder automatically within 20 sec after sensing loss of power on the preferred line. This time interval permits normal utility switching operations and minimizes transfer to the standby service during minor service difficulties.

At the request of the owner, the standby service was designed to carry about 50% of the building's load, the remainder being dropped by load-shedding equipment. This technique provides an orderly phasing out of machinery to avoid damage.

Each of the two lines enters the vault from two physically distinct directions as further protection against damage or outage at the facility. Two other service conduits were installed to handle future loads or as possible backup lines. Installation of a 60-kw gas generator was completed recently as protection against complete loss of utility power.

Multiple sets of 500MCM cables carry power from the transformer secondaries through a pull-box in the basement wall to switchgear bus.

This duality and division for reliability is also basic to the distribution scheme. A split-bus configuration within the switchgear provides isolation in case of switchgear damage by dividing the critical loads equally into two sections. A fault in one section would not likely affect the operation of the other, and each bus section carries its own disconnect switch to facilitate necessary repairs, changes, and maintenance chores. Aluminum bus connects the two free-standing, metal-enclosed switchboard sections on either side of the CT cabinet and load-shedding gear.

The molded-case circuit breakers protecting the feeders are supplemented by integral current-limiting fuses to handle high-capacity, short-circuit currents that could reach 100,000 amps. Space for additional breakers in each switchboard section (one section has two spares, the other three) anticipates a 50% expansion in building floor area. Spare conduits are already in place from the switchgear to the opposite side of the building,

CABLE RISERS, supported in orderly bundles by metal framing sections from computer room to reservations desks on floor above, are part of the vital electronic network operating around the clock. Charles Kelly, general foreman of Beach Electric Co., discusses details with consulting engineer Howard Lemelson.

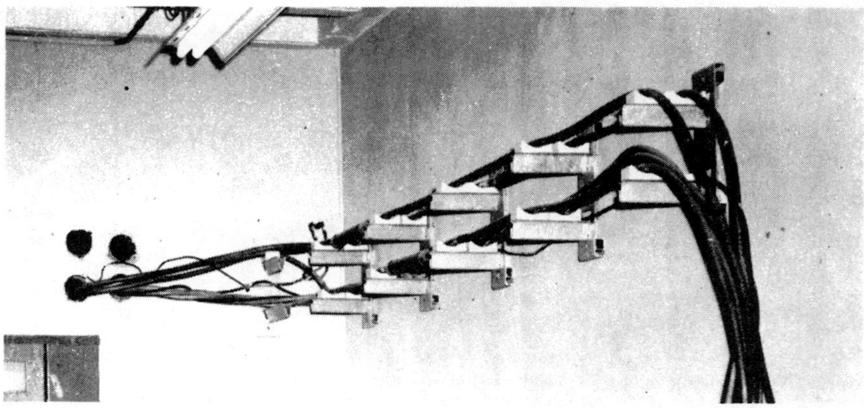

PRIMARY POWER CONTINUITY is assured by two underground 13.6-kv utility mains (one normal, one standby) emerging into vault from bank of four 4-in. steel conduits on 9-in. horizontal and vertical centers about 30 in. below grade. No. 2/0 conductors are spliced between wall-mounted rack supports. Two vertical conduit sections carry the conductors beneath concrete floor to throwover switch enclosure at left.

where westward expansion is planned.

Critical loads are equally divided on each section of the switchgear for diversification and reliability. The motor control center serving the comfort-conditioning equipment and plumbing network also is divided into two equal parts, each part fed from a breaker on opposite sides of the switchgear bus.

To supply the operational loads of the building, two breakers on each switchgear bus section feed power to the first and second floors. Each of the four feeder circuits consists of three 500MCM cables in 3-in. rigid metal conduit and supplies a 460-208/120-volt delta-wye transformer. The two transformers on each floor are located at diagonally opposite corners, on either side of the building's long axis.

Located in a generously sized electrical closet, each transformer serves about half of the floor area. Lengths of circuit homeruns are thus kept to a minimum. These transformers are identical in size, making emergency substitution possible. If a transformer serving a critical function were to fail, another unit (such as the one supplying the less-critical kitchen and office areas) could be taken out and substituted for the failed unit until a new one could be obtained.

Computer equipment on the first floor, vital to the reservation function, was protected against voltage, frequency, harmonic disturbances, and total power loss, as follows:

1. A transformer secondary interlock tie extends from the other first-floor transformer to the computer panels, making possible a transfer of power with minimum delay in case of feeder or transformer failure.
2. To prevent line disturbances caused by motorized kitchen equipment from being imposed on the computer power lines, each of the three kitchen distribution panels are fed from a different electrical closet, one feeder coming from the second floor.
3. Capped conduits were installed to facilitate wiring of computer power panels into a future motor-generator set that could be used to override voltage and/or frequency spikes coming from the utility system if experience indicates the need for it.

Power to the reservation desk, where each clerk uses a cathode-ray display and optical card display unit, is facilitated by steel underfloor ducts extending from each electrical closet.

Fluorescent lamps in fixtures using 277-volt ballasts are the major interior light source, the 460/265-volt lighting panels being fed directly from the main switchgear bus.

Kallen & Lemelson, consulting engineers, New York City, did the electrical design, working closely with architects Frank Grad & Sons of Newark, N.J. and J. N. Pease Associates of Charlotte, N.C. Electrical contractor was Beach Electric Co., Inc., East Orange, N.J.

Chapter 4

Application and Installation Methods

CONTRACTOR SPECIALIZES IN COMPUTER POWER

Robert J. Lawrie

Here's how an electrical contractor developed a specialized organization to design, build, and maintain computer power systems.

Computer power specialists for Data Site Constructors review drawings of a new construction project. From left are John Lucas, project manager; Joseph Koch, president (standing); Peter Bergman, project manager; and Alan Zwiebel, design electrical engineer.

ABOUT FIFTEEN YEARS AGO, Allan Electric Co., a successful, well-established electrical contractor located in Totowa, NJ, recognized that computer power work offered a great opportunity for growth. Noting that this kind of installation demanded special requirements and capabilities, the company began to accumulate information, specifications, and data needed to design and install computer power systems. As a variety of computer facilities were installed, the firm acquired experience, know-how, and expertise in the construction of these facilities. As a result, a subsidiary company, Data Site Constructors, was formed to provide a total capability for the design and construction of computer power systems. The firm also offers complete turnkey construction of all phases of data-processing facilities.

Services offered

Data Site's total effort is aimed exclusively at the computer power and data-processing construction market. The company presently offers a number of design/construction approaches such as:

Total turnkey. This concept offers complete design, construction, and maintenance of *all* functions, such as civil, mechanical, electrical, structural, finish work, and HVAC.

Design and build. With this approach, the firm provides total design and construction of all *electrical* facilities.

Total bid. The firm submits a bid on all phases and types of construction of the data-processing facility.

Bid—electrical. A bid is submitted, based on supplied drawings and specifications, on only the electrical systems.

Negotiated project. Data Site provides a computer power system, sometimes based on submitted designs or proposals, that is tailored to meet objectives at the best possible price.

Time and material. When performed under these terms, the computer power project is usually modernization or alteration work and is completed on the basis of materials used and work required.

Maintenance contract. An annual contract provides for regular maintenance of UPS system, batteries, generator, load tests, fire alarm, security, and other related items.

Typical projects, which range from one-half-million to five-million dollars, include computer facilities for national broadcast companies; communications firms; banks, accounting, insurance, and other large financial companies; hospitals; research and development facilities; and similar operations.

In these applications, work may include design and installation of uninterruptible power supplies (UPS); standby power generators; power distribution for the data-processing center; HVAC; fire, halon-suppression, and security systems; raised floors; general interior construction; site development; and preventive maintenance.

Because the company concentrates solely on

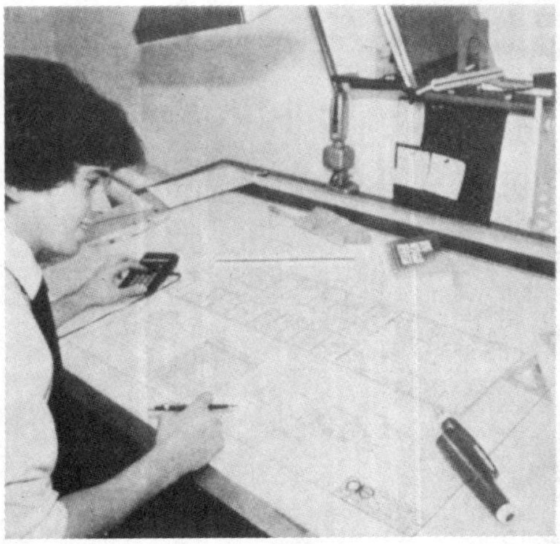

Design details for a new computer power system are being incorporated in the system concept by Michael Messina, designer/draftsman.

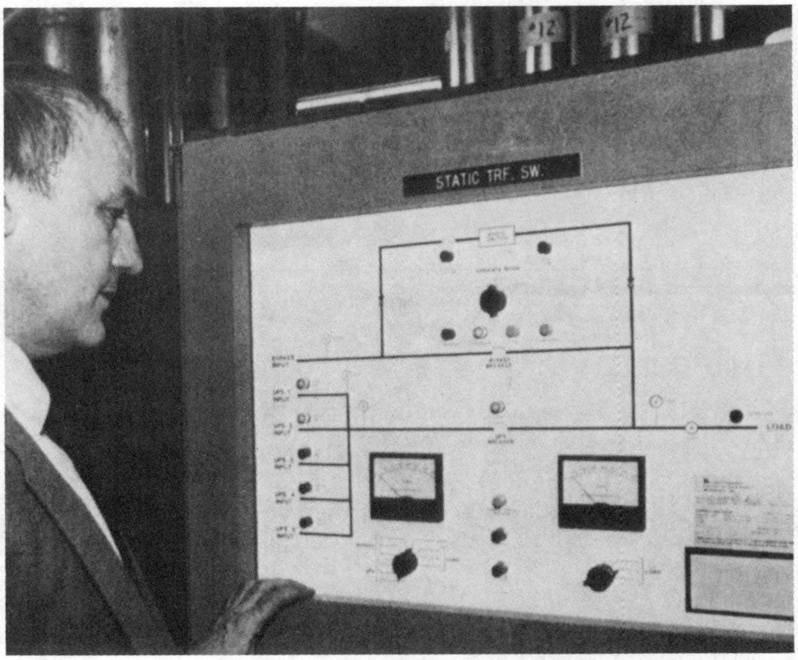

Mimic diagram on front of static switch panel, being inspected by project manager Peter Bergman, shows the arrangement of major components of the UPS system.

computer power and data-processing centers, numerous benefits are obtained for the customer as well as for Data Site. Major benefits for the customer are:

- Single-source responsibility. On most projects, architects, consulting engineers, various contractors, and a number of vendors may be involved. Data Site takes full responsibility for all of these.
- Reduced costs. Because of inherent close coordination, tight control, close cooperation and long-time experience, the firm usually presents the lowest cost.
- Guaranteed lump-sum price. For the same reasons, the firm can guarantee its quoted price. Close relationships with subcontractors, vendors, and consultants make this possible.
- Speed. Data Site "in-house" engineers, project managers, and others who have been with the firm for many years have the experience and know-how to design and construct all types of computer power, assuring fast job completion.
- High-quality design, components, and installation. Data Site can combine the practical experience and engineering know-how of their people with those of Allan Electric to obtain the necessary expertise.

Many other important advantages accrue because of these special capabilities. Custom design of systems or design changes can be easily accomplished. Maximum dependability of system operation results not only because of the firm's capabilities, but also because the subcontractors (on turnkey projects) and vendors are required to offer the highest level of service. This speeds the entire construction process, delivery of components, start-up, debugging, troubleshooting, and preventive maintenance.

The benefits for the contractor are due to the advantages of specialization, resulting in increased efficiency, more capabilities, and focused operations, all of which permit the firm to provide better work at lower costs.

Team approach is key to success

Data Site facilities are in the same building as Allan Electric; however, its offices are physically separated. With this arrangement, the company is able to maintain its independence, yet have ready access to the resources of the parent firm.

The key to the success of their operation is a coordinated in-house/subcontractor team approach. Joseph Koch, president, in charge of company operations, coordinates this design and construction team. The team consists of Alan Zweibel, design engineer, who is also a professional engineer; Michael Messina, a designer/draftsman; two project managers—John Lucas, who is responsible for technical data, equipment, and estimating; and Peter Bergman, who is in charge of construction/scheduling. Expediters, clerical, and other office workers provide assistance.

At the initial stage of a project, Joseph Koch confers with the customer to set the general needs and objectives for the computer power system. From these Alan Zweibel, working with John Lucas, develops a design concept. While John is investigating vendors, technical details, and checking designs, Pete reviews all aspects of the job from a construction point of view. He checks for location of UPS equipment, switchgear, and HVAC; decides which

Huge computer room for a world-wide accounting firm required extensive electrification. Installed under a raised floor are power, control, and signal cables for hundreds of computers, related apparatus, computer power centers, HVAC units, and peripheral equipment.

wiring methods will be most effective; checks on wire management, availability of labor, and later manages the progress of construction; and interfaces with the customer.

Equipment vendors are vital to the success of every project. Data Site has found that the vendors' performance can mean the difference between a profit and loss on these specialized computer power projects, and poor vendor performance has a negative effect on their reputation. As a result, they select vendors only after careful analysis and comparison. Initial cost of the equipment means nothing if the supplier does not provide a quality product and good, dependable service. Any savings easily will be lost if the service is not there for start-up, special assistance, troubleshooting, or on-time delivery of new components or replacement parts.

Typical case history

A recent project illustrates the effectiveness of this kind of electrical construction operation. A world-wide accounting firm decided to move major operations into four large office buildings in northern New Jersey. The contract, for complete computer facilities, included computer power, electric-service expansion, electrical-system alterations, and HVAC and certain mechanical aspects. Electrical systems for computer power comprised the major portion of the work.

The project was about three-million dollars, and the bid had to be submitted in two weeks. Furthermore, the work had to be done in three months.

The new service includes an underground duct bank consisting of 10 4-in. PVC conduits encased in concrete, carrying multiple 600MCM conductors. Supply voltage is 480V.

A 900kW, diesel-generator set, installed in a new building nearby, serves as an alternate supply. From the generator, standby power supply conductors are carried underground in a concrete-encased PVC conduit bank to a main switchboard. A 2000-gal tank provides a two-day fuel supply for the diesel generator.

The main switchboard houses a 4000A main power circuit breaker and seven feeder breakers ranging in size from 800 to 1600A. All circuit breakers have ground-fault protection.

A UPS system consists of two 200kVA modules, that supply 60-Hz power, and two 125kVA modules that supply 415-Hz power for the project. The 60-Hz supply is expandable to 1000kVA. The UPS modules are supplied from circuit breakers in a UPS distribution board that is fed from both the utility and generator supply.

Three automatic transfer switches provide an alternate supply to vital equipment, such as UPS systems, maintenance bypass circuit, and standby power panel, when the primary source is interrupted.

A static switch is included in the UPS to assure the high-speed switching necessary to avoid disturbance to computers in the event of a power transfer. The static switch is rated lower than its corresponding circuit breaker because it carries power for only milliseconds while the main transfer switches are operating.

Four battery banks are installed in a separate battery room, one for each UPS module. Batteries are lead-calcium, and each bank is protected by a fused 600A disconnect switch.

Two battery banks (foreground) for 60-Hz system are made up of 180 cells each, providing two supplies at 405VDC (2.25V per cell). Two battery banks (background) being checked by Al Fiore, foreman, Data Site Constructors, supply 270VDC (120 cells) for 415-Hz system.

Computer power center rated 100kVA is one of nine that supplies 208V power to computers and related equipment. Jay Gronbeck, foreman, Data Site, checks circuit breaker assignments.

Geometrical jungle of EMT, in 3- and 4-in. sizes, is only a small portion of raceways installed in the UPS room. At left is main switchboard; three automatic transfer switches and generator panel are in center; and a UPS distribution board is in background.

HVAC equipment includes numerous heating and cooling units ranging in size from 10 to 20 tons. A large switchboard for this equipment is installed in the UPS room.

Other notable work included a complex fire-protection system that incorporates main and remote annunciator panels and ionization detectors that sound an alarm or release halon on a multiple trip system. Detectors are installed overhead and under the raised floor. An unusual feature was the installation of lighting systems under raised floors to provide ready visibility for planned expansions, alterations, and maintenance. Switches and signal lamps showing the ON/OFF status of this lighting are provided on area annunciators. Also, extensive control, telecommunications, and data-transmission systems were installed.

Several contractors including Data Site bid on the job. The Data Site team studied the bid documents, planned the project, obtained all costs from subcontractors and vendors, and submitted the quote on time. The bid was accepted by the customer; and Data Site accepted the job and finished it in three months, as required.

Dual UPS powers world communications center

Twin uninterruptible power supplies, huge turbine-generator sets, custom-designed computer power centers, and a secondary-selective distribution system assure maximum reliability for computerized communications.

By ROBERT J. LAWRIE

POWER system reliability is of paramount importance at the new ITT World Communications Center, Whippany, NJ. Each day, many thousands of messages are channeled to all parts of the world through this modern, high-technology facility. Operating day and night, computerized communications equipment demands total reliability as well as transient-free power. To assure this high level of performance, company engineers specified and developed a total power plan designed to provide the highest degree of power system dependability and the best possible service for customers.

The recently completed complex operates "in parallel" with a similar facility in New York City. This arrangement emphasizes the value the company places on service continuity for customers, since any malfunctioning lines at one location can be immediately transferred to the other. Pointing up the vital nature of service continuity is the fact that the company serves the Washington/Moscow hotline as well as a huge network of international Telex/telegraph services, satellite data transmission, video, high-speed data, leased channels, and many specialized message services.

LOAD-BREAK SWITCH rated 200 A at 15 kV dwarfs ITT electrical engineer Bill Pirrone. Switch, transformers, and other important equipment are installed in outdoor yard surrounded by a concrete block wall.

TWO 2500-kVA TRANSFORMERS operate in parallel, stepping down incoming 12.5-kV utility supply to 480/277 V for distribution. Either transformer can be removed from service at any time, since the other can carry the entire facility load.

ITT electrical engineers played a major role in the design of the power systems because of their long experience with "total reliability" designs and expertise in UPS systems. The plan developed had to dovetail the power-distribution scheme with the UPS arrangement. Major elements include the normal incoming power supply; an alternate supply consisting of two 2800-kW jet-fuel turbine-generator sets; secondary-selective switchgear; dual UPS systems, each rated 500 kVA and each having expandability to 1500 kVA; solid-state UPS bypass switches; maintenance bypass circuits; a UPS isolation scheme that permits one UPS system to be isolated for maintenance while the other carries the entire critical load; a UPS output tie CB; and numerous computer power centers that supply 277/120 V to computer/communications equipment.

System operation

During normal operation, the local utility delivers 12.5 kV to a service complex that includes the high-voltage yard, primary switchgear, turbine generators, and related components installed in a large facilities-service structure adjacent to the main building. As shown in the one-line diagram, power is fed through a 200-A fused load-break switch and primary circuit breakers to two 2500-kVA, oil-filled transformers that step down the primary voltage to 480/277 V. Secondary power carried by multiple 500MCM conductors extend underground to transition points in the main building where conversion is made to two 4000-A busways.

The dual busways carry 480/277 V overhead in the basement to the main secondary switchboard. Each busway supplies a 4000-A circuit breaker at each end of the switchboard. A 3200-A tie CB, which is operated normally open, can be closed to permit supply of the entire switchboard from either busway. Loads supplied are equally divided between both halves of the switchboard. These loads include motor control centers, pumps, HVAC equipment, lighting, and the UPS systems.

The dual uninterruptible power sources are each supplied from a 3200-A CB fed from each side of the switchboard. Both UPS systems supply transient-free, continuous power to computerized communications equipment. In addition, dual bypass schemes for each UPS are provided. One bypass circuit is for maintenance purposes, permitting complete isolation of the associated UPS; the other bypass arrangement includes a static switch for assured continuity without transients during transfers of power. At the out-

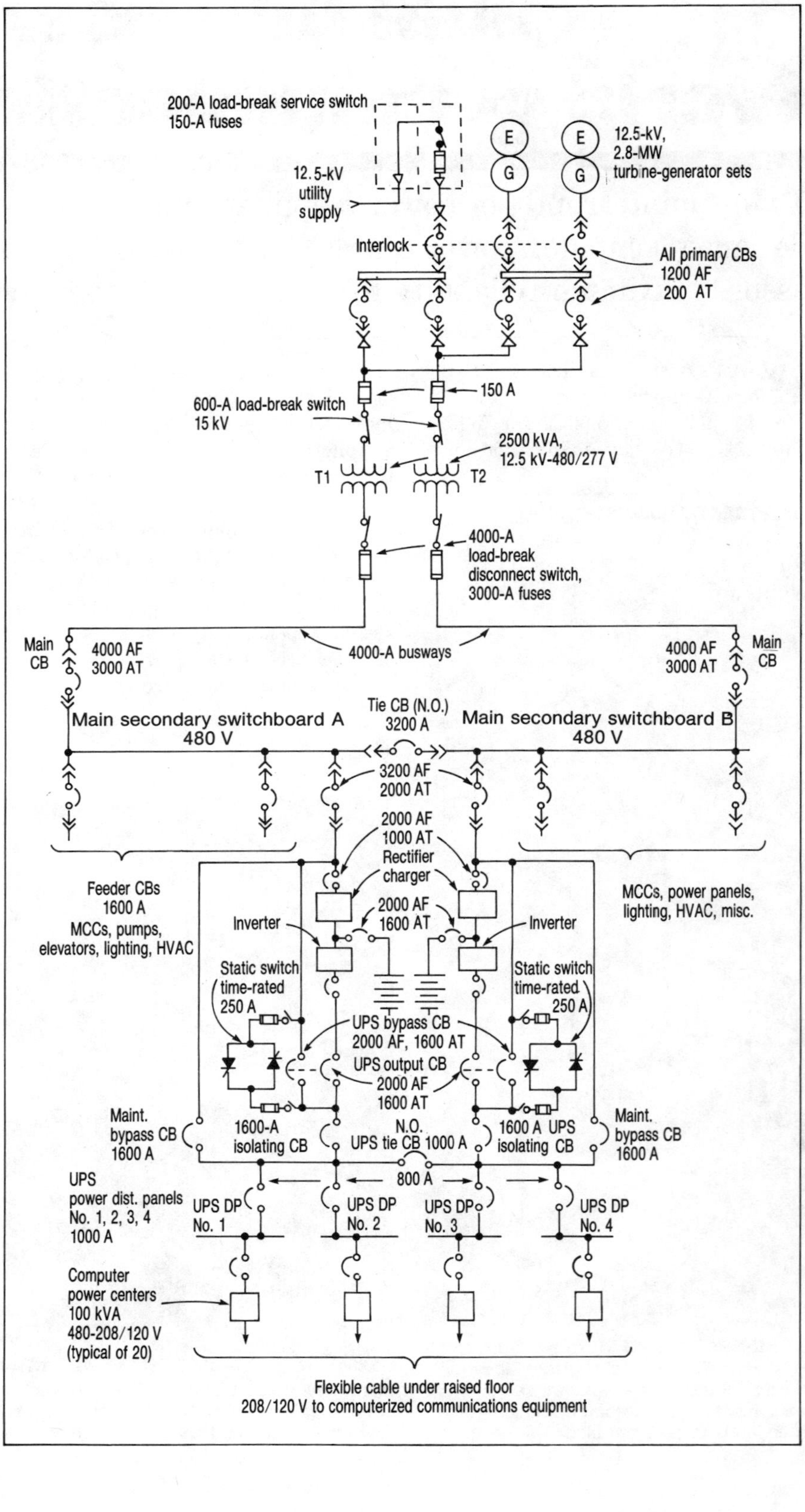

ONE-LINE DIAGRAM shows major elements and pertinent details of power distribution and UPS system serving modern communications complex.

put of the UPS systems is a tie CB that permits either or both of the UPS systems to supply four 1000-A panels designated as UPS distribution panels. These panels supply 480/277 V to computer power centers located close to the computerized communications equipment on the first and second floors of the two-story building. Rated 100 kVA, the computer power centers are furnished with dual distribution panels that protect 208/120-V circuits extending under a raised floor to the communications equipment. Presently there are nearly a dozen power centers; provisions have been made for up to about 24 such units as required in the future.

HUGE TURBINE-GENERATOR SET rated 2800 kW, 12.5 kV is one of two installed in services building adjacent to the main building. Jeff Katz and Bill Pirrone, both engineers with ITT, inspect air intake silencer and hydraulic-start motor for main turbine.

UPS systems

The purpose of the UPS systems is to furnish a regulated output voltage with frequency stability. The systems must be unaffected by transients in the utility supply and must continue to supply filtered, uninterrupted power in the event of a utility power failure. Battery banks are provided to assure power to critical loads during activation of the generator systems.

Specifically, the UPS system draws all power from the utility line during normal periods; turbine-generator sets supply power whenever utility power is not available; and the battery banks supply power during transfer of power sources. The UPS system isolates the critical loads from the utility line, keeping all transients, via inherent filtering systems, from reaching the loads. The UPS also supplies power during momentary power dips, surges, or internal faults, while permitting only conditioned, regulated voltage and frequency to be supplied to the computerized communications equipment and related items.

SECONDARY-SELECTIVE distribution is possible using this main secondary 480-V switchboard. Power enters the switchboard at 4000-A CBs at each end of the board (arrows 1 and 2). Tie CB is at center (arrow 3); George Carey, group leader engineer, inspects one of two 3200-A power circuit breakers serving UPS systems (arrows 4 and 5).

The twin UPS systems are installed in the basement of the building, each capable of powering the entire critical load in the facility. Major components of each UPS system include an incoming section with a 2000-A input circuit breaker and termination space and a rectifier-charger to supply DC power to the inverter and charging power for the battery bank. The lead-acid battery bank, floating on line, provides power to the inverter instantaneously in the event of utility power failure; the inverter and output filter assure that only transient-free power is supplied through the output circuit breaker to the loads. Other key components

UPS MODULE rated 500 kVA (450 kW) is one of two such modules presently installed. Director of engineering Joseph Weingord inspects thyristors in inverter section. Behind open door are sections housing the system rectifier/charger, input circuit breaker, and battery and DC bus sections. At right is metering and monitoring panel with mimic diagram and module output circuit breaker at end.

include a bypass static switch that operates extremely fast, high-capacity bypass circuit breakers, and a main control panel to provide for operation, metering, and monitoring of the system.

The rectifier-charger consists primarily of an input isolation transformer, SCR phase-controlled rectifier bridges, a series DC output filter choke, and a DC bus filter capacitor bank. Firing of the SCRs is adjusted continuously to a regulated DC at the bridge output. The output is 6-pulse DC with a ripple frequency that is six times the input frequency. This pulsed DC output is smoothed by a conventional series choke and is further filtered by a large capacitor filter bank connected across the DC positive and negative lines. The gate drive to the SCRs in the rectifier bridge is controlled by logic circuitry that senses the DC bus voltage and current.

Two 3-phase bridge inverters consist of SCRs controlled by digital logic. The arrangement results in a 6-phase circuit with 12 switching elements. The square-wave output is fed through a filter to obtain optimum wave shape and lowest harmonic distortion.

Installed in a separate lineup adjacent to each UPS module are the main control panel, bypass circuit breakers, the static switch, and output circuit breakers. The control console contains metering, alarm LEDs, mimic bus, controls for the UPS system, and the static switch and its controls.

BATTERY BANKS are installed adjacent to UPS room. Each bank consists of 186 lead-acid cells, each rated 2.25 V. Bill Weber, technician, checks terminations where DC power extends up into cable tray for routing to UPS modules.

CONTROL EQUIPMENT is inspected by Stephen Peters, engineering manager. Section at left contains static switch control at top, bypass circuit breaker in center, and UPS output CB at bottom. In the center is the metering, monitoring, and alarm panel. The static switch is installed directly beneath this panel. At right are output circuit breakers to circuit loads.

Static interrupter

A static interrupter bypass switch that operates in conjunction with a paralleled power circuit breaker allows instantaneous load transfer from the UPS output to the utility power bypass circuit in the event of a system failure. Normal fault-isolating devices such as circuit breakers are not fast enough to effect a transfer without causing adverse transients on the critical load supply. This switch also functions to remove a faulted UPS module, permitting the other module (or utility bypass power) to supply the load.

The solid-state switch consists of SCR circuitry controlled by fast-acting sensing devices and electronic logic circuitry. Both UPS systems are provided with such a switch, each rated at 250 A for 150 millisec. When an automatic high-speed transfer is required, or an internal UPS fault occurs, the static switch carries current for approximately 40 millisec while the paralleled bypass circuit breaker is closing. When the bypass CB has assumed the load current, the static switch will automatically open.

In the event of a UPS module failure, a second, separate bypass circuit breaker can be closed to provide a utility circuit around the faulted module, and isolation circuit breakers can be opened to competely isolate the unit for maintenance and testing.

Battery banks and generators

Two large banks of lead-acid batteries are installed in a separate room adjacent to the UPS equipment. Each bank consists of 186 cells, each rated 2.25 V. Circuits from the battery banks consist of THW conductors carried in cable tray overhead to each UPS module. Each bank is protected by a 2000-AF/1600-AT circuit breaker. In the event of normal utility power failure, each battery bank can carry its entire assigned load for at least 15 min.

Two 2800-kW turbine-generator sets are on standby in the event of normal power failure. If normal power fails, the batteries, which are already on-line, supply power to the critical load. Startup of both turbine generators is initiated. The huge, locomotive-sized units, which supply 12.5 kV alternate supply voltage to the system, require about 2 min to reach proper rpm. Upon reaching required speed, utility-power feeder CBs open and generator-power CBs close. Only one generator is required to carry the entire system load; the first one that comes up to speed will be placed on line to supply the load.

The generator sets are installed in the facilities building near the high-voltage yard. High-voltage switchgear and generator switchgear, control panels, and synchronizing equipment are also housed at this location. Installed underground nearby are two 20,000-gal tanks of high-grade liquid fuel—enough to drive the generators for approximately five days.

COMPUTER POWER CENTER is one of many located near computerized communications equipment. These centers are furnished with a 100-kVA isolation transformer that steps down 480 V to 208/120 V, which is supplied beneath the raised floor to equipment.

MODEM CABINETS are typical of the complex and highly sophisticated telecommunications installed. Most messages are computerized, on "real time" operation, demanding the ultimate in service continuity.

Computer power centers

Computer power centers are compact units specifically designed to simplify circuiting to computerized communications equipment. These power centers are located close to loads served throughout the building. The power centers were manufactured to meet the design requirements specified by ITT engineers. Each unit consists of a 100-kVA transformer that steps 480-V power to 208/120 V for distribution to equipment. Transformers are of shielded isolation-type design to minimize transmission of spikes and other transients to the sensitive communications equipment. All units are furnished with dual 42-circuit panels installed back-to-back within the unit cabinet. From the cabinet, circuits extend under the raised floor via flexible multiconductor cables to loads. All panels are provided with monitor lights, alarms, metering, and controls.

The UPS system and requirements for the power distribution system were specified by a team of ITT engineers headed by Joseph Weingord, director. Other members of the team included George Carey, J. Katz, Stephen Peters, and W. Pirrone. Consulting engineers were Jaros, Baum and Bolles, New York City. The electrical contractor selected for the project was Electrical Installations, Inc., Cranford, NJ.

Powering up for CAD/CAM

Increased power capability at the Martin Marietta plant in New Orleans, LA, will supply Computer Aided Design/Computer-Aided Manufacturing systems to speed production of NASA Space Shuttle propellant tanks.

THE SOPHISTICATED technology that is second nature at the Martin Marietta Aerospace Corp., New Orleans, LA, is being augmented by computer-aided design and manufacturing facilities that required additional power. This sprawling complex is dedicated to the production of the external tank (ET) that carries fuel for NASA's space shuttle. Manufacture of this huge space-age tank is complicated and time-consuming; however, implementation of CAD/CAM (computer-aided design and computer-aided manufacturing) is expected to simplify and speed its production. Availability of the CAD system was given first priority; therefore its power supply had to be designed and installed as quickly as possible.

The design of the CAD power source was governed by the computer manufacturer's specifications, the time available, the existing equipment and environment, and budget requirements.

The computer manufacturer's primary requirements were as follows:

1. The power source was not to be shared with large motors or other intermittent loads that could cause voltage dips, transient spikes, or other sudden fluctuations.

2. The average voltage was to remain within ±5% of nominal.

3. Noise spikes could not be greater than 1.5 times nominal voltage.

4. If the power source did not inherently satisfy the regulation and/or noise requirements, or where brownouts were anticipated, special conditioning was to be provided. Power mod-

By S. RAJ SUNDARRAJAN

CAD/CAM and ET production

In 1973, Martin Marietta was awarded the NASA contract for production of the Space Shuttle external tank (ET), which carries liquid (cryogenic) oxygen and hydrogen fuel and is the only major nonreusable element of the Space Shuttle transportation system. Since the ET is not recovered for reuse after a shuttle flight, this means that one ET will be required for each flight.

The ET is an assembly of fusion-welded aluminum alloy components that are preformed or chemically milled. It measures 154.2 ft in length, 27.6 ft in dia and carries 1.6-million lbs of cryogenic propellant. The exterior of the ET is covered with a multilayered thermal protection system (TPS) coating to provide protection against extreme temperatures during launch and flight of the Space Shuttle. This protection system covers the entire tank.

Computer-Aided Design (CAD)

The Cad System will provide the support essential for performing all required design graphics and analysis for the ET in a more-efficient and timely manner than any other means. In a preliminary evaluation an increase in design productivity of up to 5:1 was projected for complex, symmetrical structures such as the tank. This increase in productivity, coupled with the integrated data base, is expected to provide three areas of payback: the lapsed time in responding to production problems can be significantly reduced; errors resulting from poor visualization and interpretation by engineering, production, and quality personnel will be greatly lessened because of the system's ability to display realistic assembly and blowup views; and utilization of a common design database increases productivity throughout the entire design and production cycle.

Computer-Aided Manufacturing (CAM)

In the construction of the ET, many individual components must be fastened or welded together to form a single, strong, yet light-weight vehicle. Each tank contains over one-half mile of welded seams. To facilitate this process a CAM system is being planned to speed and add reliability to the welding and other procedures. Typical welding functions and parameters to be controlled and carried out automatically include arc voltage, travel speed, alignment, prevention of fluctuations, and creation of a recording of exactly what takes place during the weld.

CAD/CAM hardware

The CAD/CAM system selected will have two main processing centers, one in the main engineering building and the other in the manufacturing building. The entire system will consist of two 32-bit arithmetic processing units and seven color graphics processors, each of which incorporates high-speed memory, and 300 MB disk and magnetic tape drives. Peripherals include 18 color and 27 monochromatic workstations that will be dispersed throughout the entire facility. Drawing output will be provided by centralized large plotters and/or by small hard-copy plotters at the design site.

ules and line filters to achieve voltage stabilization and isolation were recommended.

5. Lightning protection was to be provided for the system in accordance with NFPA Vol. 1, Article 18-52, Section B-1.

To meet these criteria, it seemed judicious to derive the computer power source from a substation other than the one that was to feed the associated large HVAC load (two 25-ton units, one 10-ton unit). Careful investigation located a substation feeding a predominant existing computer load plus small distribution panels. There were no intermittent loads that could cause

RAJ SUNDARRAJAN, an electrical design engineer for Martin Marietta Aerospace Co. in New Orleans, LA, holds a Masters Degree in electrical engineering from the University of Mississippi, where he worked as a research and teaching assistant for the Electrical Engineering and Mathematics Department.

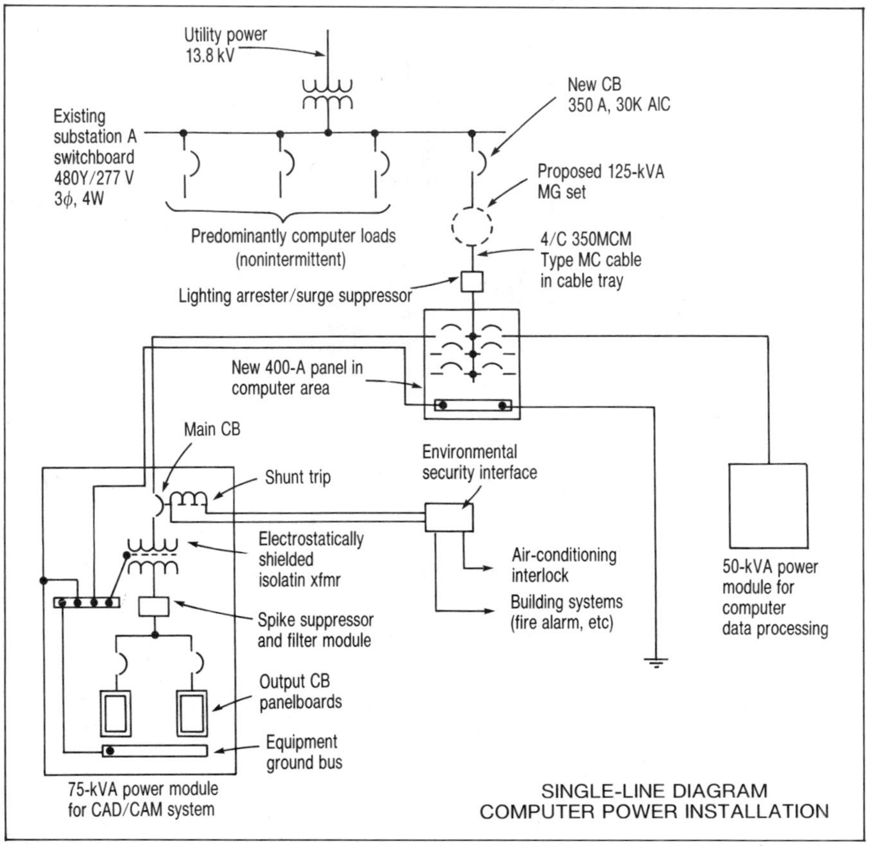

SINGLE-LINE DIAGRAM
COMPUTER POWER INSTALLATION

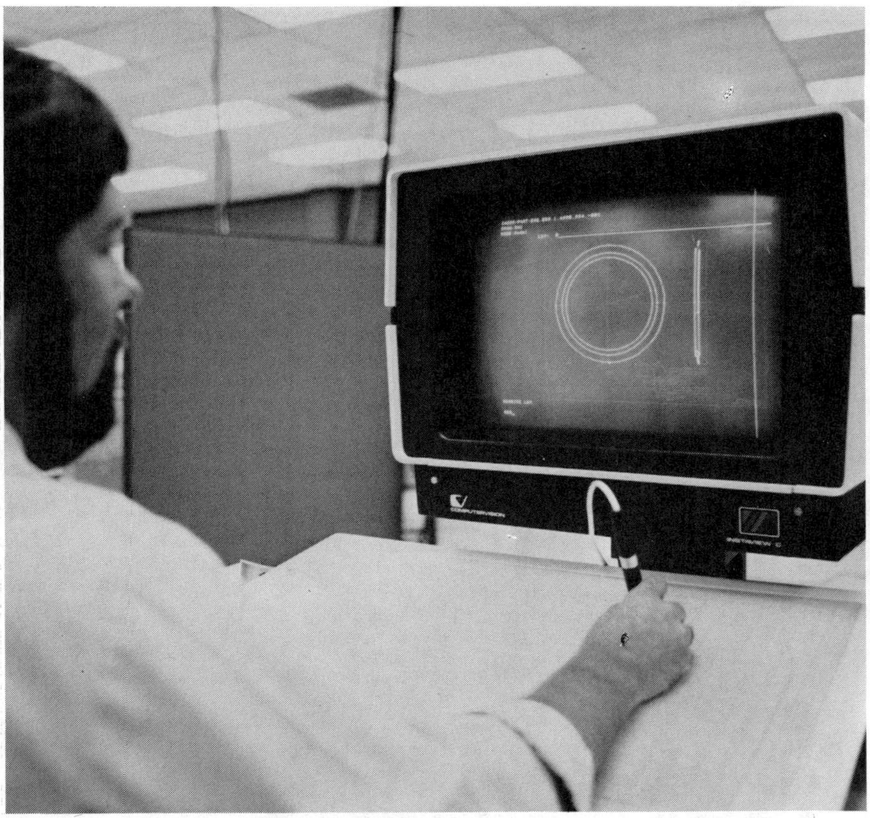

COMPLEX DESIGN is being developed by engineer in full color on CRT screen of CAD/CAM system. Computer-aided design (CAD) is expected to speed the design process and reduce errors because the subject will be created graphically for thorough analysis.

TYPE MC CABLE is in process of being installed in cable tray to supply power to CAD/CAM equipment. Cable is 350MCM, 4/C with XHHW insulation on conductors. Grounding requirements were facilitated by grounding the armor at both ends using armor clamps with integral connectors, then running a No. 3 grounding conductor to the computer room distribution panel ground bus. Jacks support huge reel during pulling operation.

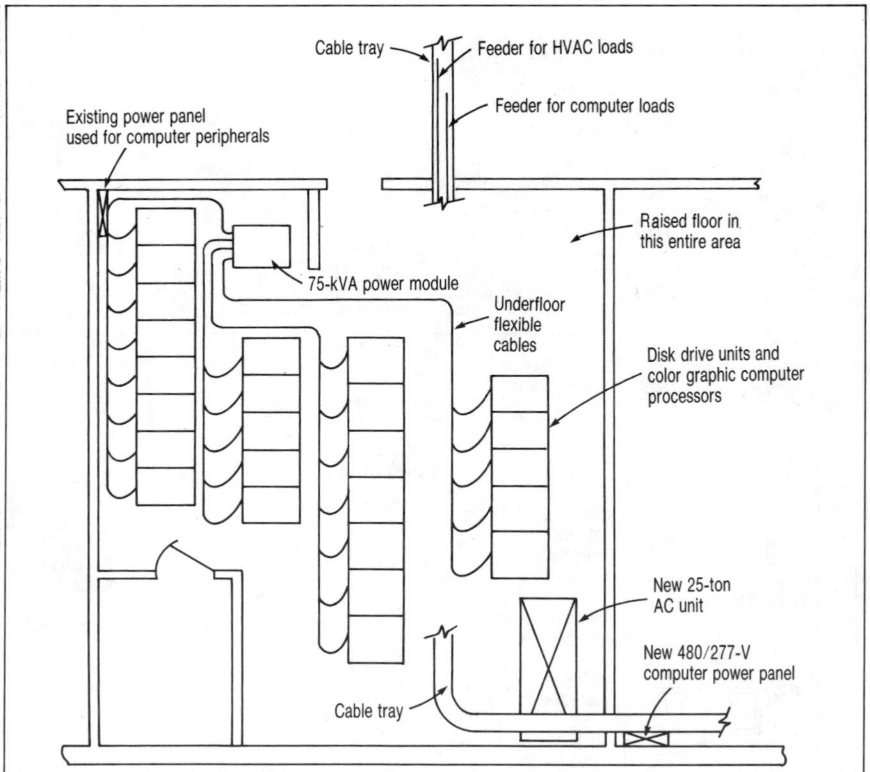

PLAN VIEW—COMPUTER ROOM

sudden voltage fluctuations.

The accompanying one-line diagram shows that power was carried from the 1500-kVA, 13.8-kV-480/277-V substation through a new 350-A molded-case circuit breaker via cable-tray distribution over hung ceilings to a new 400-A power distribution panel in the computer room. From this panel, two 2-in. flexible metal conduits, each containing four 1/0 THHN/THWN conductors, were routed under the computer-room floor to two power modules especially selected to refine power supplied to the computers.

The power modules—one rated 75 kVA for CAD equipment and the other 50 kVA for the company's existing data-processing computers—were installed without a backup M-G set or UPS mainly because of budget limitations. Features of the modules include a stepdown/isolation transformer, spike and transient suppression, output power distribution panels equipped with main and branch circuit breakers, output power cables installed under the raised computer floor, a system status and fault monitor panel, and an environmental system interface that per-

TWO EXTERNAL TANKS (ET) are in final stages of production. The huge tanks carry fuel for launch of NASA's Space Shuttle. Fabrication of the tanks will make use of computer-aided manufacturing.

mits the system user to interlock the power module with air-conditioning, fire-alarm and other building systems.

The entire project is noteworthy because of the small amount of elapsed time between inception and completion. Design was initiated on November 4, 1982 and completed on November 18; the system was on line by January 31, 1983.

The installing electrical contractor was W. J. Barnes Electric Co., New Orleans; George Hastings, construction supervisor, coordinated the work for Martin Marietta Corp.

CABLE TRAY above dropped ceilings carries metal-clad cable (Type MC) from unit substation to computer room power distribution panel. Cable tray is 12 in. wide, 5 in. deep. Note use of trapeze hangers and steel "U" struts to support cable tray.

INSTALLATION (left) of Type MC cable onto 12-in. wide, ladder-type cable tray is implemented by pulling wheel (arrow). Cable armor and cable tray were bonded and grounded to isolated, computer-grade grounding system.

POWER MODULE rated 75 kVA, which supplies 240/120-V power to CAD equipment in computer room, includes an isolation transformer (±3% output voltage regulation), a spike and transient suppression module (300-V clipping level, peak line voltage 294 V, peak current 15 kA), and an environmental system interface.

PRIMARY ELEMENTS (left) of a CAD/CAM system are the color graphics processor, the arithmetic processing unit, and the storage module (300 MB disk drive).

Computer installation calls for contractor versatility

Computer/data-processing projects offer a multitude of opportunities for alert electrical contractors. Here's how one contractor installed power, control, signal and numerous other support systems in a major banking facility.

AT THE Chase Manhattan Bank headquarters in New York City, a complex and extensive computer/data-processing system, assures accurate handling of volumes of vital statistics and related paperwork. Throughout 21 floors of a 54-story skyscraper are hundreds of minicomputers, terminals, data-processing machines, and several strategically located central processing units (CPUs).

The electrical contractor on this project, L. K. Comstock, New York City, is well aware of the many aspects of a sophisticated computer installation. This project includes a large main computer area incorporating a raised floor; extensive power-supply equipment and computer interface cabling; an uninterruptible power supply (UPS); HVAC and humidity-control systems; complex interconnect computer circuits that must extend throughout various floors as well as rise through 21 floors; special equipment, such as high-frequency generators, surge suppressors, voltage regulators, and control systems; and a variety of signal systems, such as fire-protection, alarms, security equipment, and annunciation equipment.

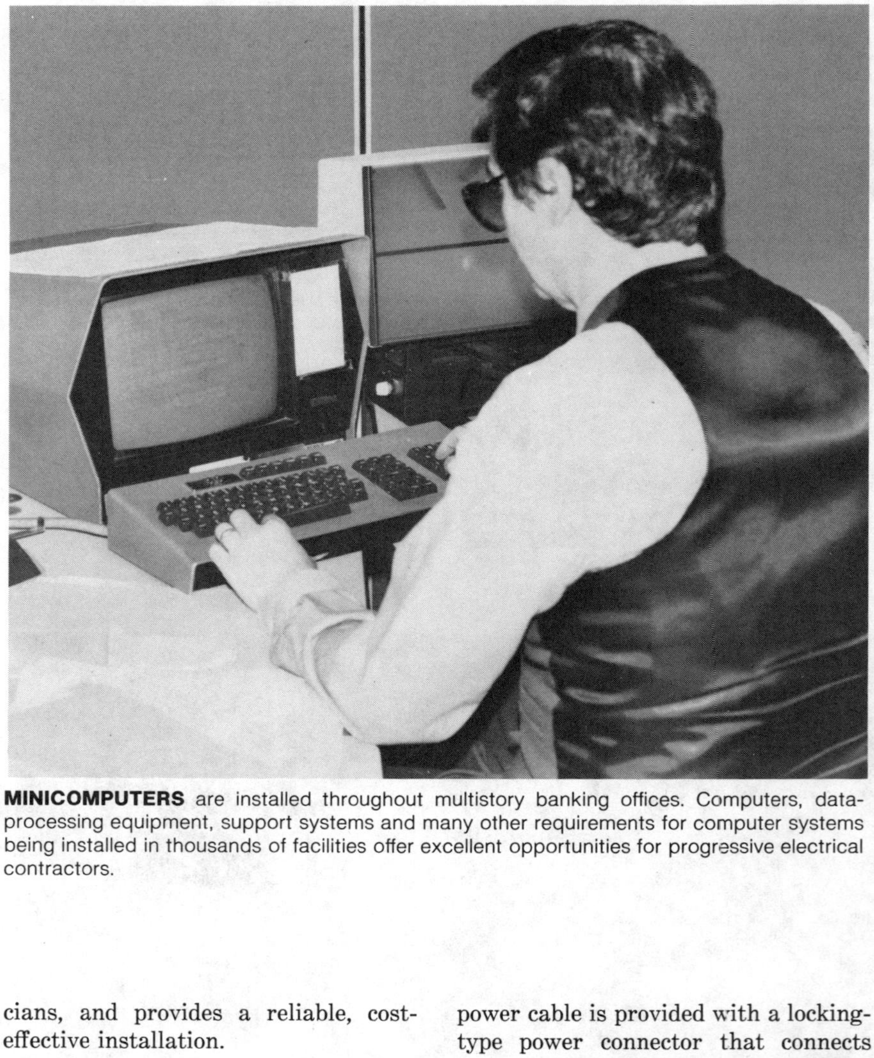

MINICOMPUTERS are installed throughout multistory banking offices. Computers, data-processing equipment, support systems and many other requirements for computer systems being installed in thousands of facilities offer excellent opportunities for progressive electrical contractors.

Computer installations not only require such additional systems; they also call for such special skills as knowledge of telephone and communication systems, shielding techniques, handling of multiconductor cables, and methods of termination or connector make-up, start-up, test and trouble-shooting ability. The profit-minded electrical contractor can take advantage of the opportunities presented if he has developed the necessary technical know-how, uses competent electricians, and provides a reliable, cost-effective installation.

By ARTHUR J. MAIHACK

Computer wiring methods

In main computer rooms, power from a large distribution panel is carried by cable in conduit under a raised floor. Each machine in the room is provided with a power cord that extends down through a specially prepared slot in a floor tile. Each supply power cable is provided with a locking-type power connector that connects with each machine power cord. Supply voltages range from 120 to 460 V. Computer interconnect wiring is handled in

ARTHUR J. MAIHACK has worked as a journeyman, foreman, general foreman, and shop steward for Local Union No. 3, IBEW, in New York City.

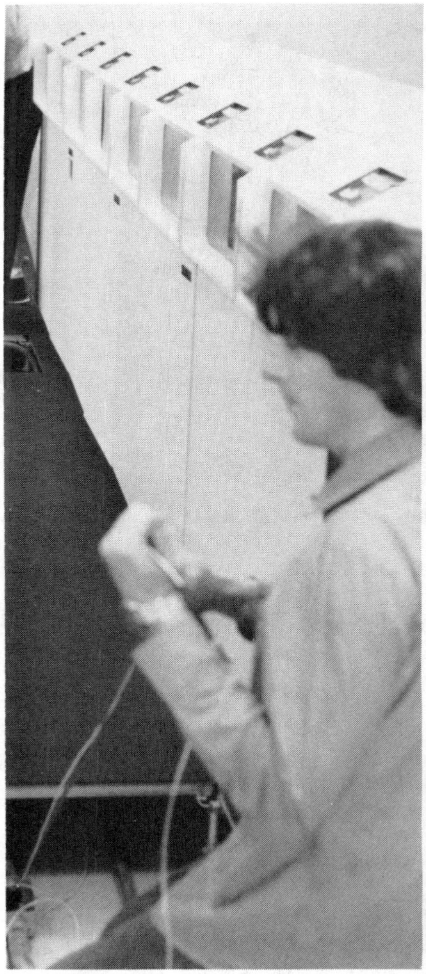

PULLING UNDERFLOOR fire-detection wiring beneath raised floor in computer room is done by electricians Peter Grillo (rear) and Dennis Caufield. The fire-detection system uses ionization detectors beneath the floor as well as overhead to provide early warning of fire during incipient stages. Moisture and water-detection sensors are also installed under the floor to protect wiring for computer equipment.

RELOCATION of computer equipment, such as the printer being moved by Phil Grippaldi, Charlie Hoffman and Bob Ragoneses, all Comstock electricians, is typical of contractor work done during system changes. Electricians moved all computer equipment and made changes to all types of wiring systems in this banking headquarters.

a similar manner, using coaxial cables and multiconductor control cables.

The raised floors provide flexibility for cabling between computers and periferal equipment and provide means for expansion and changes and room for other services. As an example, smoke and alarm systems are provided for protection of personnel and computers. Ionization sensors are installed both overhead and under the raised floor to assure early detection of any incipient fire.

It is important to know how to prepare the cable passage holes in the raised floor tiles. These holes are cut on the jobsite using a large band saw, preferably with an adjustable-speed belt and clutch that will stop the blade when it becomes twisted or jammed.

To make cuts that are extremely difficult to accomplish on the band saw, a hand-held reciprocating saw is used. The electrician is always careful not to place the cable passage hole in a location that would weaken the strength of the floor or lessen the support of the tile. To protect the cabling from the sharp edges left by saws on the tile, hard molded-vinyl coping is then cut and riveted into place. This assures a neat and reliable opening.

System changes

When changes and rearrangement of equipment in the computer rooms are required, the raised floor is a big advantage, according to John Tortorella, foreman for L. K. Comstock at the Chase Manhattan project. Comstock electricians physically move all computer and peripheral equipment as required, set them in place, level the equipment at its new location, and bolt together cabinets and multiple assemblies as necessary. Computer cabling is rerouted or added and pulled by Comstock electricians to the newly placed equipment. Frequently, specialized cables and other circuits must be pulled to remote parts of the building where related equipment is located.

To simplify routing cables up and down 21 floors in the banking facility,

400-HZ MOTOR-GENERATOR sets (above) supply power to modern high-speed computer equipment. Eddie Riccardelli, foreman for L. K. Comstock, supervises expansion of generator system as new unit is installed. At right, electricians Phil Scandaglia and Rene Rieder are installing a new wire trough for the system expansion.

VOLTAGE SUPPRESSION equipment (right) is being installed in panel by Barry Gluberman, electrician. This equipment eliminates or minimizes voltage spikes and transient voltages that could interfere with data processing.

COAXIAL CABLE, (below) which carries signals from computer banks to CRT, will make additional information available at this work station. The cable is pulled from beneath the raised floor to final placement location by electrician Barney Conroy. The contractor's electricians also put all connectors on cables and cut and bush floor tiles for cable routing.

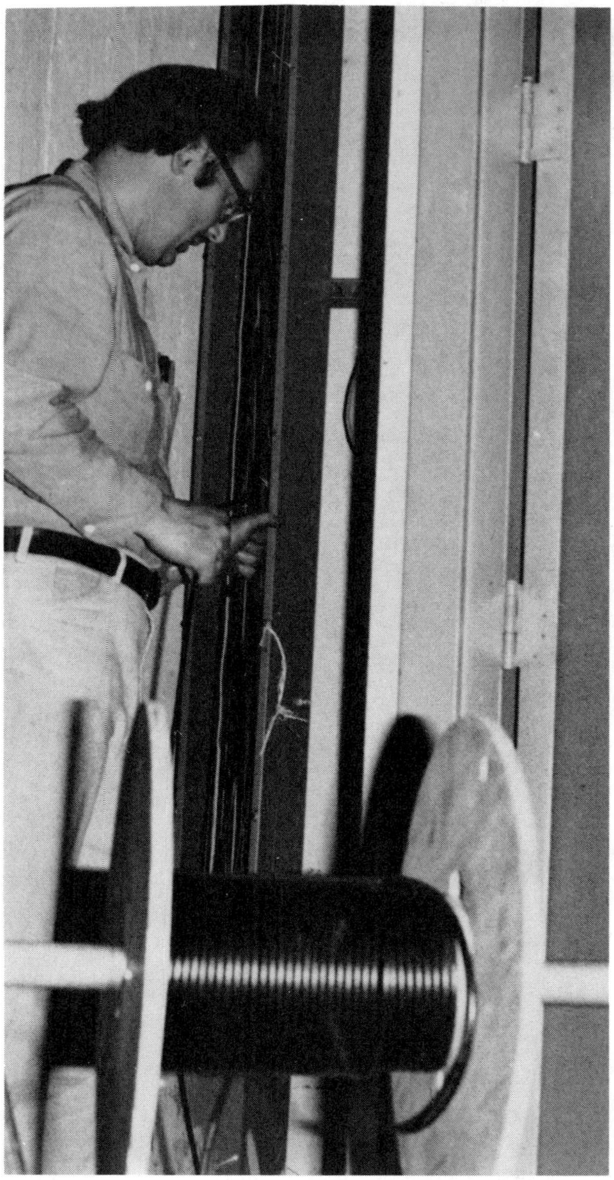

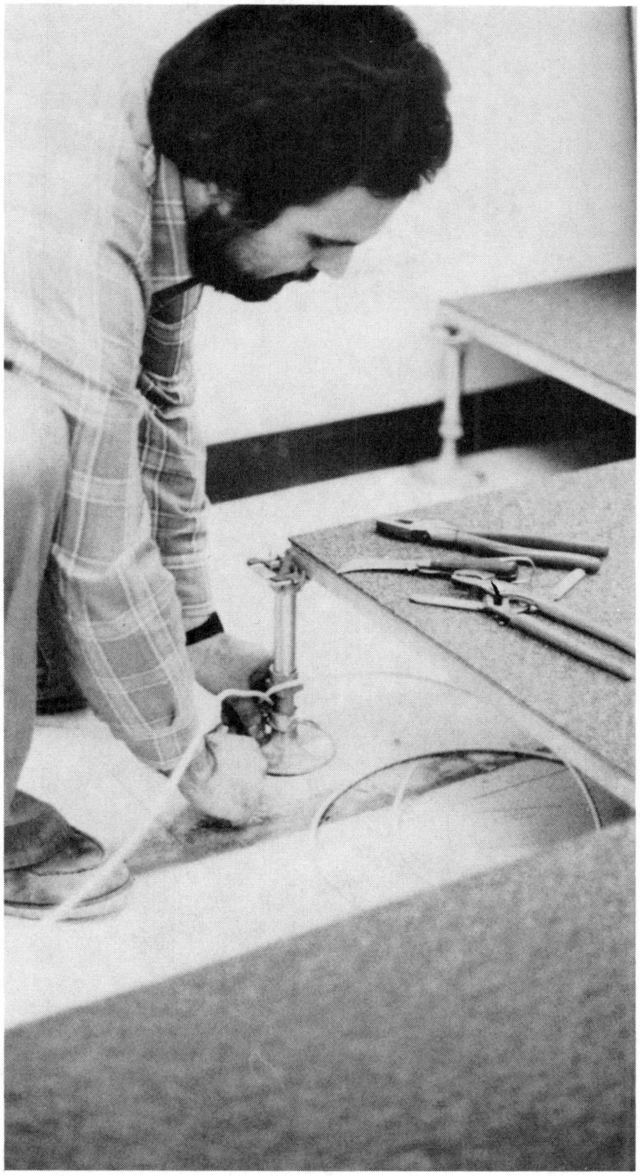

VERTICAL TROUGH (above, left) simplifies routing of computer cabling between floors. Electrician Bob Ragonese feeds CRT coaxial cable from reel into the 8- by 8-in. trough. Troughs, installed in closets, have removable covers for easy access and fast alterations.

COMPUTER MONITORING changes (left) can be made at terminal blocks in this and similar panels. The electrical contractor installed the monitoring system and, because of his extensive knowledge of the system, maintains it and is responsible for changes. John Tortorella, foreman, directs wiring changes.

GROUNDING CONDUCTOR (above) is being attached to stanchion that is part of the raised floor system. The grounding network, which is connected to a grounding bar in power panels, also connects all steel and enclosures in the computer area to maintain a common ground potential. Connection is being made by electrician Phil Grippaldi.

CABLE PASSAGE In floor tile used on raised floor is being prepared by electrician Rudy Cassano. This effective technique is popularly used. At left, cuts are made using large band saw set at proper speed for metal cutting. The saw has a clutch that automatically stops the blade if it becomes jammed. This increases the safety of the tool and saves the cost of replacement blades. At right, the edge of the cut metal is lined with molded vinyl coping to provide for protection of cables as they pass through the slot.

an 8-in.-sq vertical raceway was installed in electrical closets on each floor. Cables are run directly to equipment, or in some instances, conductors are protected in conduit or raceway to equipment.

Specialized terminating and connector techniques have been developed by the contractor and some of the more-experienced, computer-oriented electricians. These methods incorporate modern, customized terminating tools that provide fast, dependable connections. A quick return on investment of such tools is realized because of the labor and time saved.

In addition, computer manufacturers have found that by having the electrical contractor handle computer cable installation requirements, critical lengths are installed properly, excessive cable lengths are avoided, costs are lowered, and reliability is increased.

Support systems

Other important systems and features installed by the contractor include a signal and communications system that monitors the status and operating mode of all computers and terminals. Systems installed for computer safety include an underfloor-moisture and accumulated-water alarm; voltage suppression and regulation systems to eliminate or minimize transient voltages and spikes that could cause equipment malfunction or even damage to the sophisticated and sensitive electronic equipment; and a complete grounding system that keeps the raised floor at the same ground potential as computer equipment, power, etc. This helps to minimize static and spurious noise in the computer area.

Additional work available to the contractor on many computer projects includes installation of 400-Hz motor-generator sets. Many of the more modern data-processing systems use high frequencies for high-speed operation and other special requirements. Some systems use several such generating units, plus associated controls, protection and extensive related equipment. Also included is the separate wiring of the high-frequency circuits.

UPS system serves luxury hotel computers

A half-dozen small computers forming the nerve center of a Florida vacation facility are assured of reliable power.

THE ENJOYMENT, safety and convenience of guests in an 814-room luxury hotel is enhanced by an array of computers served through a reliable power system. A 15-kVA, single-phase uninterruptible power supply (UPS) provides a critical link in this reliability scheme at the Hilton Hotel, Lake Buena Vista, FL. Six computer consoles or central processing units (CPUs) form the backbone of a host of high-technology features described in the accompanying box.

By JOSEPH KNISLEY

FIG. 1. Schematic diagram shows the placement of the uninterruptible power system (UPS) in the reliability power scheme for the six computer systems used at the hotel.

These services, which meet the varied needs of business travelers at conventions, vacationing families, and tour groups, are carried out through the following facilities.

1. A national reservations computer system
2. A point of sales (POS) terminal system (cost-accounting use)
3. A key-card cutter system (a cardboard card that a guest uses instead of a metal key is encoded with a nonduplicated pattern of small holes)
4. A property management system (front desk CRT terminals)
5. A private multipurpose telephone branch exchange system (data bank)
6. An accounting system.

Assured power need

Because reliability and total availability of power is so vital to the communications, security, and safety aspects of the building, these sensitive computers need a pure 60-Hz sinewave input without interruptions. Even if electric power is interrupted for a fraction of a cycle, this condition could cause loss of vital data and/or errors in computation of the stored information. Thus, a 120-V UPS unit is provided within the power distribution scheme, as shown in Fig. 1, to serve a load panel (LP-EUPSN) directly. This load panel feeds all of the CPUs, data storage drives, printers, etc., in the computer room, as shown in Fig. 2. It

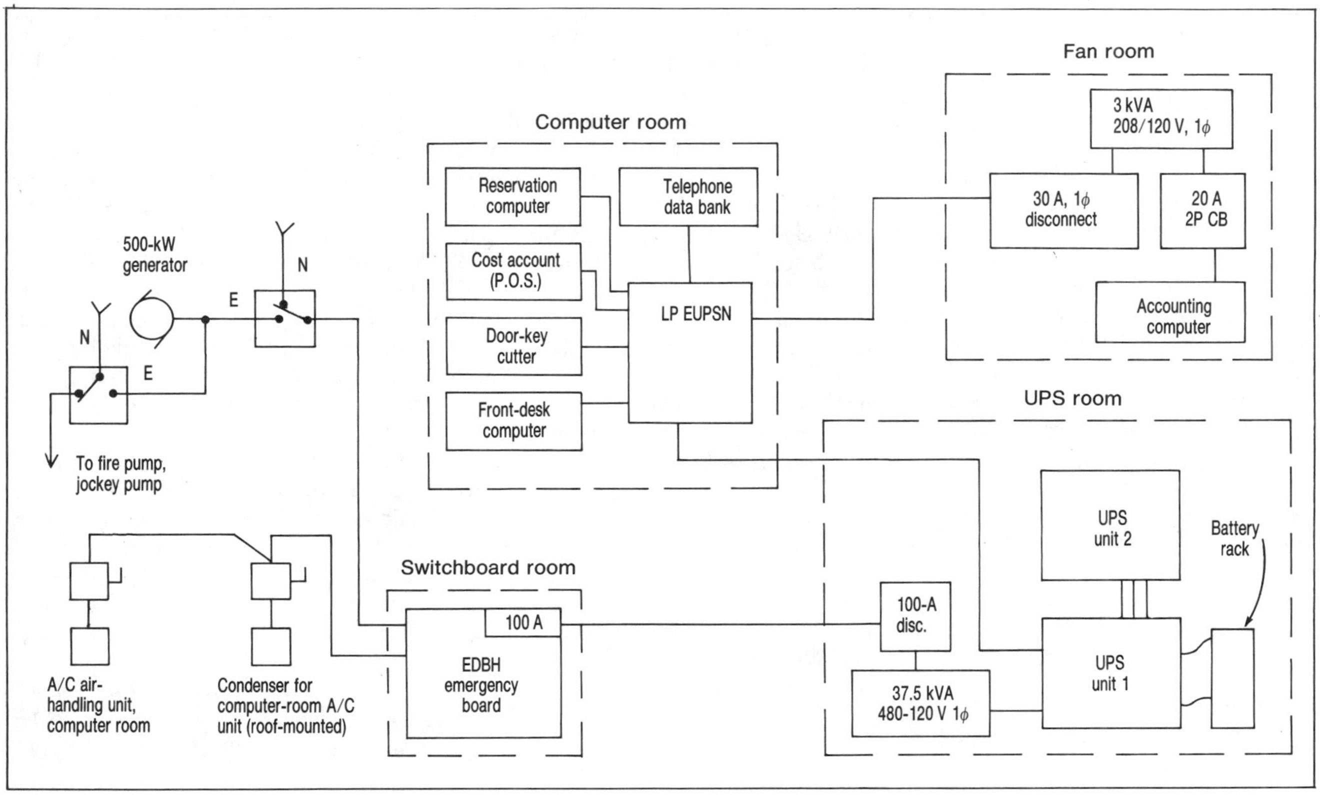

High-tech computer

COMPLETED in record time in November, 1983, the three wing, 10-story facility at Walt Disney World Village is designated as Hilton's Central-Florida flagship. Viewed as establishing a "new age of hotels," the facility offers guests an unusually broad complement of amenities and conveniences, many of which are computer-based—to sense, schedule, or interpret information for the room occupants.

Owned and developed by Tishman Realty & Construction Corp., the hotel has a fully integrated communications and control system that operates the life-safety, energy-management and security programs. The system's heart is a digitized microprocessor-based telephone in each guest room connected to a private branch exchange.

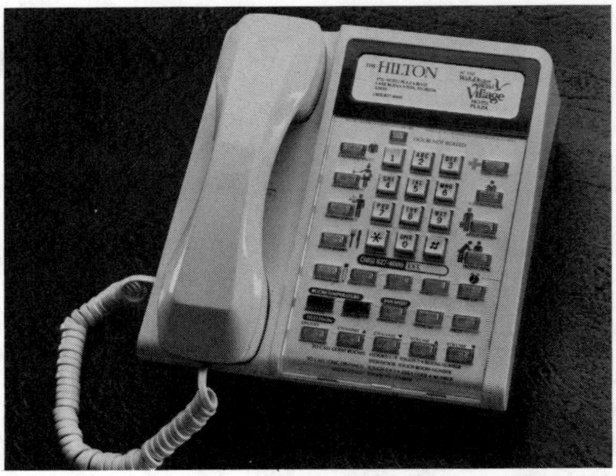

When a guest enters a room and closes the throwbolt on the door, the system is energized, and HVAC and television are controlled by individual telephone buttons. Smoke detectors in the room are also linked by the phone to central CRT monitoring screens. All phones have on-hook speakers through which the hotel management can immediately contact occupants in case of an emergency.

With a touch of a single key, guests can directly access the front desk, room service, valet, wake-up call, correct time, and a message center. As another convenience, the phone has two lines.

A tamper-proof, coded magnetic card replaces the traditional key. Created at check-in time, each guest's card bears a unique, randomly selected punched-hole code generated by a computer.

All shops and restaurants have a point-of-purchase microprocessor-based cashier register to record guest expenditures.

Designed by the New York architectural firm of Alan Lapidus Associates, the 625,000-sq-ft property also features two banks of elevators that are microcomputer-controlled to "speak" to the passengers with a synthesized voice, defining floor location, direction of travel, and general safety measures.

also feeds the 208-V accounting system console in an adjacent room.

The EUPSN panel, a second panel called ECPUE (also serving the computer room), and air-conditioning equipment for the room are all fed from an emergency switchboard bus EDBH. This bus in turn is wired back to an automatic transfer switch, served by a 500-kW standby generator.

In the event the utility line is lost, power for the EUPSN panel remains fully continuous, and all other loads on the emergency circuits are picked up within seven seconds by the standby power of the generator.

Thus, the reliability factor for the computers has two components: the immediate remedy of the UPS if commercial power is interrupted even momentarily, and the generator if the interruption is prolonged and the UPS battery is becoming depleted.

System components

The UPS system is housed in two identically sized cabinets, each 50 in. high, 96 in. wide and 20 in. deep. Unit 1 contains the rectifier charger and unit 2, mounted above, contains the inverter. The cabinets are connected by three short conduit runs containing power and control circuits. Both cabinets, as well as a two-tier storage-battery rack, are housed in a service area room on the first floor.

As Fig. 1 shows, power to the UPS is delivered at 480 V from a 100-A CB on switchboard EDBH via a pair of No. 2 THHN conductors and a No. 8 grounding conductor to a disconnect switch that serves a 37.5-kVA, single-phase, shielded transformer providing step-down from 480 to 120 V.

The 15-kVA UPS unit chosen is available in either a 120-V or 240-V rating, and the 120-V model selected is conveniently matched to the nominal 120-V output of the battery bank that was specified. In turn, the transformer secondary output is connected to input circuit-breaker CB1 in UPS unit 1 via two 250MCM THHN conductors, and the No. 8 grounding conductor is tied into a ground reference point in the UPS room.

Throughout the rest of the UPS power system an isolated grounding conductor is maintained in conjuction with all power conductor runs. Transmitted electromagnetic radiation waves from other electrical equipment could induce interferring ground-path noise on the equipment grounding conductor; but the isolated grounding procedure prevents this occurrence.

Regulated "clean" power from the UPS is delivered to lighting panel EUPSN in the computer room via a pair of 1/0 THHN conductors and a No. 6 grounding conductor in a 1¼-in. conduit. A raised floor is provided in the computer room, which facilitates distribution of branch circuits from the EUPSN panel via MC cable to receptacle outlets located in the raised floor space. All consoles, drives, and printers, in turn, are plugged into the receptacle outlets. The computer room layout is shown in Fig. 2.

UPS characteristics

Under normal conditions, the commercial AC line delivers power to the UPS rectifier/charger section, which charges the battery bank, keeping it at float voltage and also supplying DC power to the inverter section. The inverter reforms, or synthesizes, and AC waveform, and the output of the inverter section is sinusoidal AC power derived from a constant-voltage transformer. The regulation is ±5% when operating from the power line and +5 to −8% when operating from the battery.

The output ferroresonant, constant-voltage transformer and a two-coil input transformer totally isolate the input line, the battery, and the load.

A sealed lead-calcium battery group, consisting of 60 cells connected in series to produce a nominal discharge level of 120 V, are rack-mounted on two tiers a few feet from the UPS. Because of its sealed construction, the battery does not require water to be added during its 5-year rated life. Self-sealing vents eliminate corrosion problems in the adjacent sensitive equipment.

These gel-cell batteries use an immobilized electrolyte instead of a wet electrolyte, and they don't have to be put in a separate ventilated room as do the wet-cell types, which emit hydrogen. At the gel-cell battery float-charge voltage, over 95% of the gas is recombined within the cell, greatly reducing ventilation requirements.

While these high-capacity batteries could be called maintenance-free, the manufacturer recommends that voltage measurements be made on individual batteries every three months to verify that voltage balance is being maintained.

If the batteries vary by more than ±1.5 V (13.8±.15 V), they should be equalized for 48 hours and then rechecked after 24 hours on float charge.

The size of the conductor from the battery terminals to the UPS unit was selected to keep the voltage drop in the cable connections to 2% of rated DC input voltage. The battery is completely isolated from the UPS unit when the battery power breaker is OFF except for the polarity-indicating lamp circuitry.

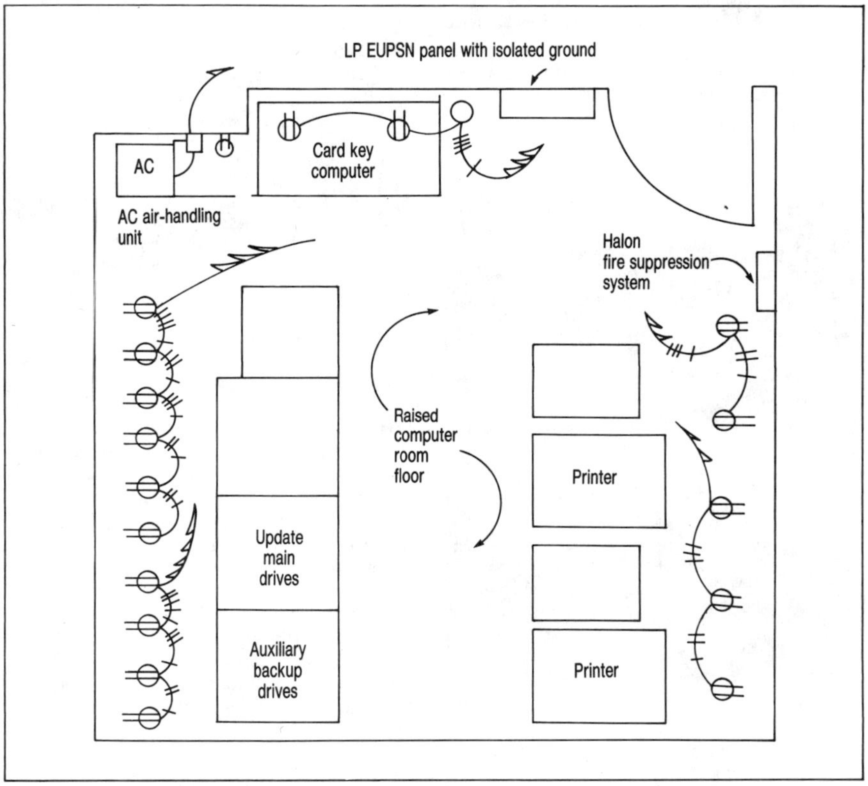

FIG. 2. Computer-room layout allows convenient access to the free-standing equipment that provides around-the-clock guest conveniences. Telephone circuits and other non-UPS 120-V receptacles are not shown.

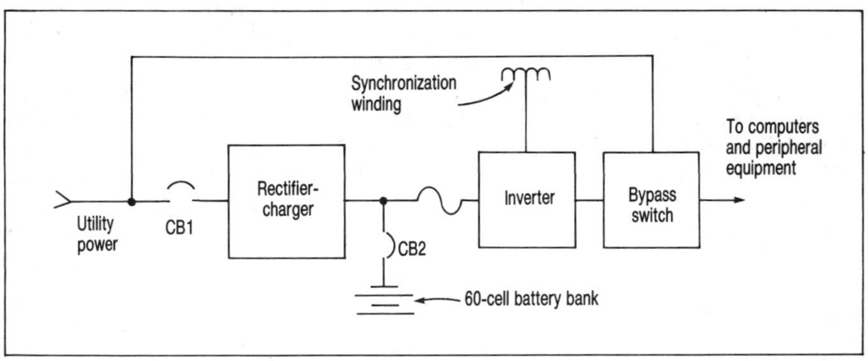

FIG. 3. One-line diagram of the UPS circuitry shows the commercial AC line bypass that can be tied into the load panel by a fast-response, solid-state bypass switch. The associated trickle-charged battery bank is always on line (i.e., connected to the rectifier output, so that whenever utility power drops below acceptable levels, the UPS unit automatically switches to the battery reserve without interruption.

Added protection

As a safeguard against malfunction within the UPS, such as an inverter failure or inverter overload caused by startup surges, the unit has an AC bypass line connected through the solid-state transfer switch. The transfer switch senses any incipient excessive change in the UPS output and responds by instantaneously switching to the AC bypass line. The load is then connected directly to the commercial AC service. Synchronous circuitry in the transfer switch monitors the AC waveform matched to it, so a transfer is done without a break in the output waveform. The bypass connection is made before the inverter output is opened.

Because the bypass feeds raw (unfiltered and unregulated) power to the load panel, it is considered a temporary emergency or maintenance condition only. Alarms are activated to notify personnel of the problem component so

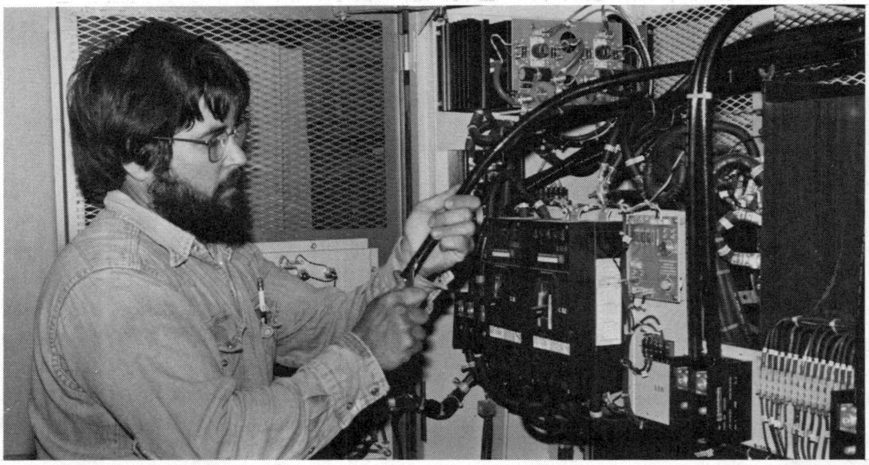

UPS UNIT NO. 1 wiring is completed by E. C. Ernst journeyman electrician. Insulation on the 2/0 conductor from the battery bank terminal is being stripped for insertion in the battery source circuit breaker in foreground. The adjacent CB (incoming power circuit) to the left has already been wired.

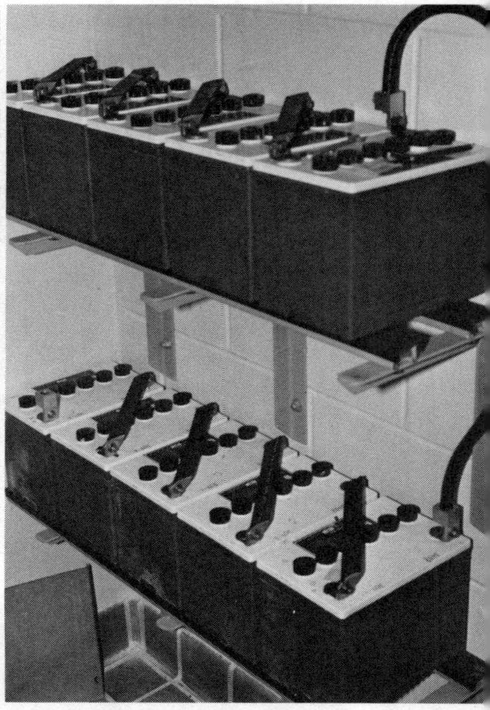

CANTILEVERED RACKS next to the UPS unit hold two sets of sealed lead-calcium batteries connected in series to form a 60-cell, 120-V DC system. Interconnect hardware from one battery unit to another on the same rack consists of ½-in.-thick copper busbar with an insulating coating to minimize the possibility of short circuit. Each busbar from the free positive terminal of one unit to the free negative terminal of the next chassis crosses over the center caps of a battery; however, the cells are sealed and do not require the addition of water. Note also that the batteries in the top tier are installed with the terminals facing the opposite direction from the batteries on the bottom tier for proper series intertie cabling with 2/0 conductors. Specifications call for torquing of all terminal connections to 50 in.-lbs.

ENGINE GENERATOR with a 500-kW standby rating is the backup reliability component for serving the computers. Equipped with a jacket water heater as a starting aid, the unit also uses a residential-type silencer in the dry-type exhaust gas manifold to minimize noise in the immediate area.

that restoration measures can be started immediately.

Switching back from the bypass to the inverter is also accomplished without power interruption. Upon retransfer, the rectifier/charger resumes its DC output to the inverter, while simultaneously recharging the battery array if needed.

UPS exercise

The need for UPS support of computers was recognized as a reality in this hotel, since an interruption of the commercial power could be expected at least once a year during a regular safety inspection. Such a full examination and under-load checkout of the emergency system would be made periodically by the inspecting authority, the Department of Buildings and Safety of the Reedy Creek Improvement District (RCID). These inspections, which include looking for construction and modification done without a permit, blocked equipment, etc., are performed at all hotels in this special-use district of Central Florida, which consists solely of the 43-sq-mi Disney property.

R.C. (Bob) Duncan, deputy manager for building and safety, who supervises the deenergizing of utility power in a hotel facility test (typically at 3 a.m.), believes that, while all hotels have emergency generators, the maintenance department usually cranks up the generator and runs it 30 min without a load, without checking the transfer switch, which is the source of many problems.

It is clearly evident that this hotel is under the jurisdiction of an inspecting and code-making authority having an enviable record and a high regard for guest safety.

The hotel's electrical system design was by Lehr Associates, consulting engineers in New York and Tampa, with Vincent Greico the project engineer. The installing electrical contractor was E. C. Ernst, Orlando, FL, with Curtis Duffield as project manager and Keith Johnson as general foreman.

Tishman Construction Corp. of Florida served as construction manager, with George Hughes as the project manager. The UPS design layout was done by Maurice Aubry and Mike Sincavage, both with Tishman.

Newspaper meets deadlines with UPS protection

Computers and peripherals perform without interruption despite momentary power outages.

INSTALLATION of a 30-kW uninterruptible power supply at the Santa Rosa (California) *Press Democrat* has kept that computerized newspaper free of power problems for the past six years. All vital data of the editorial, business and production operations are stored on disk packs, and loss of just one pack could require from 24 hrs to a week to restore the information.

The area experiences intermittent power outages, especially during the winter season, which have lasted from 2 sec to 1½ hrs. The early-morning publishing deadline leaves no time during the day to cope with a power outage, so when the newspaper's operations were redesigned six years ago, inclusion of a UPS had high priority.

UPS SETTINGS are checked by electronic services manager Larry Fans. The 30-kW unit has kept newspaper production in continuous operation for six years.

Three mainframe computers were installed. One handles the production end of the paper—news writing, editing, wire services, advertising, and typesetting. An identical setup handles the business functions—payroll, billing, and circulation—and serves as a backup to the production computer. The third system handles a large data base—information on subscribers and on nonsubscribers within a 3-county area plus programs to keep track of home delivery routes and mail subscribers.

The most significant segment of the redesign was the switch from outdated hot type to a modern, computerized photo-composition unit to set type. Since type is the heart of a newspaper, avoiding disruption during typesetting from a power outage was perhaps the strongest argument for the UPS.

VIDEO DISPLAY TERMINALS are part of a wide variety of typesetting-related equipment protected against data loss by the UPS.

UPS operations

Used with a standby battery plant to provide emergency power, the 30-kW UPS consists of a solid-state inverter, battery charger/rectifier, and an automatic static switch. It is specifically designed to isolate the critical loads from all types of potential power disruptions, from line transients and voltage fluctuations to brownouts and blackouts.

Under normal conditions, the UPS battery-charger section is powered from the commercial AC line and supplies regulated DC power to the inverter, while simultaneously float-charging the batteries. The critical load is continuously supplied clean AC power by the inverter section of the UPS, effectively isolating the computer systems and peripheral devices from line transients or fluctuations.

Should a failure occur in the commercial AC power, the critical loads are supplied by the inverter section of the UPS which, without switching, obtains power from the standby battery plant. The UPS battery plant continues to supply power to the inverter until commercial power is restored, or until the various systems can be systematically shut down.

The system uses a totally solid-state transfer switch through which power flows to the critical loads. This switch protects the load from interruption by continuously sensing both inverter and load parameters. If an overload, inrush surge, load short circuit, or even inverter failure is sensed, the static switch effects a zero-break transfer to

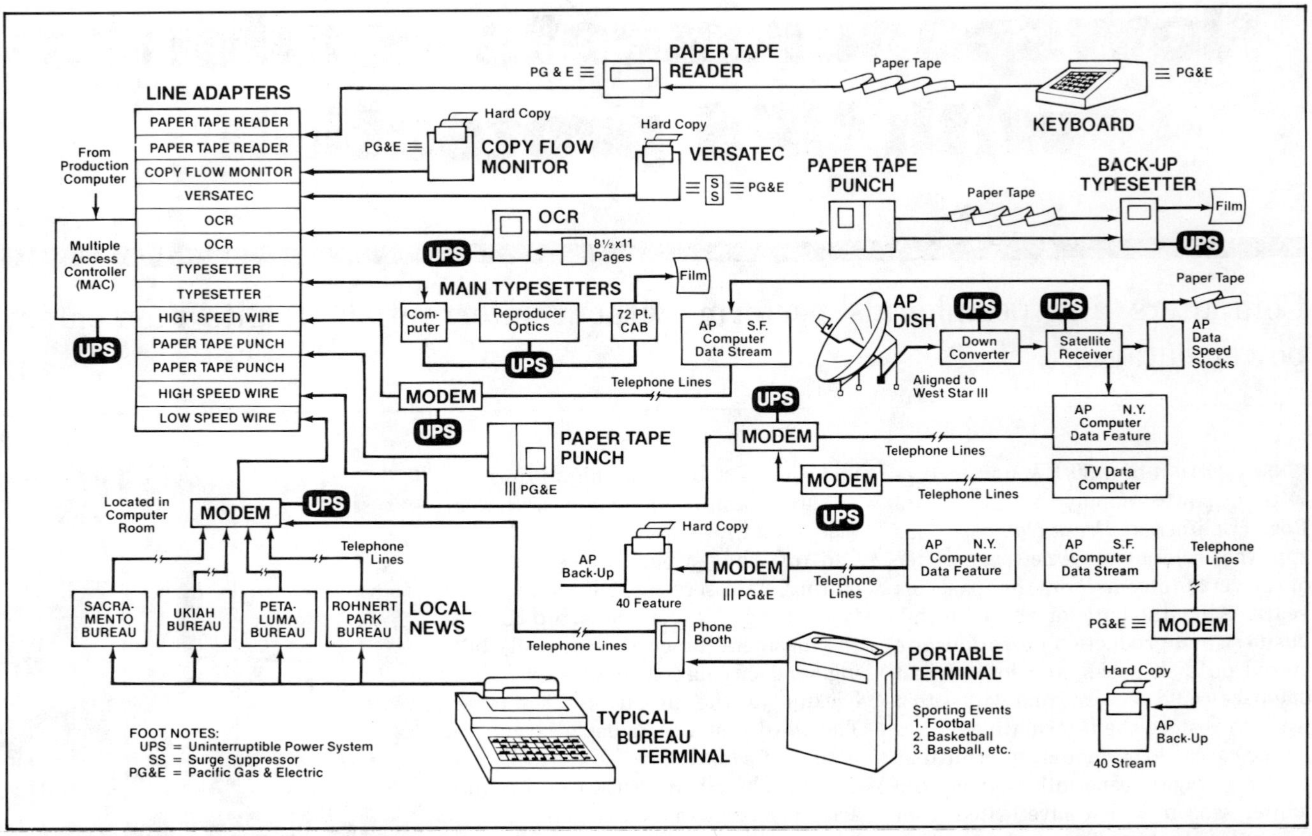

BLOCK DIAGRAM shows equipment connected to mainframe production computer. Sensitive peripherals that could suffer loss of data in the event of a power interruption are connected to the uninterruptible power supply, as shown by the UPS symbols.

a bypass line source.

As redesign of the *Press Democrat* progressed, the paper's production equipment was tied to one of the computers through a multi-access controller or "MAC." Basically, a MAC is a set of connectors, called line adapters, for input/output devices. Each line adapter connects some production function or piece of equipment to the computer. And most of the equipment, including the MAC, was hooked into the UPS. The two primary typesetters, for example, were tied to the UPS because they are directly on-line to the computer and they have cathode ray tubes (CRTs) in them that must be protected.

Also plugged into the UPS were the modems for the *Press Democrat's* remote bureaus in Sacramento, Ukiah, Petaluma, and Rohnert Park. These bureaus are the chief sources of area news for the *Press Democrat,* and power problems do not fit into the hierarchy of deadlines. A provision was also made so that a portable terminal for sporting events (football, basketball, and baseball games) could also be connected to the UPS through the modems used by the local bureaus. This insured that the latest information on these games would not be interrupted by a power failure. Finally, the wire service lines to the computer, as well as the receiving units for the wire services satellite "dish" (aligned to West Star III), were also attached to the UPS.

Connection of additional critical loads progressed in two phases. First, the paging system, phone system, and fire alarm and control systems were hooked up to the UPS. Communications, both internally and externally, would be critical during an outage; and notification and control of a fire occurring during that time certainly could not be sacrificed. The second phase involved connecting all the editorial VDTs, the composing room, the backup typesetter, and the optical character readers (OCRs) to the UPS.

The editorial and typesetting loads had one thing in common—variable amounts of data were always in volatile memory, and any power interruption would cause loss of the vital data.

Choice of equipment to be connected to the UPS had to be selective, however, to ensure that maximum power would be available when needed. Therefore, some equipment is not fed by the UPS. The copy flow monitor was not included, for example. Its function is to monitor what the production system does. If the power goes off, the monitor stops. When the power picks up, it will continue, right where it left off. No data is lost. The keyboards were not connected for the same reason: they are input by paper tape, and if power fails, they stop—but no data is lost.

Other equipment not connected to the UPS includes the hardcopy printer, which produces copy for proofreading; the backup wire-service systems; and the backup paper-tape punch, which can be used with the main typesetters. None of this equipment stores data in memory.

The insight that included the UPS in its planning six years ago has put the *Press Democrat* in position to deal with all types of potential power problems. The 1 hour, 20 minutes of reserve power available from the battery bank has been sufficient so far to avoid interruptions.

Installing computer network circuits in a New York City skyscraper

The complex cabling system at this modern facility has numerous specialized signal cables and electronic solid-state circuits that were installed rapidly with a high degree of reliability.

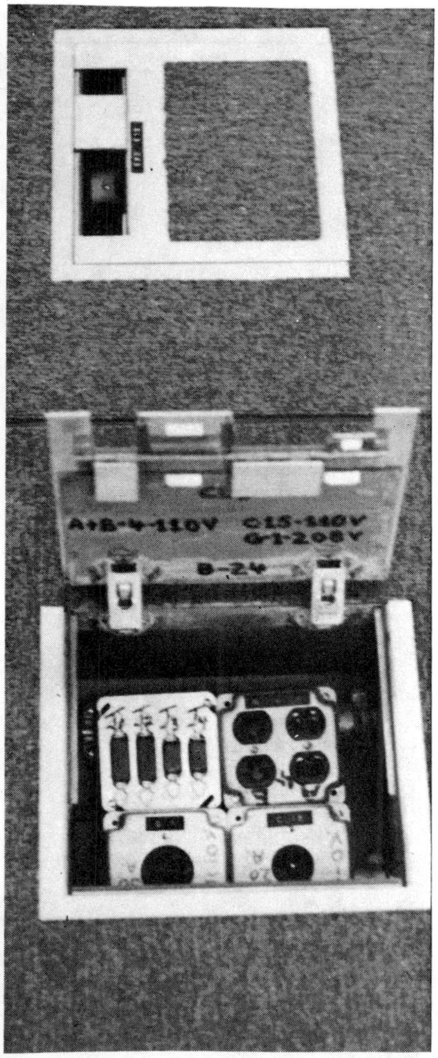

COMPUTER STATION HATCH is installed flush in section of raised floor. Hatch at top of photo is closed and will support substantial weight. Open cover at bottom reveals telephone and computer connectors at top left, 115-volt receptacles at top right, a 230-volt straight-blade receptacle at bottom right, and a 3-phase locking-type receptacle at bottom left.

MINICOMPUTERS are installed throughout the banking offices to help simplify and speed paperwork. System is typical of hundreds being installed or planned to expedite office operations.

By **ARTHUR J. MAIHACK**
IBEW Local 3
New York, N.Y.

A N EXTENSIVE computer system is the key to accurate completion of a high volume of paper work at Citicorp, Inc., a large banking firm in downtown Manhattan. The system consists of scores of minicomputers located throughout the 25-story office building plus numerous specialized computers, memory banks, and other business machines.

The minicomputers, comprised of a keyboard, controls, and a cathode-ray tube (CRT) for immediate readout, are interconnected with other computers and devices on all floors. Thus, an operator can call for immediate review of any information held in the computer storage.

Essential design requirements, as spelled out by Bolt, Barenak, and Newman, consulting engineers, Cambridge, Mass., called for maximum flexibility and reliability. This was accomplished by providing modular offices with removable walls, task lighting with capability to relocate lighting fixtures as needed, raised floors on every level to accommodate changes in location of minicomputers and arrangement of power and signal circuits, vertical trunk cables between floors and horizontal trunk lines running under raised floors, a carefully designed system of patch cables in computer closets, and power and signal "hatches" installed in the raised floors.

On each floor there are four computer termination patching closets. Installed flush in floor sections, approximately 10 ft on centers, are power and signal "hatches." Computer signal cables extend between the closets and hatches, and power circuits run between panelboards and hatches.

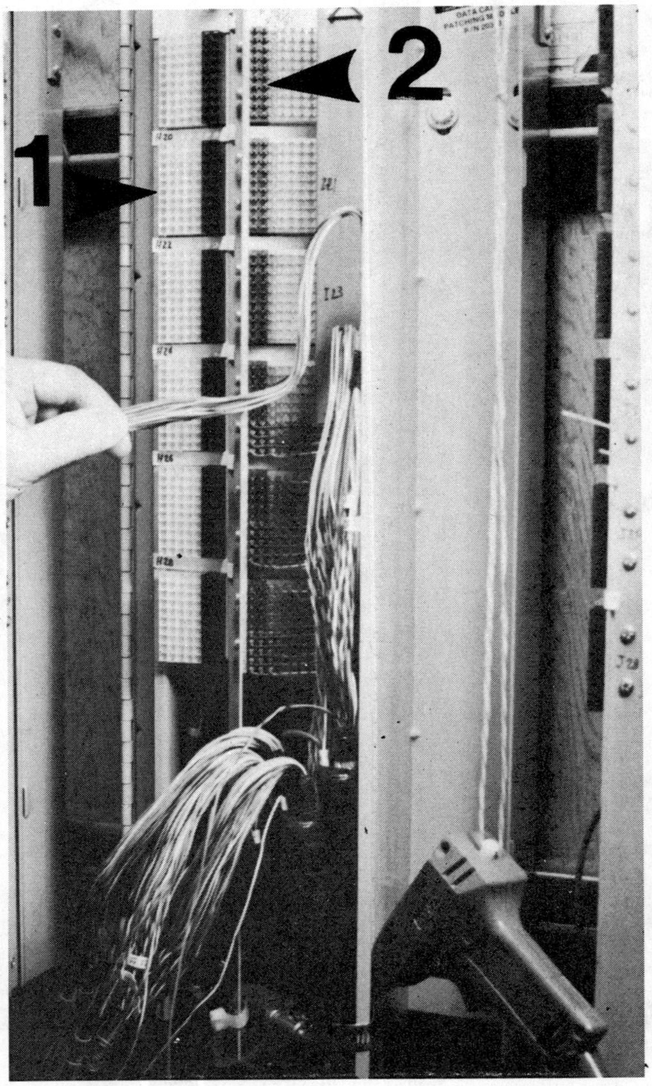

CONNECTION MODULES in computer closets each consist of six 50-post terminal blocks. Modules are mounted vertically on hinged panels so that both front and back are easily accessible. Arrow 1 points out front of terminal block that accepts plug-in connection of patch cords. Arrow 2 indicates posts on back of terminal block that accept permanent wire connections.

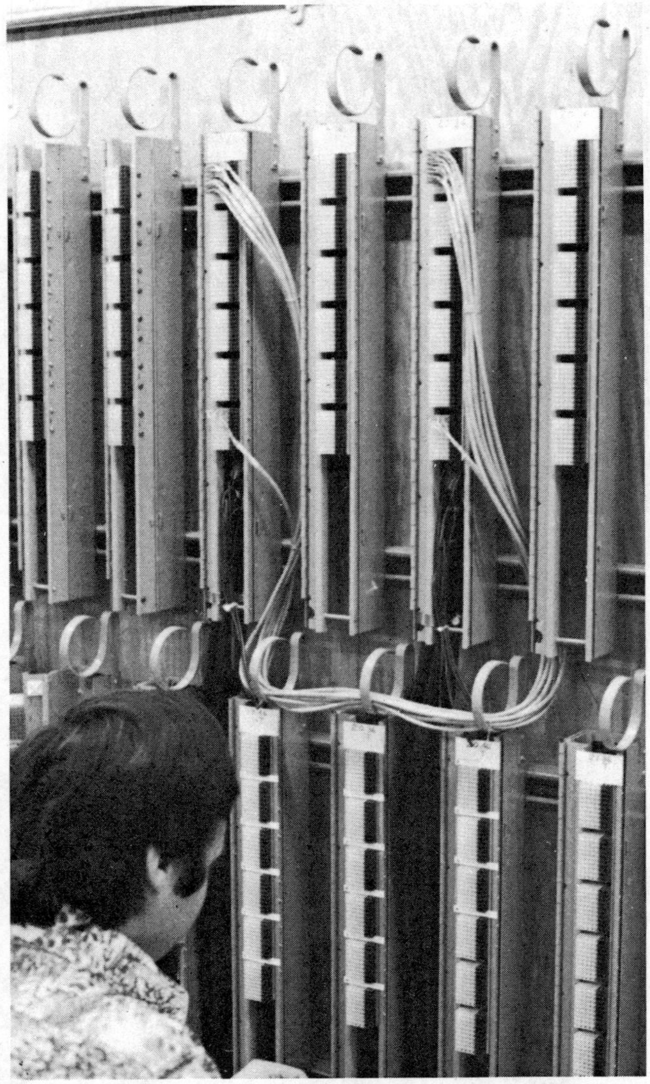

PATCH CORDS, made up by electricians on the project, are plugged into terminal blocks in computer closet. Note hook-type cord supports installed above each vertical terminal module. The patch-through scheme permits great versatility in use and application of minicomputers and other equipment installed throughout the 25-story building.

The hatches, consisting of heavy-gauge steel boxes measuring approximately 1 ft sq and 8 in. deep, provide all signal and power supplies for a minicomputer station. They contain telephone connectors, computer signal connectors, four 115-volt straight-blade receptacles, a 20-amp, 230-volt receptacle, and a locking-type receptacle rated either 30 amps, 230 volts, single-phase or 208 volts, 3-phase as required. When hatch covers are closed, they are sufficiently strong to support heavy personnel traffic and machine loads.

Flexible cables supply hatches, providing mobility, so that the hatches can be relocated within a radius of approximately 15 ft. Power conductors are carried in flexible metal conduit. Computer signal cable, specially made and provided with a neoprene jacket, contains 25 pairs of conductors. It has an aluminum shield and a special "T" shield that separates 12 pairs of conductors from the remaining 13 pairs. This feature permits "high-noise" circuits to be isolated from one another.

All installation work, which included pulling in hundreds of computer signal cables and making up thousands of connections, was done by journeyman electricians. The men were trained by foremen on the project in accordance with procedures developed by the engineers and were supervised by Arc Electric Construction Co., New York, N.Y.

In computer closets, dozens of terminal modules were installed, each with six multiwire terminal blocks. Some closets have up to 10,000 terminal posts. Each cable is assigned to a terminal block by code number, and each wire is likewise identified. All wires in the cable are attached to terminal posts using a wire-wrapping tool. This gun-shaped device spins the conductor around the terminal post very quickly and securely. The other end of the cable is similarly attached to

THOUSANDS OF CONNECTIONS were made up quickly and securely with the aid of a special wire-wrapping tool. Martin Siegal, electrician, Arc Electric Construction Co., New York, uses tool to speed connection of 25-pair cables in computer termination patching closets.

a 50-pin (25-pair) terminator in the floor hatch.

Patching cables, with special terminating pins on the ends of the conductors, were made up in a similar manner and then plugged into designated receptacles on the terminal blocks in the computer closets. If changes in the routing of computer signals or other special hookups are required, the patching cables provide a high degree of flexibility.

After the computer signal wiring was completed, electricians were trained to carry out thorough tests to assure the reliability of the system. Tests were performed with a specially designed digital testing instrument. After connection to a terminal block, the instrument scans the pairs of conductors, and if a shorted or miswired pair is detected, the faulty circuit is identified by circuit number on the instrument digital readout. Accompanying photos show additional installation details.

Access flooring provides increased flexibility

Renovation of existing office areas is making increased use of raised flooring to provide underfloor space for electrical, telephone, communications, and mechanical services, including heating and air-conditioning ducts. Resting on a building's concrete floor slab, a network of steel pedestals supports a raised floor of modular sheet-steel panels, providing 6 to 15 in. of space below. Any panel can be lifted in seconds to provide instant access to any of the services underneath.

The concept can be likened to the reverse of a ceiling plenum or as a miniature version of the interstitial technique frequently used in hospitals to alternate mechanical services with patient floors. The installed cost in offices varies from $3 to $4 per sq ft, plus $1 to $2 for carpet or other surfacing.

Originally developed to meet the exacting needs of the computer room, access flooring is now being applied to all types of office areas, including those in manufacturing plants. Components vary, but all systems have elements in common. Pedestals normally average 8 in. in height and are vertically adjustable so that the access floor is dead level regardless of subfloor irregularities. The standard 2-by-2-ft painted floor panel generally consists of a steel bottom pan and a flat top steel sheet in a welded construction. The pan, about 1½-in. deep, is the strength member, carrying the load and distributing it to the pedestals. In a typical system, four pedestals support the corners of each panel. There is also a variation, in which bolted or snap-on steel "stringers" lock all four pedestals together for greater lateral stability and additional panel edge strength. Office floors generally have the capability of supporting 150 lbs/sq ft, with a 400-lb concentrated load possible. A recent development is an electrified panel having built-in, flush-mounted service outlets for electrical and telephone service.

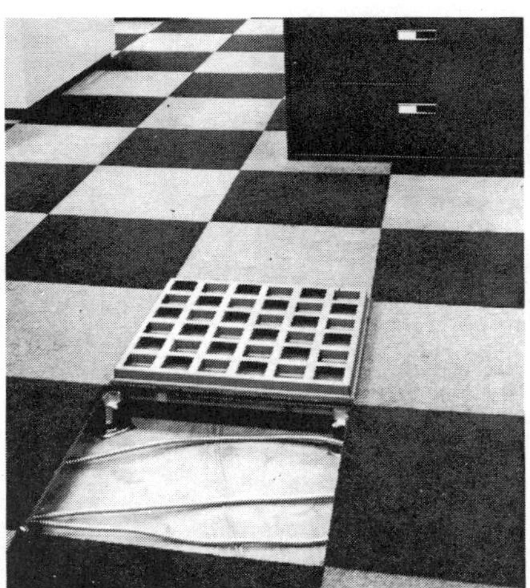

TYPICAL OFFICE LAYOUT shows a carpeted access floor with BX cable in underfloor space.

UNDERSIDE of electrified floor panel shows cable-fed, built-in service outlets.

Carpet, adhesively applied, is by far the most popular surface. Also available are high-pressure laminates, vinyl asbestos, parquet flooring, terrazzo, and ceramic tile. One innovation is a 2-ft-sq magnetic carpet tile that may be attached without adhesive after all construction is complete, assuring cleanliness.

Steel access floors are most frequently specified when there is a fairly high density of desk workers, frequent changes in space functions and in locations of needed services, and when the open landscape office is used. Exact personnel locations and service accommodations do not have to be predetermined before construction; services can be made available at any point on the floor surface at any time before and after occupancy. The system has a built-in provision for unlimited change. It can handle any volume need for phones, electric typewriters, electronic business machines, calculators, CRT terminals, minicomputers and microprocessors, intercoms, and security and fire signaling devices.

Studies show that the average office floor gets 80% of its wear on only 30% of its area. Since carpeted access floor panels are removable, those in high traffic sites can be relocated and replaced with panels from other areas on a regular rotation plan. Such a program can stretch out schedules of carpet cleaning and extend useful carpet life as much as three times.

Remodeling can be done at minimum expense simply by utilizing the existing slab floor as a pedestal base and putting all services atop the old floor and under the new one. Old, resilient floor covering and existing underfloor wiring need not be removed. Access floor versatility can even be used to bring substandard buildings up to code, particularly where task lighting is introduced to reduce energy needs. Since access floor systems can supply electricity exactly where needed, they can be the means to provide lighting exactly where needed, independent of any ceiling lighting.

Computer UPS system economically upgraded

An uninterruptible power system (UPS) serving the Navy Finance Center in Cleveland, Ohio was upgraded from 150-kva capacity, using a dual-redundant configuration, to 300-kva capacity, using a modified static-switching bypass arrangement.

For economy reasons, rather than duplicating all the equipment in order to double the capacity, the Dept. of Navy engineers used a cost-saving design. They specified the installation of another 150-kva transformer and a rectifier inverter of the same capacity. The manufacturer added one pole per phase to the transfer switch array, modifying it to create a pair of two-source switches as shown in the accompanying diagram. Except for the loss of full redundancy, the scheme has the same system capability as originally provided.

If commercial power is lost, the UPS system takes over as the power source, using the 20-min capacity available from the battery bank to serve four separate central-processing units, until a diesel-driven generator comes on line to handle the total building computer load. Under standard operating conditions, the UPS functions continuously to deliver a pure sine wave component to the computers—eliminating transients, harmonics, spikes, and dips in the "commercial-grade power."

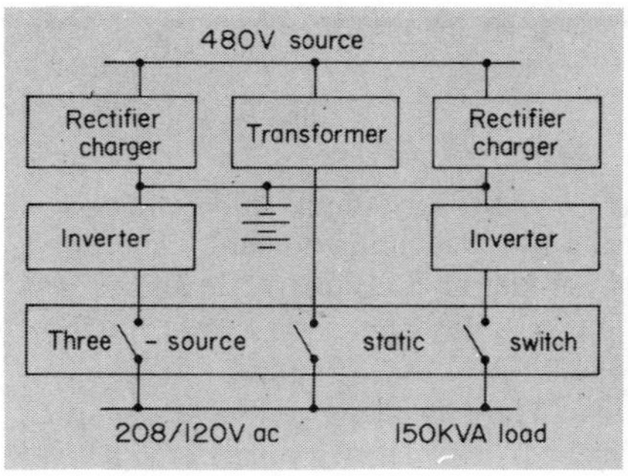

OLD UPS SYSTEM (above) was replaced by enlarged UPS system arrangement with 300-kva capacity (below).

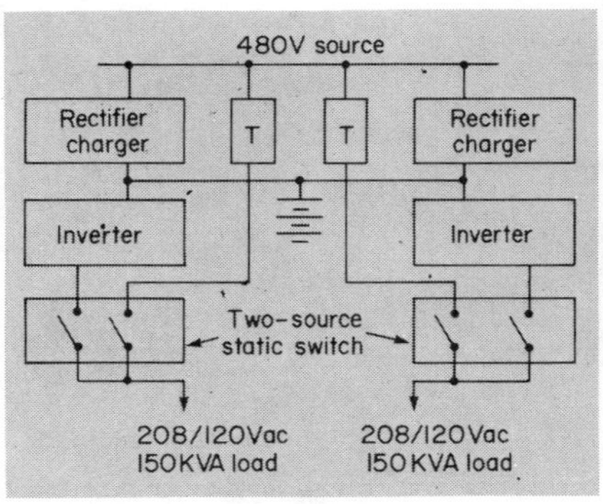

RECTIFIER/CHARGER and inverter, with 150-kva capacity, duplicates the existing 150-kva unit it faces across the aisle at the Rear Admiral Issac Campbell Kidd computer center in Cleveland.

Updating computer wiring at the NY Stock Exchange

NY Stock Exchange facilities upgrading had to be carried out without interrupting trading on the Exchange floor. Many circuit routing and logistic problems had to be solved as the job progressed.

NEW COMPUTER and communications equipment was needed as part of a comprehensive upgrading of the New York Stock Exchange facilities to handle sharply increasing transactions. Making the necessary modifications, which had to take place after the close of the trading day or on weekends, was a challenge to Albin Gustafson Co. of New York, the electrical contractor on the project.

Engineering that involves changes within existing structures is always tricky; it is even more so when no downtime is permitted. New equipment must be installed, but this requires removal of old equipment for which often no adequate electrical drawings exist. Disconnecting the wrong wire conceivably could shut down the whole operation. Exact routings of new cables must also be worked out. All these elements—plus manpower, equipment and tooling scheduling—can often be accomplished most efficiently by the man on the job. For the NY Stock Exchange project, these items were an essential part of the constructon services supplied by Gustafson.

By ALFRED BERUTTI

Engineering approach

As shown in Fig. 1, the actual trading floor of the Exchange is located within several buildings. The Main Room is the most familiar because it has often been photographed; but the Garage and the Blue Room also contain trading posts. It was principally these trading posts—at which the transactions are made and recorded—that required updating as part of the project. The 22 old posts were to be replaced by 14 modern units.

The new computer and communication equipment was to be installed on the fourth floor. From there, it was necessary to run a massive number of cables to the trading posts and other peripheral equipment. The existing posts were fed from cable that ran below the floor and up into the posts. Because of the no-downtime rule and the complexity of the wiring, it was determined that only one post could be changed over at one time. This meant that all wiring would have to remain in place except that for the one post. There was insufficient space under the floor for both new and old cables, except at the Blue Room. An alternative means of feeding the new posts from overhead had to be devised.

Architects Bonsignore Brignati and Mazzotta and electrical consultants Jaros, Baum & Bolles devised a perimeter raceway for the Main Room that was to be supported by an open structural grid called a "space frame." The perimeter raceway (Fig. 2) was to carry all power, computer, and communication cables for the posts in the Main Room as well as provide access to an open cable-tray system that was to be run above the hung ceiling of the Garage.

SPACE FRAME FLOATS above trading posts on the Main Room floor of the Exchange. These posts were to be removed and replaced with new modern units without disrupting normal trading activity. Because of this limitation, only one post could be changed over at one time, and all work had to be planned very carefully.

Contractor's role

Specifics had to be worked out. Laborious tracing and ringing out of existing wiring had to precede any disconnecting. Exact methods of running cables from the computer room

PERIMETER RACEWAY encircles the Main Room. A temporary work platform allows the electrical contractor to pull cable into the trays in advance of the changeover of the posts. The cables for the vertical drops are coiled below the raceway in preparation for the next phase of construction.

through existing structures and to the perimeter raceway and from the raceway down to the trading posts also presented problems. How much pulling of cable could be done prior to the start of post changeover? And then there was the determination of the exact manning required to make the changeover within the time allotted (Friday, 4 pm to Sunday, 8 am). Because of the nature of the project, these items could best be solved by the electrical contractor.

A typical new trading post consists of a lower section in the form of a double horseshoe that contains card readers, printers, time stamps, keyboards, telephones, and other units. Suspended by cables from the space frame is an upper section, also in the form of a double horseshoe. Typically, the upper section supports 48 23-in. CRTs and 22 9-in. CRTs mounted on outrigger arms. HVAC units are also located in the overhead section.

Determining the exact number of cables that are required to drop down to the equipment in both the upper and lower posts was a major undertaking because equipment of many different manufacturers was involved. Gustafson's Borio Gatto and John Andreala, Jr. had to work closely with Doug Grafflin of Securities Industry Automation Corp. (SIAC)—who specified the equipment—to achieve an accurate count. An average drop, it was determined, would consist of 208 coaxial cables for video equipment, 59 12-conductor computer cables, a run of No. 2 AWG cable in flexible conduit as a power feed—plus chilled-water piping for the HVAC units. Extensive mockups had to be made to come up with the best method of supporting and bundling the cables and piping in the vertical drop of approximately 30 ft. The final result had to allot room for future expansion; it also had to be esthetically acceptable.

Each trading post has specific equipment associated with it. Cables from these units were routed via the perimeter raceway to the computer room. Once there, cables associated with one post had to be spread out to different computer and communications racks. Since many cable crossings are required, there was danger of overloading the raised floor space allotted for wiring at these transition points.

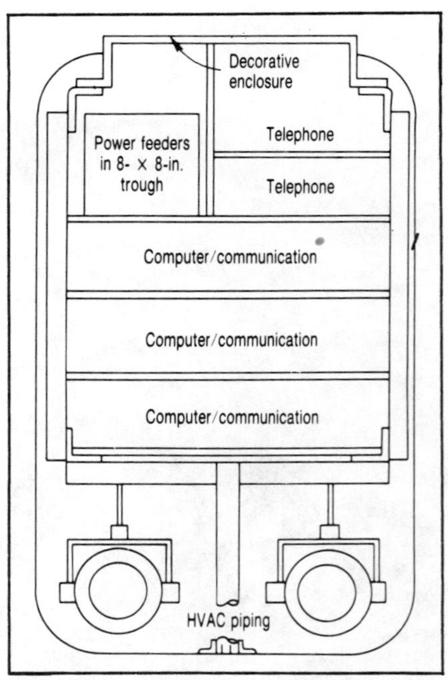

FIG. 2. Section through the perimeter raceway shows segregation of power and communications circuits. A catwalk alongside the enclosure permits access for servicing and installing cable.

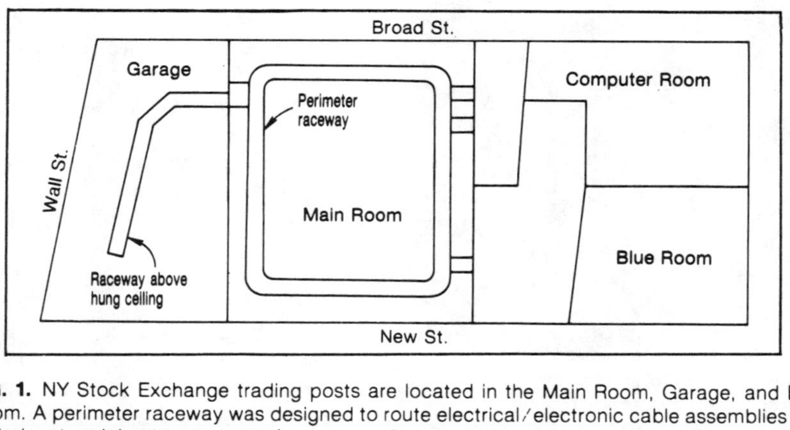

FIG. 1. NY Stock Exchange trading posts are located in the Main Room, Garage, and Blue Room. A perimeter raceway was designed to route electrical/electronic cable assemblies and chilled-water piping to new posts from an overhead position while maintaining the esthetics of the Main Room.

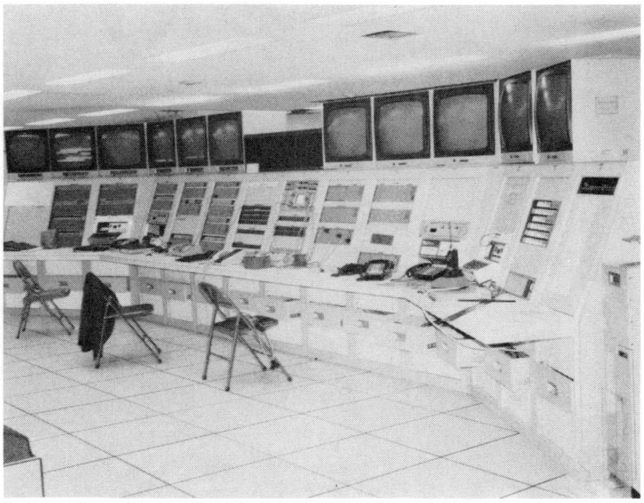

COMMUNICATION AND COMPUTER CABLES (top, left) from the posts exit from the perimeter raceway and enter the computer room through conduit sleeves. The cables drop down and are then run under the raised floor (top, right) of the computer room to their final destination. In this run, transitions must be made from groupings according to posts to groupings according to function. Raceway space at crossover points tend to be overcrowded (bottom, left). The electrical contractor therefore chose to have all communication and computer cables twisted into bundles with allover jacket to reduce space requirements. In addition, detailed layouts of cable runs were made so that crossovers could be minimized. This resulted in a neat installation, as seen in the photo of the main console (bottom, right).

MOCKUPS were made of the vertical drop from the perimeter raceway to the trading posts. A typical drop included 208 coaxial cables for video equipment and 59 computer cables, power feeder cable, and chilled water piping for HVAC units that were mounted in the upper section of the post.

The overcrowding solution was twofold. Coaxial and computer cable were bundled into cables made up of 19 individual multiconductor cables twisted together. Even with their overall extruded jacket, these bundles occupy less space than the individual cables laid at random. By making exact layouts of cable runs, Gustafson additionally was able to develop routing that would minimize crossovers.

Exact measurements had to be made of the length of each cable from the initiating point in the computer room to its final proposed termination at the trading post. Many of the runs had lengths approaching 550 ft, which was close to the maximum length to which the cable could be bundled. Therefore, precise measurement of the proposed

WORK PLATFORM suspended from the space frame (center of photo) is used to assemble the upper section of the new trading post. The platform is located directly over the post to be replaced. Note the vertical drops that already have been made to new posts on either side of the old post. A decorative shroud was later placed around the cable and piping bundles.

cable runs at the jobsite was critical.

Logistics

Each trading post consists of custom-made cabinetry. All equipment that was to be located in the upper and lower sections arrived at the jobsite separately. Gustafson had to mount and wire all electrical/electronic units.

A room at the trading-floor level was used to work on the lower sections. Each lower section consists of two independent 14- by 16-ft horseshoes and a central module. These could be worked on during normal working hours because they did not interfere with Stock-Exchange activities. The upper section of the posts, however, was a single structure measuring roughly 13 by 27 ft. It was too large to assemble in the same location. Partial assembly and then final assembly and hoisting into place once the old post had been removed was not feasible because of the time limits involved. The only solution was to assemble the structure overhead and then lower it into place during changeover.

To accomplish the task, a wooden platform was suspended from cables from the space frame exactly over the spot the new post would occupy. The upper frame was assembled and the units mounted on it and wired. This approach had an additional benefit; it permitted all wiring to the upper units to be permanently connected to the computer room. Checkout prior to the actual changeover day also allowed defective units or other bugs to be caught. This greatly reduced the amount of work that had to be accomplished in the limited time allowed for the installation. Once the upper sections were completed, all cables that had to drop to the lower sections were coiled on the platform, the jacket stripped, and cable connectors installed on the ends of the individual cables where appropriate. The only drawback of this system of overhead assembly was that the work could only take place after the Exchange was closed for the day, weekend, or holiday.

Changeover periods involved intense activity by the electrical contractor. All wiring to existing posts had previously been identified and tagged. All obsoleted wire was disconnected and pulled back into the space below. Other trades

CHANGEOVER BEGINS (photo, left) with the demolition of the existing post. The upper section of the post—which had been assembled and wired on a temporary platform—is lowered (photo, right) to its final working height. Vertical drop cables that are intended for equipment of the lower section of the post are trained to one side.

then removed the posts. A trench was then cut into the floor to accommodate a wiring trough used for routing vertical-drop cables to the two horseshoes. Once the floor was patched, the lower sections were rolled into place and all cables connected.

Meanwhile, the working platform was being disassembled, and the upper section—supported by cables from the space frame—was lowered to its final working height. Once everything was in place and connected, SIAC put all equipment through its operating paces in preparation for use on the next trading day. All the posts, other than the three in the Blue Room which were fed from below, were handled in this fashion. All were completed within the time scheduled, and no interruption of Exchange activities occurred.

Many details

Installing ties between computer and posts was the major part of the project; but power, lighting, and other modifications were also made. Each new post had HVAC and lighting requirements that exceeded those of the posts they replaced. New feeders were installed from existing switchgear to a new electrical closet on the space-frame level. From there, seven sets of 500MCM cable were routed to different centers of load. In the perimeter raceway, the cables were installed in an 8-in. by 8-in. trough. The cables were tapped and a circuit breaker was installed at each post vertical drop location. No. 2 AWG cable run in flex conduit was then run down to the posts.

Temporary lighting was also required to make up for obscuring of existing ceiling-mounted fixtures by temporary platforms and catwalks. These were constantly being moved as work progressed to different areas. Once the overhead work was completed, new lighting for visibility on the floor and for architectural highlighting was installed.

LOWER SECTIONS of trading post are rolled into place. The two approximately 14- by 16-ft horseshoe-shaped units were assembled in a room off the Exchange floor. Mounting of electrical/electronic equipment and internal wiring was done prior to shutdown. Vertical-drop cables are then routed via a trough in the floor to both horseshoes.

MAIN POWER TRANSFORMER is rated 1500 kva, with a 12.5-kv primary and a 480/277-volt, 3-phase, 4-wire secondary. The transformer is connected to the main switchboard by a service lateral with six 500MCM THW copper conductors per phase and neutral. These 24 conductors are in four 4-in. PVC conduits.

Converting a plant to a computer center

New service and complete rewiring were needed to make an existing manufacturing plant suitable for housing computers.

When a former manufacturing plant in Rockaway, N.J., was purchased by the International Paper Co. for a new computer center, an entirely new electrical system had to be provided. The lighting system was outdated, the old cooling system was inadequate, and extra power was needed for computer operations.

The old power system had two separate services, one 4160-volt system for air cooling, and one 208/120-volt system for lighting and general receptacle power. These were removed and a single 480/277-volt, 2000-amp service was installed, including a 1500-kva, 12 kv-480/277-volt transformer.

A 320-ton air chiller from another International Paper plant was used to replace the old cooling system to keep down the cost of new equipment, and a new starter/control panel for this 244-kv air unit was purchased. A separate air-conditioning system was installed in the computer room to provide the extra cooling necessary for the computer system's critical temperature requirements. Three 15-ton units are supplied from a separate breaker on the main panel to provide independence from the main cooling system, in case of main chiller failure.

The old, stem-mounted, two-lamp 120-volt fluorescent fixtures were replaced by 1000 new 277-volt, two- and four-lamp, recessed, air-handling troffers for better efficiency and tem-

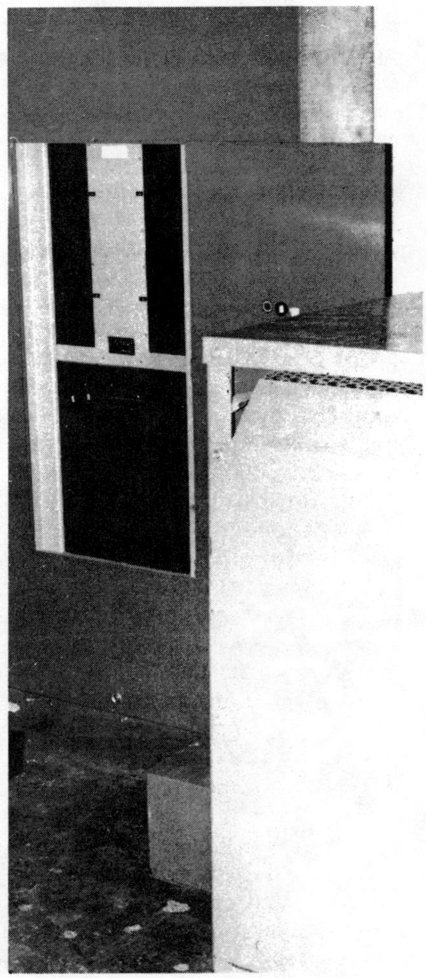

TRANSFORMER rated 300 kva and low-voltage distribution panel supply 208/120-volt general receptacle power and computer-room power. The distribution panel has two 600-amp breakers, one for the receptacles and one for the computer room.

MOTOR CONTROL CENTER serves 14 motors from ¼ to 30 hp used in the air-handling system for blowers, pumps, and exhaust fans. Three 500MCM cables in an overhead duct to the control center are protected by a 400-amp CB on the main switchboard.

perature control. A motor control center replaced individual starters for the air-system motors. This system includes 14 motors from ¼ to 30 hp, including three 1-hp motors added for exhausting the computer and tape storage areas within a set time should the fire system operate, to minimize damage from extinguishing gases.

The new power distribution system includes a 2000-amp, 480/277-volt main switchboard, one 300-kva and one 115-kva, 480-208/120-volt transformer for receptacle power—the smaller unit serving a separate office section of the building.

All existing 208/120-volt receptacle and lighting wiring was removed, and only the panel feeders and boxes were used in the new system. New breaker panel interiors were installed for receptacle wiring, and the new 480/277-volt lighting system was provided with separate panels and wiring.

An emergency power generator was installed to provide power for emergency and exit lighting in case of normal-power failure. A transfer switch automatically switches those lights to the generator, and the system is in full operation in less than 90 sec.

The new electric system was installed by Devon Electric, Inc., of Rockaway, N.J.

GENERAL LIGHTING is provided by 277-volt, 2- and 4-lamp recessed air-handling fixtures with 40-watt, cool-white fluorescent lamps.

EXISTING 208/120-VOLT PANEL BOXES (right) were retained and fitted with new interiors. No. 1/0 feeders connect these 225-amp panels to the main switchboard.

COMPUTER ROOM PANEL (below) is fed by eight 350MCM cables (two per phase and neutral). Three 125-amp breakers in this panel feed subpanels, each with 16 20-amp and one 30-amp branch circuits.

Installing a Computer-Assisted Fire Dispatching System

Upgrading of a 50-year-old audible-signal alarm system for the New York City Fire Department with a mini/microcomputer-based system obsoletes a time-consuming, eight-step manual procedure.

A COMMUNICATIONS system consisting of minicomputers, microprocessors, and a network of teleprinter terminals—one for each of 300 firehouses—provides the New York City Fire Department (FDNY) with a space-age procedure for rapid deployment of men and equipment in response to alarms.

The new system replaces a manual dispatch system that was set up 50 years ago with the establishment of a communications office (CO) in each of the city's five boroughs—Manhattan, Brooklyn, Bronx, Queens and Staten Island. Each CO dispatches firefighters within that borough.

With the old manual setup, it could take four men on duty in the CO, working in sequence and using written tickets, rotary files and voice-radio calls as long as eight minutes to assign and dispatch units, as the alarm arrival rate went above 30 an hour.

Basically, the dispatcher must assume that all alarms are real (although false alarms account for about 50% of the yearly total), so at least two pieces of equipment are usually sent—one engine (or pumper) and one ladder truck. Since the alarms also include emergency situations other than structural fires, such as someone trapped in a stalled elevator, the demands for FDNY service have been increasing at a rapid rate. Alarms on a yearly basis totaled 130,000 in 1965, 400,000 in 1975 and 473,000 in 1978.

Along with the startling and constantly rising number of total alarms being experienced, two other conditions emerge; a reduction in availability of units to respond and unequal work loads placed on busy companies.

Thus, FDNY management realized the need to use modern electronic and electrical methods to assign units, to track the status of 350 pieces of equipment, and to maintain records of incidents. Toward that end, a contracted management study in 1969 recommended that an automated procedure to receive and process alarms be set up

MODERNIZED Manhattan Communications Office employs a variety of consoles and display equipment. Alarm incidents are displayed on a pair of Manhattan Island boundry maps (1). Engine companies are shown on left map, ladder companies on right map. Alarm data is displayed on CRT terminals located at dispatcher watch station (2). Chipboard consists of an array of transilluminated slots (3). Radio dispatcher station (4) receives reports and talks directly with units in the field. Raised floor is used throughout.

By JOSEPH R. KNISLEY

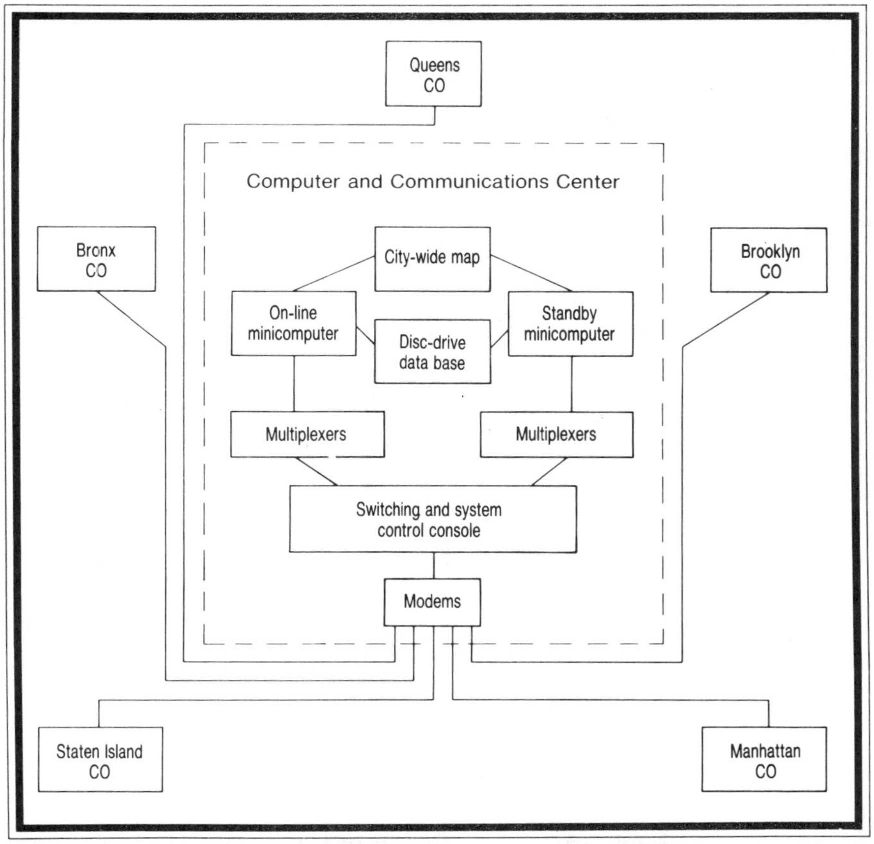

FIG. 1. Computer communications network carries the designation "STARFIRE" because the five borough communication offices (CO), each representing a node of a five-pointed star, collect or concentrate the alarms received in its geographic area and send them along to the Computer and Communications Center (CCC), whose computer makes decisions on dispatching of equipment. Triplexed phone lines connect each CO to the CCC, and duplexed phone lines provide separate control circuits.

with a real-time link back to a central deployment center. When an elaborate system was developed under this concept and released for bids in 1972, two responses were received—one for $40 million and another for $38 million.

Since this cost was more than twice the estimated budget, the project was temporarily postponed. Then, in 1972, FDNY management prepared internally a more cost-effective system redesign. Unlike a fixed assignment card, this system is able to select those units to provide the quickest and most effective response based on knowing which units are actually available at the time of the alarm.

Termed STARFIRE, the $15 million network consists of 14 minicomputers, 12 microcomputers and 500 computer terminals. The logo designation, STARFIRE graphically depicts the system's operation. A central computer located at the Compute and Communications Center (CCC) holds all software programming and decision-making intelligence, and it processes all alarms collected at each node of a five-pointed star. Each of the star points represents a borough CO, where a smaller local computer acts solely as a concentrator of all alarms received throughout that borough and sends them to the CCC for processing. After a deployment recommendation is de-

WIRING CHECKUP at 400-A vertical distribution board shows lineup of 200-A disconnect switch assemblies employing 125, 150, and 200-A fuses. Note the ½- by 2-in. copper ground bus at bottom of enclosure.

NEW SERVICE SWITCH and distribution board assembly in Manhattan CO (left), a free-standing, 3-phase, 4-wire, 208/120-V unit, is located at right angles to rear wall, on which the existing service switch is mounted. Uncovered wiring box shows two sets of 500MCM copper conductors leaving CT cabinet and extending to transfer switch associated with the standby diesel generator. Completed changeover to new service equipment (right)

termined, it is sent back to the borough CO for the dispatcher's approval before being forwarded to the responding fire station. Fig. 1 shows that arrangement.

Redundancy for reliability

The entire system is fully redundant for high reliability, with a backup computer at the CCC and at each borough CO. At a CO, in case both minicomputers fail, concentrator functions can be taken over by the microprocessor that normally controls status entry panels, a wall-mounted map with transilluminated location lights, and an automated chipboard. This processor also has a backup unit, as shown in Fig. 2

The chipboard consists of an array of slots, also transilluminated, with color-coded indicator lights, into which plastic chips representing fire companies are inserted. The indicator lights graphically identify the company's availability. If the entire computer system fails or if the dedicated telephone communication lines from CCC are down, the CO dispatcher can revert back to manual operation using the chipboard. Manual proficiency is maintained by periodic exercises.

Alarm arrival/notification

Alarms received from mechanical street boxes, from newer type electron-

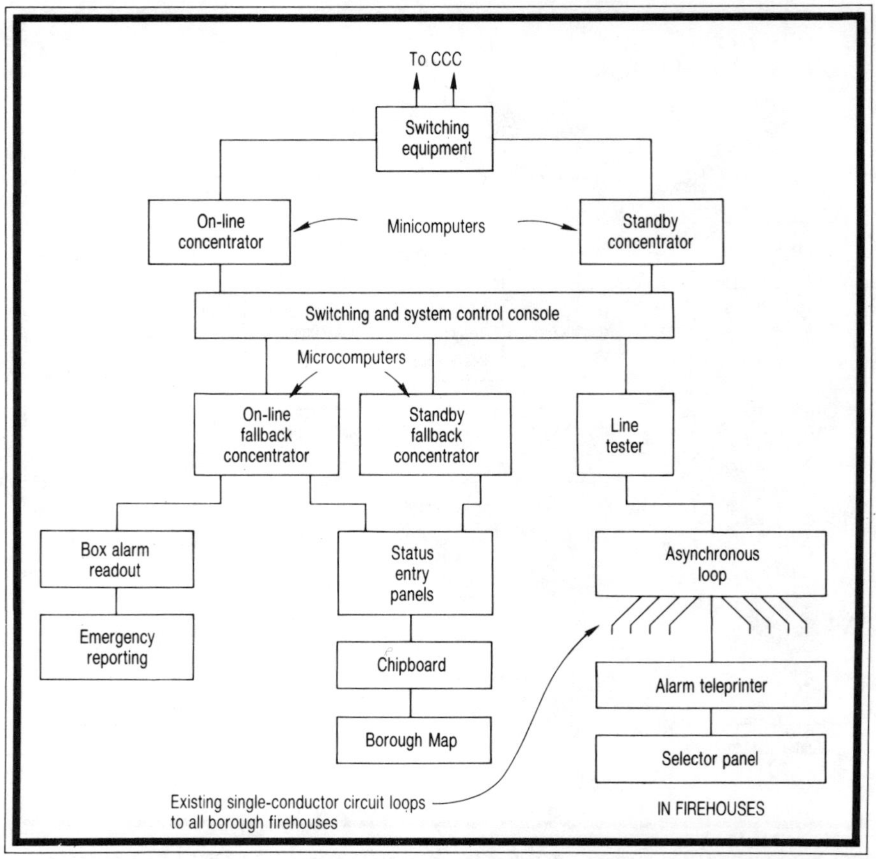

FIG. 2. Borough communications office has a standby unit for the minicomputer that concentrates and organizes all incoming alarms for transmission to the CCC plus a standby unit for the microcomputer that controls status-entry panels and graphic displays. Single-conductor, FDNY-owned loop circuits connect a CO to the alarm teleprinter at each of its station houses, thus providing a hard copy of all alarms and equipment dispatch orders.

shows two sets of rerouted 500MCM copper conductors leaving CT cabinet and descending to trough directly below where they run to service switch at right. The wiring troughs below each carry a single set of 500MCM copper conductors, with one set going to, and the other set returning from, the transfer switch. All multiple conductors were cut to exactly the same length to prevent unbalanced loading (and potential overloading) on any of them.

SERVICE END BOX on basement wall of Manhattan OC receives an extension section to provide adequate bending room for new sets of 500MCM copper service cables extending to existing CT cabinet on wall at left.

ic street boxes (that permit two-way voice communication as well as transmission of digital signals) and from direct telephone contact with a citizen are processed through the CCC computer.

Since STARFIRE assists rather than controls, the computer recommendation is displayed for review and approval of the dispatcher before further action is taken. Units can be added or deleted from the recommendation, or any of several other actions can be substituted.

LOAD PANEL in computer room uses a wall and floor trough assembly for circuit distribution. Note transition along the floor from wiring trough to conduit runs and the location of box-enclosed, single locking-type receptacles. Raised floor will cover all of the conduit and wireway runs.

ISOLATION TRANSFORMER, rated 75 kVA, 3 phase, receives a checkout prior to startup. The unit, serving all sensitive computer/signal transmission equipment, prevents transients caused by a variety of stray signals and spikes in the electrical distribution system from being coupled onto the secondary winding.

After reviewing the computer's assignment, a dispatcher validates the decision by transmitting a message to the firehouse using a circuit that loops back through the CCC computer.

Each fire station is equipped with an alarm teleprinter and a microprocessor-based selector panel that also functions as a controller for the teleprinter. When an alarm arrives at the station, it is printed out on the teleprinter for the fireman on house watch to read. This hard-copy message shows units scheduled to respond and a time-stamped summary of all data that is known about the alarm.

The fireman reads the alarm message and responds rapidly by pressing three buttons on the selector panel—first his company identification code button, then a 10-4 (able to respond) button and finally the SEND button to transmit the response back to the CO. At that point, the fireman tears off the dispatch to take it aboard the apparatus that is turning out.

On average, the lapsed time from alarm receipt to watch-desk response is less than 20 sec. If the company is at another location, the order is sent by radio and a hard copy record printed on the firehouse teleprinter. Radio contact is constantly maintained with apparatus in the field.

Since all of the 10,000 firemen in the department are assigned to house watch on a rotating basis, extensive training in computer operations for all personnel was unthinkable. Thus, the response panel configuration was kept simple, so that its functions could be easily explained and used. When a fireman presses a panel button, a signal light directly above the button is illuminated, indicating that it has been pressed. Thus, any incorrectly pushed button can be canceled out and the correct one pressed.

At the CCC, video screen displays and printers provide a review of current and past alarm activity in any and all boroughs, providing vital information in the event of a city-wide emergency.

A variety of management support functions are also provided by the computer data bank, including data recording, activity reporting, information retrieval, and load generation of artificial alarms for performance monitoring of CO personnel. (Although uniquely suited for the FDNY application, the software program could be applied to other municipalities with similar requirements and planning techniques.)

Installation sequence

Brooklyn, the borough with the highest alarm rate, received a CO computer first to check out system performance within the geographic confines of that single borough. When processing of the alarms was being done satisfactorily locally, installation in the other four boroughs and the CCC proceeded.

With the exception of the electrical work within the Brooklyn CO building, the city-wide STARFIRE system was installed by E-J Electrical Installation Co., Inc., New York City, under a contract from Bradford National Corp., who developed the software programs and the computer hardware specifications. This project points up how, in an increasingly electronics-oriented age, an electrical contractor's customer can be someone completely apart from the traditional architect, general contractor or plant/building owner.

E-J followed a Bradford-prepared installation schedule that incorporated a tracking procedure to monitor firm delivery dates on equipment, so that checkout of various segments of the project could be done in sequence. This construction management procedure also aided in quality-control reviews of components and construction steps, since the life safety of citizens depended on system availability following cutover from manual to computer-based operation. All of the design and installation personnel were aware that one faulty component or connection could upset the most carefully engineered plan.

The contractor installed power distribution equipment in all five CO buildings and the CCC to serve the computers and their peripheral equipment. In the Manhattan, Bronx and Queens CO, E-J put in new service-entrance equipment and feeder distribution switchboards to handle the increased power demand. Since each CO computer facility had a diesel generator to provide a redundant power source, the new distribution boards at the three locations were connected to the existing transfer switch with a new set of conductors.

An extensive number of feeder power and branch circuits were routed along the floor at these sites, anticipating placement of a raised access floor.

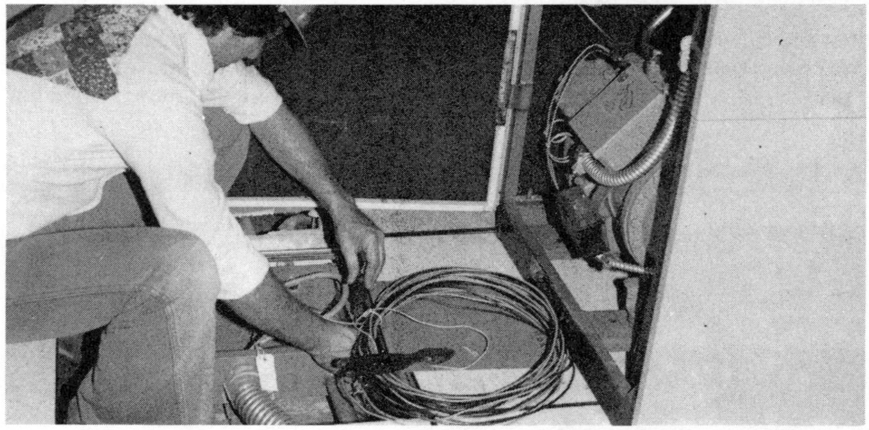

COMPUTER FLOOR, made up of 2-ft-sq removable panels, permits access to power and communication cabling and also serves as a cooling plenum for the air-conditioning unit at right, which discharges dust-free air and maintains a slight positive air pressure within the room.

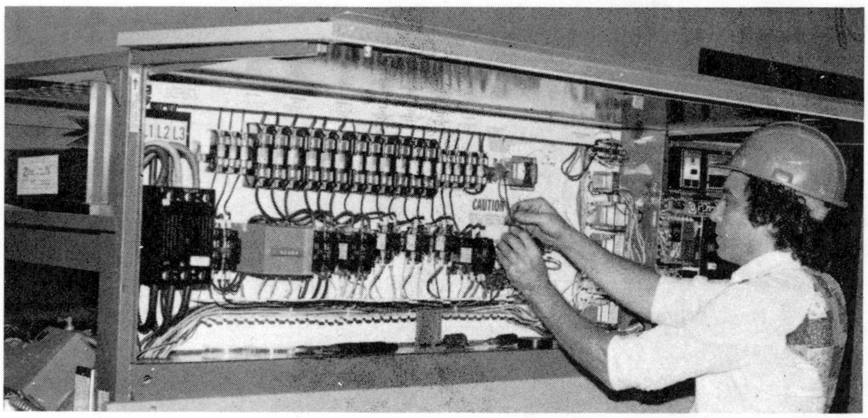

SPECIALIZED air-conditioning unit to maintain optimum temperature/humidity ambient within the computer room receives control-circuit checkout. Note ceiling-recessed wiring box for ionization-type fire detector.

STEEL BEAMS, welded to front and rear frame elements of existing signal relay racks, raised the side members enough to permit conduit runs through these free-standing racks and brought the racks to elevation of new access floor.

PATCH PANEL used to tap into circuits and monitor the data permits checking-out of system elements. Procedure is demonstrated by John J. Mohan, Director, Bureau Information & Computer Services.

TELEPRINTER, (right) employing only six moving parts and used in conjunction with a control console located on rack below, is the field terminal that provides hard copy of alarm information at a firehouse. Display board in background at the Computer and Communications Center shows the location of all engine companies on the city boundry map at left, and the location of all ladder companies on the city map at right. J. Robert Mann, Jr., vice president of E-J Electric Installation Co., reads printout containing alarm summaries.

MIMIC BOARD shows status of computer equipment, triplexed data transmission lines and duplexed control lines from the CCC to all COs. Inoperative status of loop circuits to firehouses are shown by red LED.

The NYC electrical code permits power and signal cables to be installed under raised floors that are of suitable construction and in accordance with NFPA No. 75.

Because of the critical nature of these spaces, an early-warning fire detection/suppression system was integrated into the structures. Ionization-type detectors that respond to the incipient stage of fire activity are located on the ceilings and under the access floor. The smoke detectors are cross-zoned, requiring a minimum of two different or adjacent zones to be energized before an extinguishing control circuit is energized.

The CO computers are located in a separate room outfitted with specialized air-conditioning equipment to maintain optimum temperature/humidity conditions for the solid-state gear. Because unanticipated loss of power could result in thermal, electical and mechanical stresses on the minicomputers, a UPS system was subsequently specified and placed between the distribution load bus and the computer loads at each CO, under a separate contract.

Field wiring

At each firehouse, the contractor ran a 120-V branch circuit to serve the teleprinter and response panel. Final connection of the printer to the alarm-loop circuit was done by FDNY maintenance personnel.

The FDNY-owned, single-conductor telegraph network (for the existing audible alarm) from a CO to the borough firehouses was retained to operate the teleprinters, a decision that proved cost-effective even though some of the conductors date back to the use of cotton-braid insulation within wooden conduits. The single-conductor designation means that earth is the return path of the circuit.

System designers and FDNY management recognized that with the computer hardware specified for the job, these existing circuits would be adequate for the 30-characters-per-sec (300-baud) rate of signals fed to the printer, a rate much slower that usually employed on phone-line data transmission. However, to determine if reuse was feasible, FDNY maintenance personnel tested each line for insulation value, inductance and capacitance, the last being an important consideration, since the length of some runs exceeds 20 miles.

Beefing-up computer power

Expansion of electrical distribution capacity is part of the continuing growth and alteration of the power system behind computers for the worldwide reservation system of a major airline. Here's a look at one part of the distribution system revamp.

By FRANK WILLIAMS, President
Tenafly Electric Service
Tenafly, N. J.

A MERGER of two large airline companies required a major beefing up of the data-processing center for passenger reservations. That dictated more generator capacity as emergency supply for the uninterruptible power system and more circuits for increased computer applications. As with most electrical modernization work, a vital step in this project was adding more power capacity and effectively and economically integrating the system additions and revisions into the system that already existed, with all of the work performed on a high priority, fast time schedule.

Performing electrical modernization work is usually conditioned by the owner's type of business and by past experience. Our contracting firm has been steadily involved for over five years doing the electrical alterations and additions at this computer facility. That is common for the vast majority of commercial, institutional and industrial premises, because of the high density of electrical application and the impact of steady load growth and continual development of new and better

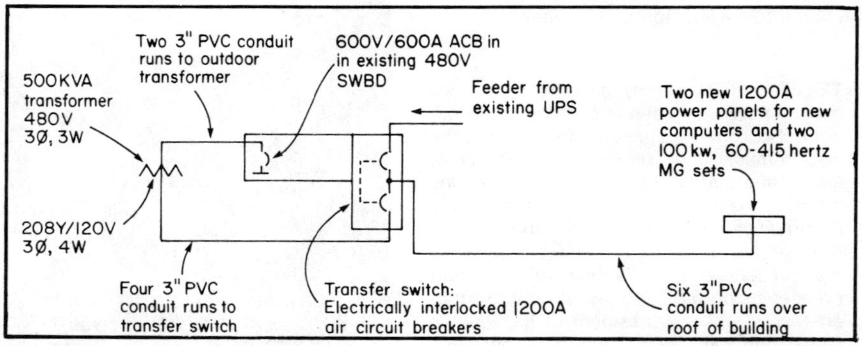

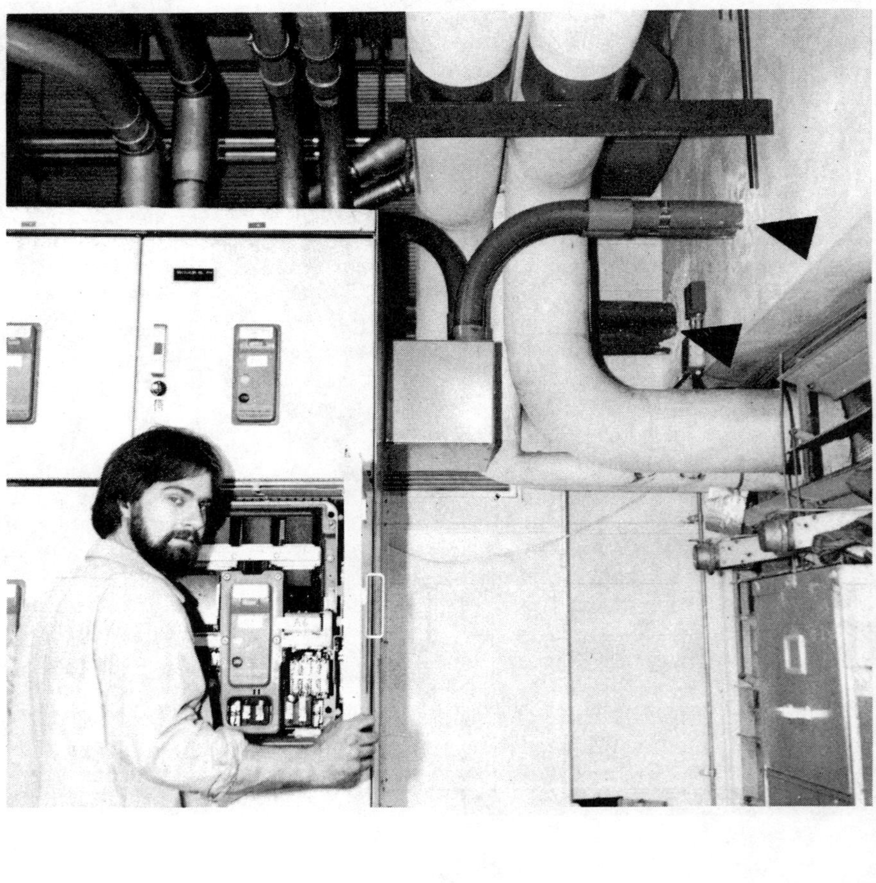

1—SPARE AIR-CIRCUIT BREAKER (ACB) in existing 480-volt switchboard was used to derive power for the new loads. The ACB is rated 600 volts, 600 amps and protects the 480-volt, 3-phase, 3-wire feeder circuit of two parallel conduits (top arrow), that runs through the wall of the electrical room to the primary of a new 500-kva transformer installed outdoors adjacent to the room. Secondary circuit from transformer (bottom arrow) carries 208Y/120-volt power back indoors to a circuit-breaker transfer assembly installed at the far end of the 480-volt switchboard. Bob Mossman, foreman for Tenafly Electric Service is at the ACB.

2—500-KVA TRANSFORMER to power new computer loads was set on roof of building just outside electrical room (which did not have space for this unit). Two 3-in. PVC conduits for primary circuit (arrow) come out of building to right side of transformer and four conduits carrying parallel makeup of eight No. 3/0 conductors per phase supply secondary output current back to circuit-breaker transfer assembly in the room.

3—CB TRANSFER BOARD installed at end of existing switchboard normally feeds the output of the UPS equipment to the two new 1200-amp power panels. When needed, this assembly provides for transferring the load from the UPS to the secondary of the new 500-kva transformer fed by utility supply to the existing switchboard. Each ACB has a trip rating of 1200 amps to protect the load feeder in either position of supply source.

types of electrically powered machines and processes used by owners. Because of the constant need for the services of an experienced, competent electrical contractor with skillful design capability, an industrial or commercial owner today wants a continuing association with such an electrical contractor, who, in effect, functions as the electrical design and construction department of the owner's overall business organization. The recent addition of more computer power capacity at this data processing center is a typical example of the kind of work our firm has been doing at this facility.

As is so often the case in modernization work, the decision to add the new power feeds was made only when it became obvious that there was an urgent need for this new capacity. As a result, a tight time schedule was set to get the capacity in as fast as possible and the pressure of time affected electrical design considerations.

The need for branch circuits to supply the new data-processing units that were added at the facility dictated the installation of two 208Y/120-volt, 1200-amp panelboards in the computer room. Analysis of the computer loads indicated that the two panels combined would have a feeder demand load of 1000 amps. Then design began on how to feed these panels and how to tie them into the UPS system, which provides the ultra-high reliability of continuous power input to the computers, whether the UPS is fed from the utility power system, from its batteries, or from emergency generators at the site.

To provide the necessary power for the two new 1200-amp power panels in the computer room, a design layout as shown in Fig. 1 was selected. For all of the circuits shown in Fig. 1, multiple-conductor circuits were made up of No. 3/0 THW copper conductors in multiple 3-in. PVC conduit runs, as follows:

1. The 500-kva, 3-phase, dry-type transformer has a rated secondary full-load current of 1380 amps, which satisfies the needed 1000-amp capacity with a good measure of spare capacity.

To handle the transformer's rated primary current of 601 amps, two 3-in. conduits are used, with six No. 3/0 THW copper conductors in each for the phase legs. Between the two conduits, the four No. 3/0 THW conductors in parallel per phase leg have an ampacity of 640 amps. With six No. 3/0 conductors in each conduit, the ampacity of each is derated to 80% of its 200-amp value as shown in NE Code Tables 310-16 through -19. Then with two No. 3/0 conductors per phase, each conduit makeup has an ampacity of 0.8 × 200 × 2, or 320 amps. And the two conduit makeups have a total ampacity of twice that value—640 amps.

An equipment grounding conductor has to be run for the primary circuit in the rigid nonmetallic conduits. NE Code Table 250-95 shows that the 600-amp ACB in the switchboard requires that a minimum of No. 1 copper be used for the equipment grounding conductor. But because of ready availability, a No. 1/0 conductor was pulled in each conduit to serve as the equipment grounding. NE Code Section 250-95 requires the same size equipment grounding conductor in each conduit. And each 3-in. conduit (with 40% of its c.s.a. equal to 2.95 sq in.) has adequate space for the six No. 3/0 THWs and the No. 1/0 contained in each (0.3288 sq in. per No. 3/0 conductor times six equals 1.9728 plus 0.2367 for the No. 1/0 equals 2.2095 sq in.).

2. The secondary 208Y/120-volt circuit from the transformer to the CB-transfer housing is made up of four 3-in. PVC conduits with seven No. 3/0 copper conductors and one No. 1/0 THW copper conductor in each. Again, there are two No. 3/0 THW conductors per phase leg in each conduit, providing a secondary ampacity of 1280 amps per phase leg for the four conduit makeups. That does not provide for the full 1388-amp secondary capacity of the transformer, but is well above the 1000-amp demand load. And the seventh No. 3/0 conductor in each conduit is an equipment ground, based on an assumed protection rating of 1200 amps and Table 250-95. The secondary circuit from the transformer to the CB-transfer is an unprotected 25-ft tap that conforms to NE Code Section 240-21, Exception No. 8.

Although the new computer load that was added does not require any 120-volt load, a neutral conductor from the transformer secondary wye was run to provide such voltage for possible future needs. Because NE Code Section

4—SIX PVC CONDUITS (right) (arrow) carry parallel feeder of twelve No. 3/0 THW copper conductors per phase from the CB-transfer unit up through ceiling of electrical room, across the roof, and then down through the roof into a junction box above the two 1200-amp panelboards in the computer room. Wireway at top right carries feeder circuit from the UPS.

5—PULLING SETUP (below, right) was arranged for making the over-the-roof pull in the six PVC conduits that run from the CB transfer to the new panelboards. Seven reels of No. 3/0 THW copper and one reel of No. 1/0 THW copper provide the eight conductors that were pulled into each of the six conduits. To assure safe and effective hookup at all points in the parallel makeup, the phase conductors were color coded (black, red, blue) for their entire lengths. No. 3/0 equipment grounding conductors were marked by green tape at all ends; No. 1/0 neutral conductors were marked by white tape at both ends.

300-20 requires a neutral in each conduit when multiple-conductor circuit makeup is used and because NE Code Section 310-4 gives No. 1/0 conductors as the minimum size that may be run in parallel, each of the four secondary conduits must contain a No. 1/0 to make up a properly sized parallel neutral.

3. From the CB-transfer board, the circuit run to the new 1200-amp panels is almost 200 ft long. To limit voltage drop in that line to an acceptable value, the feeder circuit makeup from that point consists of six parallel conduit makeups with seven No. 3/0s and one No. 1/0 conductor in each 3-in. PVC conduit. In such an arrangement the total of twelve No. 3/0 conductors per phase (two per phase per conduit) have an ampacity of 0.8 × 200 × 12, or 1920 amps. But each of the 1200-amp air-circuit breakers in the transfer assembly will provide protection at that value. And in each conduit, the No. 3/0 equipment grounding conductor satisfies NE Code Table 250-95, which calls for a minimum of No. 3/0 copper when the circuit overcurrent device is rated 1200 amps, and Section 250-95 which requires the full-size equipment grounding conductor to be run in each conduit. Again, the neutral had to consist of a No. 1/0 conductor in each conduit, for the reasons previously stated.

The sequence of elements in the whole power distribution scheme is as shown in the accompanying numbered photographs.

6—EACH PARALLEL MAKEUP (left) of seven No. 3/0 conductors and one No. 1/0 conductor was carefully pulled over a pulley from the cable reels on the ground to the roof of the building. All conductors were then carefully laid out on the roof and then cut to precisely the same length, to satisfy NE Code Section 310-4.

9—PULLBOX ABOVE COMPUTER ROOM (right) is installed with six PVC conduits stubbing down into a junction box above the two 1200-amp panels. One eight-conductor makeup is shown installed in one of the six over-roof conduits, with the come-along cable grip of a second eight-conductor makeup just coming out of another conduit at left.

10—TWO 1200-AMP PANELS in the computer room (far right) are fed by the over-roof multiple feeder circuit by means of a large custom junction box above the panels, with six conduit stub-downs feeding the junction box from the roof pullbox.

7—PULLBOX is installed on roof directly above the CB-transfer unit in the electrical room, and is fed by six PVC conduit runs up from that room. The six conduits run from this pullbox to another pullbox (almost 200 ft away) which is installed on the roof directly over the location of the two new power panels in the computer room.

8—CONDUIT SPACER-SUPPORTS were used to align and position the six PVC runs to the pullbox above the computer room.

11—EFFECTIVE TERMINATIONS for a total of 48 conductors coming down from the six roof conduits were provided by special bussing within the junction box, which provides adequate room for the necessary sweeps in the conductors. This was required because all of the 48 conductors are virtually exactly the same length, as required by NE Code Section 310-4 to prevent unbalanced loading (and potential overload) on any of the conductors. Each phase busbar shown has 12 lugs for the incoming No. 3/0 conductors and is also equipped with eight lugs for 500MCM THW taps—four 500MCMs per phase to the 1200-amp main CB in each of the two power panels. Equipment grounding conductors and neutral conductors are run down into each panel from the junction box and are connected to separate terminal blocks. △

Rooftop generators power Wall Street computers

Six huge engine-generator sets, installed on a building roof 15 floors above the streets of New York City, designed to coordinate with an existing UPS system to assure uninterrupted power for critical stock-exchange operations.

By ARTHUR J. MAIHACK
IBEW, Local No. 3
New York, N.Y.

CONTINUOUS, accurate transactions at the New York Stock Exchange and the American Stock Exchange depend, to an extremely high degree, on computer systems operated jointly by the Securities Industries Automation Corp. (SIAC) and the Depository Trust Co. (DTC). Two full floors of computers plus other office facilities serve the high-speed processing required by the world's busiest stock exchanges.

The computers and related equipment receive power from Consolidated Edison Co. of N.Y. through a parallel redundant uninterruptible power supply (UPS). In the event of power failure, three 1000-kw diesel engine-generator sets supply auxiliary power to the computers and support facilities. These three engine-generator sets were installed along with three other 1000-kw units that serve another tenant in the building to obtain the economic advantages of combining the two installations into one operation. The installation was made atop a 15-story office

12-TON ENGINE-GENERATOR SET is being lifted to roof of 15-story building (photo at right). High-capacity crane with 350-ft boom lifted six of these units to rooftop location. A flatbed trailer (above) carried the engine-generators to the jobsite. In the background is an aluminum enclosure typical of housings used to protect the standby power equipment installed on the roof.

building that is part of a two-building complex in downtown New York. The two buildings are connected by a structural crossway.

This "bridge" not only permits people to walk between the structures; it also serves to carry utilities between the buildings, including the new 480-volt standby power circuits from the generators to the loads in the 53-story main structure.

The building power supply and distribution arrangement includes (1) a utility supply to a UPS system that serves the SIAC computers; (2) a parallel path through the UPS system (redundant operating mode) in the event of trouble within the system; (3) a bypass bus circuit that supplies utility power *around* the UPS to the computers in the event that time-consuming problems occur within the UPS; (4) three 1000-kw diesel-generator sets that, in the event of normal power failure, supply 480-volt power to the computers and related loads via a system of automatic and selectable transfer switches; and (5) a second set of three 1000-kw generators that supply standby power to another tenant in the building. All six generators are installed in close "physical" arrangement to simplify operation, since a power failure affecting the total complex would require coordination of their start-up and running requirements. For maximum reliability, even though there are numerous modes of alternate power and power routes, the computers

1000-KW, 480/277-VOLT DIESEL engine-generator set (above) is one of six to be installed that forms two systems of three standby units each. Systems will supply standby power for two tenants in main 53-story building. One tenant operates computers for stock-exchange transactions. Other standby power loads include HVAC for computer rooms, emergency lighting, elevators, fire pumps, and communications systems.

TWO OF SIX engine-generator units (left) are set in position on platforms that also serve as the floor of housings that will cover the units. Each platform will support two engine-generator sets. In foreground is aluminum housing for generator controls and distribution switchgear.

PLATFORM (right) is hoisted to final position adjacent to previously installed units on roof. Each platform is 10 ft wide by 45 ft long. Note square precut conduit holes in platforms.

for the stock exchanges have additional backup resources.

Installing the generators

After the building owners were informed that two of their tenants required standby power for their computer systems, the owners called for an engineering study to evaluate the feasibility of an alternate power installation. If appropriate, the owners planned to install the required systems and set up a lease-back arrangement.

Space was not available in the basement or lower floors, and most computer loads were located on middle floors in the main building. Therefore, the study found that the most economical and practical arrangement would be to put the generators on the roof of the 15-story building and carry most of the extensive alternate circuits, consisting of cable in conduit, directly to pertinent transfer switches and computer systems.

The power source installation consists of the six diesel engine-generator sets, mounting supports and protective housings, generator control equipment and distribution switchgear installed within a common enclosure, two 20,000-gal fuel tanks, day tanks, mufflers, and other related equipment.

Engineers on the project emphasized that the superior strength of the building was a key factor permitting the rooftop installation. The weight of each engine-generator set is nearly 12 tons, and associated equipment added con-

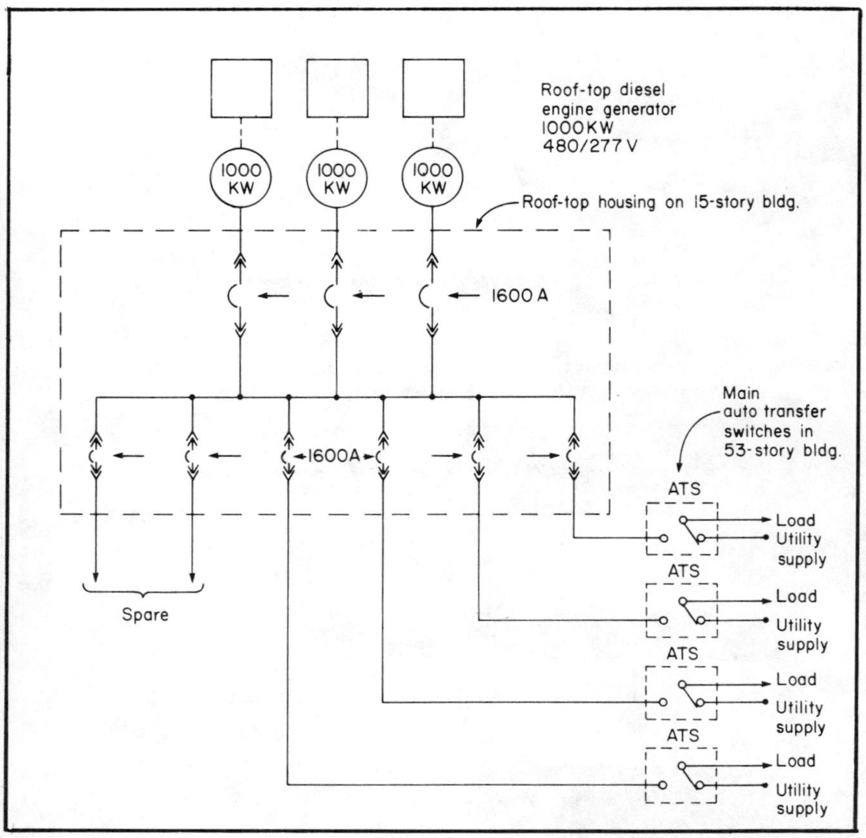

ROOFTOP ENGINE GENERATORS supply standby power to UPS system via roof-mounted circuit breakers, cable-in-conduit, and four main automatic-transfer switches. Special controls permit circuiting of standby power to selected loads through 24 transfer switches as needed.

siderably to the total weight of the installation. Because the structure was strong enough to support this equipment, only a minimal steel framing was needed. This was mounted on building steel that was extended through the roof to provide the main support.

The generating units were brought to the site on high-capacity flatbed trucks. A powerful crane rated 250 tons and outfitted with a 350-ft boom lifted the 12-ton units into place. Previously, a steel I-beam web had been constructed, and the crane was used to place huge 10-ft-wide by 45-ft-long platforms into place to serve as the base for the generating sets and as the floor for the enclosures.

Three of these platforms were installed, each providing mounting space for two engine-generator sets. After the platforms were set and the engine-generators were mounted on vibration isolators, a metallic housing was placed over each pair. In close proximity to the generators are the generator controls and standby-power distribution switchgear. All housings are constructed of aluminum with a baked-enamel finish. The housings are designed for walk-in operation and accessible maintenance of the equipment. On the generating units, enclosures are provided with sound attenuation features and silencers rated 85 db at 50 ft for quiet operation.

From the generators, cable in conduit carries 480-volt standby power to distribution switchgear on the roof. From drawout power circuit breakers, power continues via cable in conduit to main automatic transfer switches and other switching as required by each tenant.

UPS operation

Design of the SIAC system combines normal utility power through a UPS system with a power supply from three 1000-kw generators in the event of

STRUCTURAL STEEL FRAMING (left) is attached to main building columns to provide enough strength to support weight of generator sets. Steel columns were extended through roof to form solid foundation. After platforms were installed on framing, vibration isolators were provided under generator steel skids.

INTERMEDIATE METAL CONDUIT (right) furnished with spin-type couplings is mounted on 3-tier steel hanger, assembled in the field using slotted angle iron and ½-in. steel rods. Large box at right provides point for pulling or splicing of system conductors as required.

500MCM THW COPPER CONDUCTORS (below) are pulled from pull boxes directly below generators to automatic transfer switches in main building. Electrician Claude Papa feeds cable into the box.

ALUMINUM ENCLOSURES (above) protect generator controls and distribution switchgear installed on roof. Enclosures provide plenty of interior space for operation and maintenance of equipment for standby power system. Two separate enclosures were installed in line to provide for the two separate standby power systems.

GENERATOR CONTROLS (right) include a separate panel for each set of three generators. Electrical foreman Charles Kretzschmar, Arc Electric Co., New York City, operates meter selector switch on one control panel. Panel in center provides for overall system control.

power failure. The system incorporates a sophisticated generator control system, four main automatic transfer switches, and 24 selectable transfer switches that are operated according to need at the time of normal power loss. In the event of a total power failure, the system will permit plenty of time via the UPS system batteries to selectively "power-down" so that all computers can be properly de-energized without loss of information. If desired, the SIAC computers, powered by the standby generators, can be operated on an "emergency" mode for up to two days.

Loads on the standby power system include the required computers (brought on line as needed according to the operations at the time via a special control panel that permits selection of the position of each of 28 transfer switches), certain air-conditioning equipment (also selected as needed to cool operating equipment to proper temperature), elevators, emergency lighting, communications, and signaling systems.

The system as designed will start and parallel the three engine-generator sets upon receipt of a signal from any of the locations where undervoltage relays are installed.

Whenever the automatic engine start controls are placed in their automatic position, the engine-generator system is on standby in readiness for automatic starting and synchronization. In the event of a power failure, the three engine generators automatically start and come up to speed. The first generator to achieve 90% of rated voltage and 59 Hz is connected to the standby power bus. Electronic interlocks permit the connection of only one generator in the event of simultaneous generator relay operation. When the first generator comes on line, remote indications are given to an operator that priority 1, 2, and 3 loads may be transferred to the standby power bus.

Synchronizers will automatically adjust the frequency of the on-coming generators to produce synchronism with the bus. (Provisions for voltage adjustment are included in the system.) When synchronism is achieved, the on-coming generators are paralleled to the bus.

If a generator fails while operating in the automatic mode, it is disconnected from the bus and shut down. If the load is equal to or greater than 100% of the remaining on-line capacity, the priority loads are automatically dumped off the bus. Similarly, should two generators fail and the load be greater than 100% of the remaining generator's capacity, the priority 2 loads will be removed from the bus. Following a load dump, the removed load may be manually restored by use of load-dump bypass switches.

Actuation of the switches will give remote indication to an operator that load may be added to the bus. The operator at the system cubicle may supervise load-add operations and dump priority 2 or 3 loads off the bus as desired. If at any time the standby power bus becomes overloaded, the priority 3 or 2 loads will be dumped off the bus as a function of total generators on line.

Upon receipt of a signal that commercial power has been restored and that the transfer switches have retransferred the load to the normal source, the generator CBs are simultaneously opened and the engine generators run for an adjustable no-load cooldown period of one to 30 minutes. Then the controls are placed in readiness for the next power failure.

Engineers on the project were Jaros Baum & Bolles, New York City. The electrical contractor was Arc Electric Co., also of New York City.

AUTOMATIC TRANSFER SWITCH enclosures are in place and ready for wire pull. Interior components of each of the four main transfer switches will be delivered and installed during later stages of construction. Spin-type couplings simplified entrance of IMC to enclosures.

Computer reliability assured with UPS power

IN 1979, the results of a long-range planning study prompted Miles Homes, a division of Insilco Corp., located in Minneapolis, MN to design a new 60,000-sq-ft headquarters, incorporating a sophisticated computer system into the company's operations. Now nearly 50 CRTs and three Computer-Aided-Design (CAD) workstations are utilized by Miles Homes, on-line 14 hours a day, to custom-design home building plans and supply customers with current account information. But in order to alleviate problems resulting from frequent power-line disturbances in the area, Miles Homes also installed UPS power protection equipment in the new facility—assuring total power reliability essential to the computer system.

Founded in 1946, the company supplies building plans and pre-cut materials to do-it-yourselfers.

At the heart of the company's management information system operations are two computer systems. CRT operators call up information from the company's customer data base to handle customer service calls as well as taking care of all inventory and general accounting procedures. A dedicated CAD computer system services the company's home planning department in drafting designs and preparing cost estimates.

MILES HOMES customer service representative Jan Netz keys in account information at one of about 50 CRTs used by the company. All conversations with customers are logged into the company's data base for immediate retrieval by any representative, when needed.

Miles Homes' customers can choose building plans ranging from one of fifty standardized Miles models to customized homes drawn completely to customer specifications. Minor revisions such as moving an interior wall, or major modifications like adding additional footage or completely changing the interior layout are possible to fit each home builder's preferences. In addition, the company offers for sale such items as electrical and heating systems, floor coverings, plumbing, bathrooms, and kitchen cabinetry.

The CAD system also specifies the bill of materials required to complete a building plan. Estimating material is a very labor-intensive job, and calculating these figures manually leads to a high degree of error. This leads James E. Fenske, director of management information systems to note, "We help eliminate these errors by using this computerized system. We want to make sure that customers in the middle of building a home don't end up short on materials."

While planning for its on-line customer information services and CAD system, Miles Homes realized the potential for power problems in its new headquarters. The immediate geographic area has a history of power problems. About 95% of the power outages last less than a minute. However, these interruptions can cause information to be lost from computer storage discs. Since the company had some problems at their old site, they wanted to eliminate the possibility of power interruptions in the new building.

To assure smooth, continuous power for its data-processing operations, Miles Homes installed a 50-kVA UPS. When operating at full load, the unit provides up to 20 min of emergency power from its battery reserve. All computer equipment in the data-processing department is protected by the UPS. Under normal conditions, this system—known as a reverse-transfer or "true" UPS—draws power from the primary AC line and rectifies this to DC in order to power its inverter and keep the battery bank fully charged. This DC power is then reconverted to AC in the inverter stage for output. If utility power fails, no switching delay to an alternate source of power occurs. Because of this "no-break" operation, this type of UPS is ideal for protecting sensitive, electronic equipment. Once primary line power is restored, the UPS simultaneously recharges its batteries while maintaining output to the equipment.

Because the UPS is always on-line, full advantage can be taken of its voltage regulation and noise-filtering capabilities. Typically, output voltage can be maintained with 1% of nominal voltage despite input voltage swings of up to ±15%. Regulated, noise-free output is maintained even during brownouts without the need to discharge the UPS battery banks. This leaves the batteries fully charged for blackouts that often follow brownouts.

"Since our UPS installation we have had several major power interruptions, including electrical storms and a power line cut by construction workers," Fenske says. "However, we experienced no harmful effects. It's difficult to put a dollar value on our savings, but in a heavily customer-oriented business like ours, providing top-quality customer service day-in and day-out is the key. And our UPS assures us that we'll have the clean power necessary to keep our computer systems running smoothly."

UPS provides reliability at data-processing center

A 200-kva uninterruptible power supply system (UPS) provides nonfluctuating "clean" ac power to the Continental Telephone Service Corp. data processing center in Bakersfield, Calif. A 350-kw standby diesel-generator serves as a redundant power source. To offset the possible catastrophic loss of the UPS unit, a previously used motor-generator/regulator, working through a separately derived utility feeder, was retained as a further backup for the computer equipment.

The solid-state UPS system provides uninterruptible power by carrying the ac input through a rectifier (rated 500 amps dc); the output feeds a dc bus. In turn, this bus serves as a feed to the dc-to-ac inverter, while floating a fully charged storage battery bank. The rectifier/inverter serves to eliminate transients on the power system, with the battery acting as an emergency power supply.

If a complete utility outage occurs, the battery can supply the dc load to the inverter until the diesel generator starts up and assumes the load, usually within 15 sec. In the event that the engine-generator didn't start up automatically because of a malfunction, the computer operation room would receive an alarm when the battery bank had 5 min of capacity remaining to final discharge voltage. This enables the operators to make an orderly shut down and preserve data in the memory storage units. The 180-cell lead-calcium battery group has a 15-min total capacity under full draw.

As shown in the accompanying diagram, Feeder No. 2 serves a main switchgear lineup that has three fused switch outputs. Two of the three output feeders normally supply a motor-control center and the UPS equipment. The third feeder can serve semi-essential loads by means of a nonautomatic transfer switch in the event that feeder No. 1 is lost. This switch, which is remotely controlled by pushbuttons, permits shifting of the semi-essential loads from the No. 1 feeder to the diesel generator during periodic exercising. At least 50 percent loading of the prime mover for a sufficient time period is usually recommended in order to reach operating temperatures in the engine and forestall buildup of sludge in the lubricating oil.

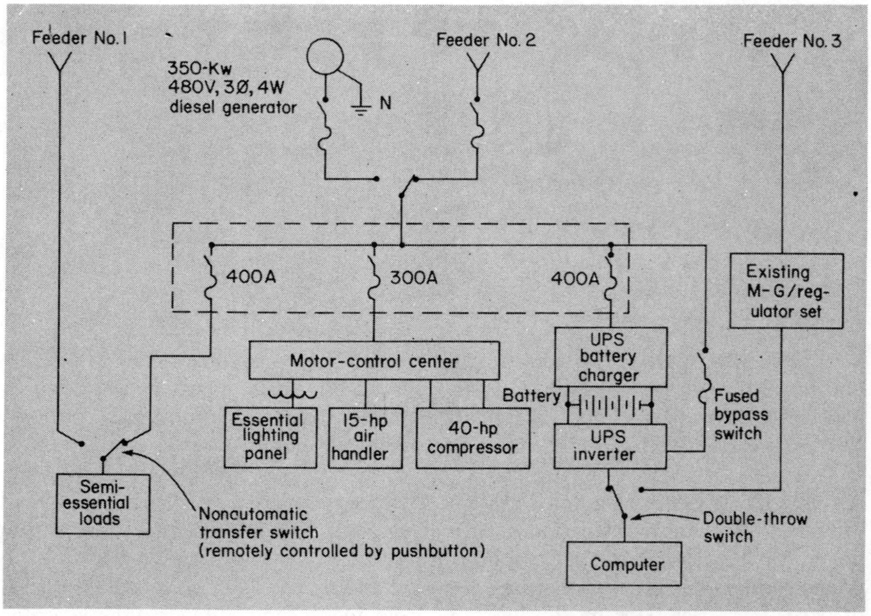

THE 200-KVA UPS ARRAY for normal support of the computer in Continental Telephone's Data Processing Center receives power through Feeder No. 2. The standby generator and Feeder No. 3 are alternate power sources. Feeder No. 1 normally handles the semi-essential loads.

Two identical engine annunciator panels (one at the automatic transfer switch and the other in the computer room) provide audible and visual signals on: generating and connection to load, failure to start, low oil pressure, high oil temperature, high or low water temperature, low fuel supply, and high or low battery voltage.

A key feature of the electrical design is the use of a manually operated, double-throw, 200-amp transfer switch to permit shifting of the computer load over to the old motor-generator/regulator set in case the UPS system fails.

Before the UPS array was designed and built, the m-g/regulator provided conditioned computer power. So, rather than scrapping this rotating gear, management decided to keep it as an additional backup. As shown in the diagram, the m/g set is served by a separately derived utility feeder (No. 3).

All switchgear and controls are located in a cool, well-ventilated basement area of the two-story structure, with the diesel generator located outside just beyond the foundation wall.

Filtering circuits within the generator block battery charger harmonics from feeding back and interfering with the electronic automatic engine controls. In addition, built-in time delay circuits on two major motor loads prevent simultaneous inrush starting current from drawing too heavy a load on the generator.

The fire suppression system for the battery room and the UPS switchgear spaces employs a halogenated hydrocarbon (Halon 1301) as the discharge agent. Besides its extremely rapid extinguishing action, the material leaves no residue and does not degrade delicate electronic components.

A fire alarm buzzer/light signal array, coupled to a time-delay circuit for operating the discharge heads, permits personnel to evacuate the area before the discharge of the colorless, odorless gas. A remote fire alarm also was installed in the computer room to warn the operators.

Chapter 5

Power Supply Problems and Solutions

Effect of power disturbances on computer operation

*Intermittent abnormalities, inherent in both commercial and industrial power systems, affect the reliability of sensitive electronic equipment depended upon for data processing, communications, and process-control operations. Here are some considerations the system designer and user should be aware of to reduce computer downtime.**

By THOMAS S. KEY, Lieutenant
Naval Facilities Engineering Command
Washington, D.C.

POOR-QUALITY power is only one of many reasons for computer operating problems. Computer system failures related to hardware malfunctions, software problems, environmental conditions, and operator errors are much more common.

Types of power-line disturbances (Table 1)

However, there are computer failures in which power definitely is to blame. The short duration dip in voltage from a power-system fault (short circuit) is a common disturbance causing lengthy downtime. That the electric utility company is not likely to accept responsibility for this is understandable, considering that these disturbances go undetected by the majority of customers. And computer manufacturers do not say that their systems will operate properly during power-line disturbances. In fact, computer input power requirements provided in manufacturers' specifications cannot be met, all of the time, by even the best utility power service. The user must face the dilemma of connecting an overly sensitive computer to a disturbance-prone power source. The only solution is to bridge the gap between power quality and equipment susceptibility.

Steady-state conditions

Just what quality can be expected from the *power source*? ANSI stan

**Adapted from IEEE Industrial and Commercial Power Systems Conference Record; paper delivered Cincinnati, Ohio, June 7, 1978.*

Table 1. Types of power-line disturbances and typical characteristics

Definition	Type I Transient and oscillatory overvoltage	Type II Momentary undervoltage or overvoltage	Type III Outage
Causes	Lighting; power network switching (particularly large capacitors or inductors); operation of other loads	Power system faults; large load changes; utility equipment malfunctions	Power system faults; unacceptable load changes; utility equipment malfunctions
Threshold level*	200 to 400% rated rms voltage or higher (peak instantaneous above or below rated rms)	Below 80 to 85% and above 110% of rated rms voltage	Below 80 to 85% of rated rms voltage
Duration	Spikes 0.5 to 200 microsecs wide and oscillatory up to 16.7 millisec at frequencies of 0.2 to 5 kHz and higher	From 4 to 60 cycles, depending on type of power system distribution equipment	From 2 to 60 sec if correction is automatic; from 15 min to 4 hrs if manual

0 — 0.5 — 120 — ∞

Duration (cycles of 60-Hz wave)

* Approximate limits beyond which disturbance is considered to be harmful to computer.

dard C84.1 lists two steady-state voltage regulation limits, Range A (+4%, −10%) and Range B (+6%, −13%). Range B should be satisfactory for modern computers and will usually be met under load variations or even brownouts. Excessive phase displacement or harmonic content are rarely found in utility power. Problems of this type are usually the result of the user's own equipment. Solid-state components of the computer itself are a major source of harmonic distortion; however, computers have been found to be relatively insensitive to their own harmonic pollution, and harmonic distortion is not a major problem. Sudden variations in frequency can cause tape and disk drive difficulties, but this is rarely a utility problem.

Transient monitoring

The major incompatibility between the computer and the power source is caused by *intermittent* or *transient* conditions on the power line. Power-system transients have been studied for many years. With a need to support sensitive electronic systems, the U.S. Navy developed 3-phase voltage monitoring equipment capable of automatically tracing the waveshapes of power-line disturbances at their time of occurrence. Beginning in 1968, data was collected at communications and computer sites. This work, which has been expanded with the advent of smaller and lower-cost commercially available monitors, is still under way.

Distinct categories of recurring power disturbances identified by the monitoring were analyzed with respect to their effect on electronic equipment, resulting in the definition of three types of disturbance (Table 1). Note from the horizontal scale that it is the duration of the disturbance, not the magnitude of the voltage, that determines the type.

Type II momentary *undervoltage* disturbances are the most common cause of power-related computer system failure. The duration of the undervoltage is measured in milliseconds (Fig. 1). While momentary *overvoltages* lasting more than ½ cycle also are considered to be Type II disturbances, they occur infrequently and rarely disrupt computer operation.

Tolerance envelope

The precise extent to which today's computers are susceptible to these disturbances is difficult to determine.

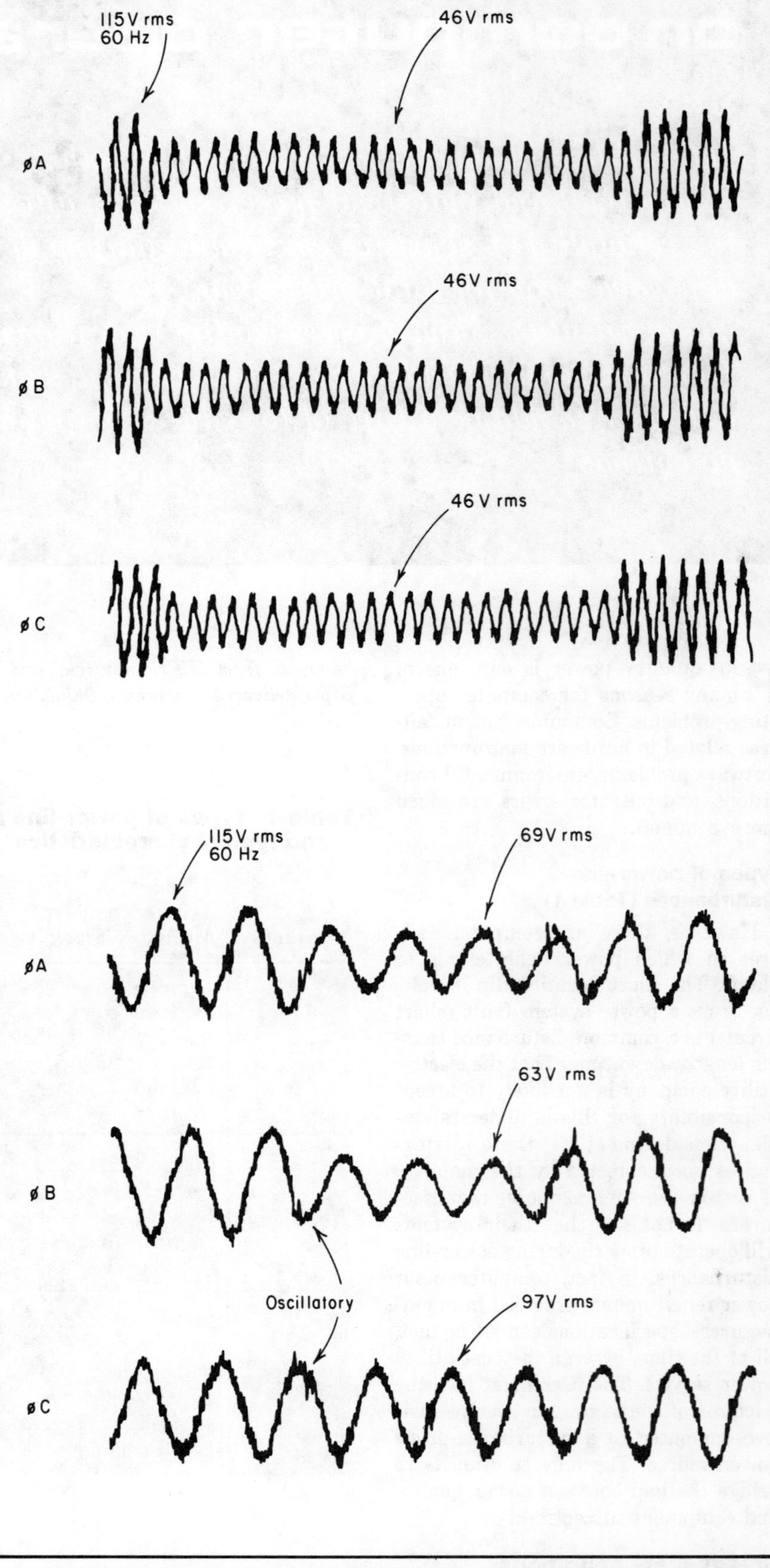

FIG. 1. Monitors recorded these examples of Type II power-line anomalies, characterized by periods of undervoltage and oscillatory disturbances. The two groups show a reduction in voltage on all three phases caused by a lightning fault (top) and undervoltage with superimposed spike and oscillatory disturbances recorded during a thunderstorm (bottom)—both in Rome, N.Y. in 1971.

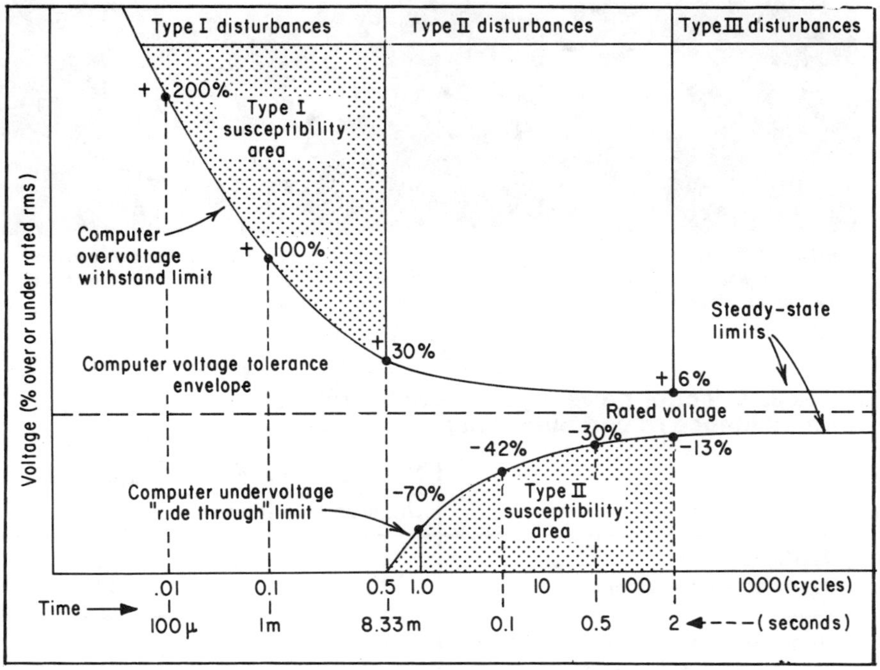

FIG. 2. U.S. Navy susceptibility tests, computer power studies, and input from computer manufacturers resulted in the development of this computer voltage tolerance "envelope" as a guide to computer design and to help in diagnosing power-related failures. Most computer manufacturers agree that they can meet these criteria.

Users are reluctant to do extensive testing because of high costs, resultant downtime, and possible violation of equipment warranty. Nevertheless, some guidelines are needed to encourage the computer designer to provide built-in protection against line disturbances and to aid in the diagnosing of power-related failures.

Accordingly, a generalized computer voltage tolerance "envelope" has been developed based on U.S. Navy tests, computer power studies, and suggestions from computer manufacturers. Fig. 2 shows the relationship of overvoltage and undervoltage with respect to time and the three types of disturbances.

Table 2. Causes of power-related computer failures, Northern Virginia, 1976

Recorded cause	Disturbance: Under-voltage	Disturbance: Outage	No. of computer failures
Wind and lightning	37	14	51
Utility equipment failure	8	0	8
Construction or traffic accident	8	2	10
Animals	5	1	6
Tree limbs	1	1	2
Unknown	21	2	23
Totals	**80**	**20**	**100**

Type I *transient overvoltage disturbances* may have an indirect effect on computer operation, resulting in "unready" or "can't print" messages, lost addresses, etc. However, the effects are unpredictable, and it is difficult to identify them with actual disturbances within the Type I range.

Type II *undervoltage disturbances,* on the other hand, have a direct and predictable effect on electronic equipment, so that the computer manufacturer must provide built-in sensors (usually on the dc side of the power-supply rectifiers) designed to power-down the equipment if voltage falls below the preset limits and yet avoid nuisance tripping as a result of otherwise harmless low-magnitude transients, etc.

In general, adequately large power supply filter capacitors can help reduce the effects of Type II *undervoltage* disturbances and momentary outages, and it costs less to provide transient *overvoltage* protection in the form of suppressors and surge arresters than to conduct a study to determine if such improvement is worthwhile.

The probability that a computer will have problems from voltages outside the tolerance envelope of Fig. 2 may be estimated if the common causes of undervoltage disturbances and the probable exposure of the power source to these causes are known.

A power-line disturbance monitor can be useful in recording out-of-tolerance voltage occurrences, and helpful data can be gathered from building outage records, computer operation logs, utility company interruption reports, weather data, and records of such potentially disturbing loads as the starting of large motors.

Combining information from such sources in Northern Virginia in 1976 enabled the author to determine the causes of 100 power-related computer failures at six computer sites (see Table 2). Since all causes were natural phenomena that can be expected to recur at approximately the same rate, future computer failures resulting from these causes can be estimated.

For every recorded failure, the specific disturbance was either voltage "sag" or an outage, and behind each of these was a power-system short-circuit fault. No transient overvoltages were found to have been at fault, and utility equipment failure accounted for 8% of the computer failures.

These results confirm that weather

is the major cause of power-related computer failure, and the incidence of thunderstorms in a given area may be used in predicting future failures. If wind and lightning caused 51 failures in Northern Virginia with an average of 40 thunderstorms per year, computers in Jacksonville, Fla., with 70 to 80 thunderstorms per year, may be expected to have twice as many weather-related failures.

The type of the utility power transmission and distribution system has a significant bearing on power-related computer failures. Table 3 shows that in one study underground systems accounted for about ⅓ as many failures as overhead lines, and that twice as many voltage "sag" disturbances were recorded on overhead spot networks as on overhead radial systems, accounting for a higher rate of computer failure.

Identifying alternatives

The installation of an interface system that reduces the incompatibilities between the computer and its power source is a common solution to power quality-related computer problems. One most-effective system, the uninterruptible power supply (UPS), is also one of the most expensive solu-

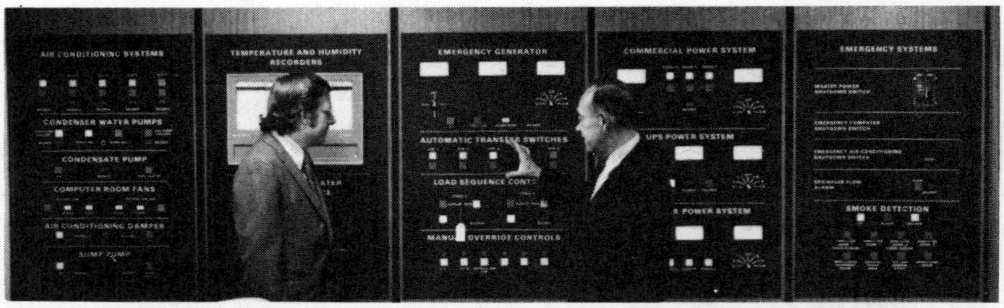

CONTROL BOARD has data inputs and switching options arranged so supervisor on duty can monitor computer power and correct many problems from this location. If problem cannot be corrected here, action can be initiated to minimize effects of power system disturbance.

Table 3. Effect of power-system configuration on incidence of computer failures

Configuration	No. of disturbances		Recorded failures
	Undervoltage	Outage	
Overhead radial	12	6	18
Overhead spot network	22	1	23
Combined overhead (weighted)*	16	4	20
Underground radial	6	4	10
Underground network	5	0	5
Combined underground (weighted)*	5	2	7

*Combined averages were weighted based on length of time monitored (30 to 53 months). Ten sites were monitored in Washington, D.C., Norfolk, Va., and Charleston, S.C.

Table 4A. Relative effectiveness of power-enhancement methods in moderating or eliminating power disturbances

Disturbance type	UPS system and standby generator	UPS system	Dual power feeders: Secondary spot network	Dual power feeders: Secondary selective*	Motor generator	Solid-state line-voltage regulator	Specialty shielded insulating xfmr	Suppressors, filters, lightning arresters	Balance computers on 3-phase; improve grounding
I	All source transients No load transients	All source transients No load transients	None	None	All source transients No load transients	Most source transients No load transients	Most source transients No load transients	Most	Some**
II	All	All	None	Most	Most	Some (depends on response time)	None	None	Some**
III	All	Only outages of duration less than the battery discharge time	Most	Most	Only "brown-out"	Only "brown-out"	None	None	None

* Includes special application of solid-state static switch between two independent sources.
**These improvements do not eliminate or moderate power-line disturbances but do reduce computer susceptibility to undervoltage and overvoltage. Computer manufacturer should be consulted for grounding assistance.

tions and is not always the best economic alternative. Tables 4A and 4B have been prepared to assist in comparing the cost and effectiveness of the available alternatives. Costs given include site preparation, operation and maintenance training, and spare parts. (These costs are meant for comparison purposes only, not for estimating.)

In summary, a reduction in power-related computer failures can be achieved by combining a knowledge of computer voltage tolerance, an understanding of power disturbances, and an appreciation of options available to diminish or eliminate the effect of power-system disturbances.

COMPUTERS (left) are expensive to own or lease, and mistakes or downtime which result from poor quality power to the computer can range from costly to disastrous. Power must be of the quality necessary for stable computer operation, without going to unnecessary extremes.

Table 4B. Relative costs of power-enhancement methods

System rating	Basis of comparison	UPS system and standby generator (diesel)	UPS system	Dual power feeders	Motor generator	Solid-state line-voltage regulator	Specialty shielded insulating xfmr	Suppressors, filters, lightning arresters	Balance computers on 3-phase; improve grounding
	Installation and equipment cost	$1600-$1900 per kva	$1200-$1500 per kva	Installation cost varies greatly from site to site.	$400 per kva	$250 per kva	$50-$150 per kva	$1-$10 per kva	Varies with installation require-ments
	Mainte-nance cost	$12,000 per year	$8000 per year		Less than $1000 per year	Less than $1000 per year	N/A	N/A	N/A
100 kva	Operating efficiency*	85%	85%		83%	90%	98%	N/A	N/A
	Energy cost**	$5300 per year plus generator exercising costs	$5300 per year		$6000 per year	$3500 per year	$700 per year	N/A	N/A
	Installation and equipment cost	$1200 per kva	$700-$800 per kva	Monthly service charge for the additional feeder is probable.	$250 per kva	$200 per kva	$25-$100 per kva	$1-$5 per kva	Varies with installation require-ments
	Maintenance cost	$15,000 per year	$10,000 per year		Less than $1000 per year	Less than $1000 per year	N/A	N/A	N/A
300 kva	Operating efficiency*	89%	89%		88%	90%	98%	N/A	N/A
	Energy cost**	$11,600 per year plus generator exercising costs	$11,600 per year		$12,600 per year	$10,500 per year	$2100 per year	N/A	N/A

*Efficiency applies to power equipment only and does not account for losses in environmental support systems.

**Based on a power cost of $.05 per kwhr and consumption at 0.8 power factor.

Utility power and your computers

Computers need steady power, at constant voltage and frequency, "clean" of transients and disturbances. Electric utility power as supplied is usually acceptable for most electric power equipment but is often unsatisfactory for electronic-based equipment, especially computers, without further conditioning.

THE SELECTION of appropriate, cost-effective power-conditioning equipment depends to a great extent on the quality of incoming power, as well as the effects of power disturbances on the computer operations.

The Bell Telephone System has always protected its telecommunications network from incoming AC power failures. Battery plants with several hours of reserve power are standard equipment in all central and transmission offices, and diesel or turbine engine alternators back up the batteries in larger offices. It became obvious several years ago that, for adequate reliability, some of the system computers would require some type of power-conditioning equipment. (At the present time there are over 10,000 computers throughout the Bell System.) Taking into account both the cost of computer errors and downtime caused by power disturbances and the cost of various types of power conditioning, what are the most economical and cost-effective means of protecting these computers?

The first requirement for answering this question is information about the quality of incoming power to the computers. From May 1977 through September 1979, the Bell System studied commercial AC power disturbances at 24 data-processing sites, for a total of 270 site months. The objective was first to collect actual field data on the type, frequency, magnitude, and duration of disturbances, and second to convert this data into a statistically correct model that could be applied to any particular installation. The 24 operations chosen for study included two AT&T Long Lines sites, one Western Electric, and the remainder at seven Bell Operating Companies. Representative samplings were obtained from locations in the metropolitan New York City area and its suburbs, Syracuse, Newark, Boston and parts of Michigan, Ohio, Maryland, Virginia, Kentucky, Tennessee, Louisiana, Missouri, and Texas.

By PAUL SPERANZA

Fig. 1. Typical AC power distribution

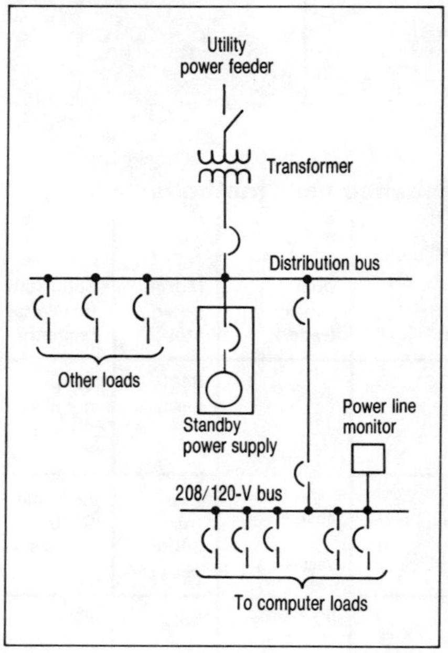

Site preparation

To insure that all results were obtained on the same basis and could be compared meaningfully, each site was inspected and monitoring devices of the same design were installed, calibrated and set to respond to the same parameters. The distribution transformer serving the computer loads was checked, visually and by ohmmeter, for proper grounding. The steady-state voltage level was checked to make sure it was well within the acceptable working limits recommended by the computer manufacturer (115-125 V). All nominal voltage levels were 208Y/120 V, 3 phase, 4 wire plus ground, 60 Hz. Commercially available recording monitors, which could be set to the values of disturbances to be measured, were used. Each monitor was checked by simulating up and down voltage changes at the settings for slow average, sags, surges, and impulses. The monitors were connected to sense on each phase of the common AC bus that fed the power data-processing centers (Fig. 1). Sensing was from each line to neutral, so the neutral (grounded) conductor would be a common reference. The monitor was purposely placed "downstream" in the electrical system, near the computer load, so that the disturbances sensed would be a realistic mix of those generated within the building power distribution system as well as outside of the building, from the utility source, under normal loading conditions. Monitors at all sites were set to the same threshold values (see box).

Analyzing the data

To develop a statistically correct model, the observed field data was analyzed carefully. Among the 24 locations studied, monitoring time ranged from a minimum of 2.1 months to a maximum of 22.4 months, with an average

PAUL SPERANZA has been with Bell Laboratories, Whippany, NJ for 28 years. He is a staff member of the Network Power Systems Engineering Group and serves on the IEEE Surge Protection Development Committee.

Parameters used

FOR PURPOSES OF THIS STUDY, to assure like interpretations of the data from all sites, the following definitions and parameters were adopted.

Steady-state voltage—rms voltage values that stay constant for 10 sec or longer. Nominal voltage level at each location monitored was 208Y/120 V.

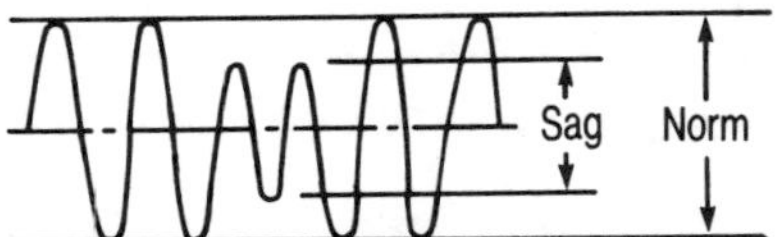

Sags—cycle-to-cycle decreases in the power-line rms voltage on any one of the three phases sensed. The duration of a sag is the number of cycles of the 60-Hz line frequency that the disturbance is below the threshold level. (Threshold: −5 V)

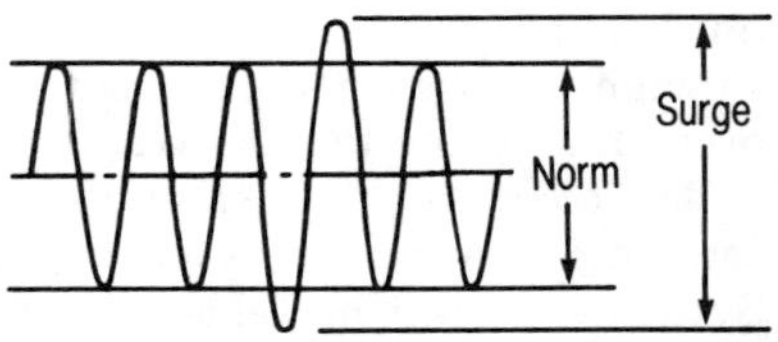

Surges—cycle-to-cycle increases in the power-line rms voltage on any one of the three phases sensed. The duration of a surge is the number of cycles of the 60-Hz line frequency that the disturbance is above the threshold level. (Threshold: +5 V)

Slow-moving average—variation of the steady-state voltage slowly from initial value for long periods of time. (Threshold: ±10 V)

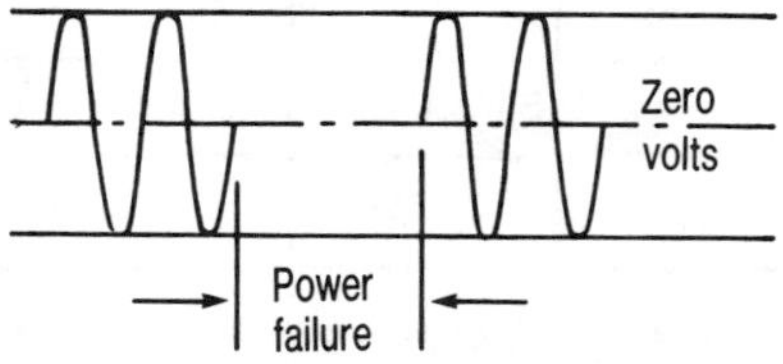

Power failure—a zero-voltage condition lasting for more than one cycle on any one of the three phases being sensed.

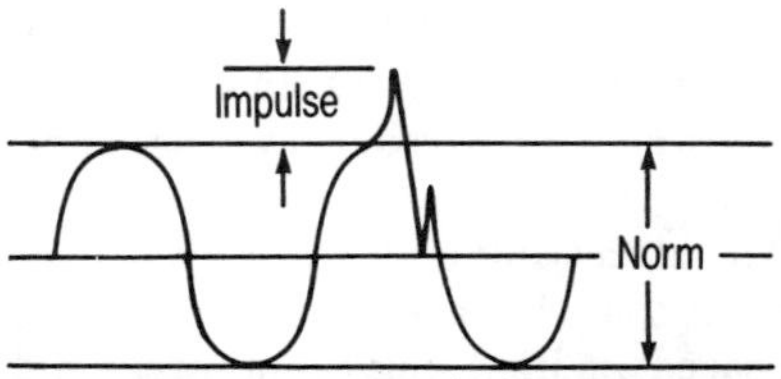

Impulse—short-term overvoltage disturbance superimposed on the 60-Hz wave and lasting between 0.5 and 100 microseconds. (Threshold: 200 V)

Frequency variation—change from the nominal 60-Hz power line frequency. (Threshold: ±½ Hz)

Fig. 2. Distribution among disturbance events

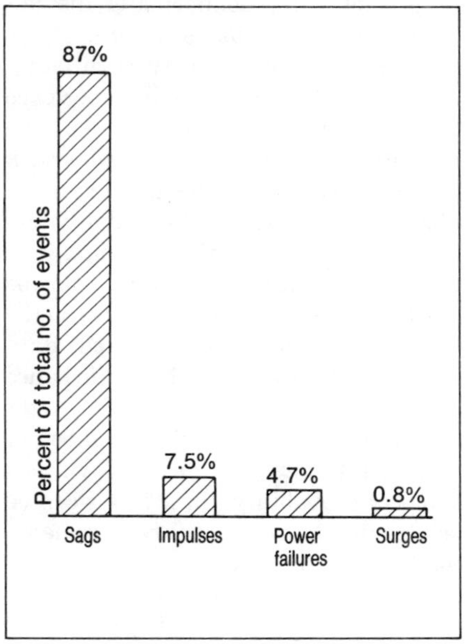

of 11.3 months and a median of 12.3 months. Sensing time at any site was long enough to prevent bias toward any particular seasonal variations.

When a set of sags or surges was recorded for each phase at a single time indication, the lowest voltage sag value or highest voltage surge value, with its associated duration, was considered to be a single event. Only the most severe sag or surge of the series, with its associated lasting time, from any one of the phases, was counted and tabulated as the single event.

When a series of power failures or sags was recorded at a single time indication, the event was counted as a single power failure or sag which lasted for the entire period. If recorded at different time indications, but within the same hour, the event was considered as a single power failure or sag with a lasting time equal to the sum of the individual lasting times. A series of events that occurred during the same day and that were known to have been caused by the same incident or condition were also considered a single power failure or sag whose duration was equal to the sum of the individual durations. By using this procedure, 113 initial events were reduced to 100 independent events. When a series of impulses ("spikes" or transients) was recorded at a single time indication on any or all of the phases, the highest impulse on any *one* of the phases was counted as the single event.

Statistical analysis of the observed data showed that there were significantly different disturbance rates from site to site. A normal distribution did not fit the data, nor could the overall cumulative data be applied to any given site. Moreover, the disturbances at any given site were rare events, when the total time of disturbances was compared to the total disturbance-free time. Since the recorded disturbances were statistically rare, a probability density function, called a Polya distribution, was used to predict the probable occurrence of these events

Table 1. Duration of power failures and sags (96 V or less)

Percentage lasting less than indicated time	Duration of event	
	Power failures	Sags
10%	0.6 sec	0.03 sec
25%	1.1 sec	0.09 sec
50%	38.0 sec	0.12 sec
75%*	40.0 min	0.24 sec
90%	4.2 hrs	0.53 sec

*75% of all power failures last less than 40 min; 75% of all sags last less than 0.24 sec.

over a set period of time. The Polya distribution has been used successfully to predict rare, nonnormally distributed events and was used here to predict what proportion of sites throughout the Bell System will experience various power line disturbances. One advantage of this statistical approach is that it enables reliable predictions of disturbances at various probability levels to be made. Another advantage is that upper statistical limits can be placed on the number of predicted events, whereas the observed data related only how often the disturbance would occur on the average. After determining the maximum number of expected events per year at various levels of probability, these levels can be related to the reliability requirements for a particular site to determine the most effective corrective action.

Table 2. Predicted overall disturbances

Sites having a maximum of "N" events	Number of disturbances (N) expected per year			
	Sags (96 volts or less)	Power failures (One cycle or more)	Impulses (Greater than 200 V peak)	Surges (Greater than 130 V)
10%	10	0	0	0
25%	16	2	0	0
50%	25	4*	2	0
75%	36	7	8	2
90%	50*	12	23	4
	Sites that will not experience the indicated disturbance within a year			
	0.05%	14.9%	43.2%	53.8%

*Of all the sites, 90% will experience less than 50 sags per year; 50% will experience less than four power failures per year.

Overall findings

The distribution of AC power-line disturbances at the 24 sites is shown in Fig. 2. Sags are by far the most frequent disturbances, followed by impulses, power failures and surges. Note that low-voltage conditions—sags and power failures—account for 91.7% of all the defined events. During the monitoring period, 56% of the sites experienced utilization voltages that fell outside the limits of ANSI utilization band B, 106 to 127 V. The steady-state line frequency was between 58.7 and 60.7 Hz at those sites (45% of the total) where line frequency was monitored.

An interesting and unexpected finding was that the rate at which impulses appeared in lightning-prone areas did not, as expected, increase dramatically over other areas. In fact, more sags than impulses (spikes) were recorded during electrical storms. The reason is that the impulse-suppression equipment employed in the power distribution system protects by momentarily shorting the power line to ground. This produces sags that are seen throughout the system, whereas impulses are quickly attenuated as they travel through the system's loads and other overvoltage protection devices.

The *duration*, or lasting time, of sags and power failures on the Bell System is shown in Table 1. Sags to 96 V or less and lasting at least 16 ms (1 cycle) are counted, because at this level computer equipment is always set to shut down—the computer will "crash." Note that the average sag lasted less than 0.12 sec, and that 90% of the sags lasted less than 0.53 sec.

ENGINEER studies printout on recording power disturbance analyzer in a computer room. Printout records time, type, and magnitude of any disturbance beyond the limits set into analyzer. Note that the small unit can easily be located inconspicuously and left in place in a computer room without disturbing normal operations for extended periods of monitoring.

Predicting disturbances

The overall predictions for the Bell System, using the statistical methods discussed, are summarized in Table 2, which indicates the *maximum* number of *each type* of disturbance that a given percentage of sites can expect. For example, the table indicates that 10% of the locations will experience 10 or fewer sags (96 V or less) per year, while the average number of sags per year, the 50% point, is 25 or less. The table indicates that a certain percentage of the sites will not experience a specific type of disturbance in a year. The data emphasizes the rarity of some types of events and the prevalence of others. For example, more than 50% of the sites will not see a single surge in a year, while practically all (99.95%) of the sites will experience at least one sag per year. Similar tables, not shown, were developed to give more detailed prediction data for each individual type of disturbance, to help in selecting the most cost-effective power conditioning equipment. However, the data varied so widely from site to site that it cannot be applied directly to any individual site. By modifying the Polya distribution equations, a statistical model was developed that could be applied to an individual site, *based on data collected at that site.* This data must be collected over a minimum 9- to 12-month period to smooth out seasonal variations and to be sufficient for statistical validity. This information is shown in Table 3. It is based on the Polya equations and can be used for any of the types of disturbances discussed. Using the data collected at any location over the required period, the average number of events per month of the type of disturbance being studied is calculated. Referring to Table 3, the maximum number of events the site can expect at various levels of probability can be determined. For example, assume that three sags were recorded in a 12-month monitoring of a particular location. That is equivalent to 0.25 sag per month. Going to Table 3, we find 0.25 in the left column and, reading across, see that the number of events (in this case sags) will be less than three per year 50% of the time, or less than seven per year 90% of the time. Put another way, the site should expect to experience three or more sags in five out of ten years, or seven or more sags in one out of ten years.

Power conditioning

A variety of power-conditioning equipment is available to correct for disturbances. It ranges from relatively inexpensive line-voltage regulators to medium-cost motor-generator (M-G) sets, to comparatively costly uninterruptible power systems (UPS). For example, a simple line-voltage regulator could intercept 73% of all power disturbances, excluding voltage dips below 96 V, surges above 144 V, and all power failures. One type of M-G set could intercept 95% of the disturbances, missing only dips below 70 V and power failures longer than 100 ms. The UPS could intercept 98% of the problems, excluding only those power failures that last longer than the battery backup, usually 15 min. In some cases, the UPS can be backed up by engine-generators, which can provide power for hours, days, or even longer.

For any particular location, the data and method shown will permit an estimation of the frequency of various types of power disturbances at various probability levels. This, along with knowledge of the effect of each type of disturbance or the data-processing operation and the possible cost, can help select the most cost-effective power-conditioning method to be used.

---

*Adapted from a paper "The Quality of U.S. Commercial AC Power," by M. Goldstein and P.D. Speranza, delivered at the 1982 IEEE International Telecommunications Energy Conference (Paper 82CH1818-4, © IEEE 1982).

Table 3. Maximum number of disturbances expected in one year at a specific site

Average number of events per month	The maximum number of events to expect per year at the indicated probability level				
	10%	25%	50%	75%	90%
0.08	0	0	1	2	4
0.09	0	0	2	3	4
0.10	0	0	2	3	4
0.15	0	1	2	4	5
0.20	1	2	3	4	6
0.25	1	2	3	5	7
0.30	1	3	4	6	8
0.35	2	3	5	7	9
0.40	2	4	5	8	10
0.45	2	4	6	8	11

Voltage requirements for computer centers

Computers and other electronic equipment are sensitive to electric voltage variations. Con Edison of New York, one of the country's largest electric utility companies, has produced these guidelines to help users of such equipment deal effectively with voltage fluctuations.*

COMPUTER equipment will, in general, operate satisfactorily using normal electric utility service; but the capability of computer systems to handle momentary transients depends on their specific design.

Typical steady-state voltage requirements for computers are 208Y/120V, ±10%. However, there are many variations. One computer manufacturer's equipment reportedly operates satisfactorily with +10% to −8% voltage, while another's can handle only +6% to −8% of nominal. Some manufacturers say that problems could be caused by transient voltage spikes or dips greater than ±20% of the steady-state voltage for durations as short as 8 milliseconds (½ cycle.)

There are three ways to improve continuity of service: ensure good electrical plant design within the building, follow an effective equipment maintenance program, and provide adequate buffering equipment.

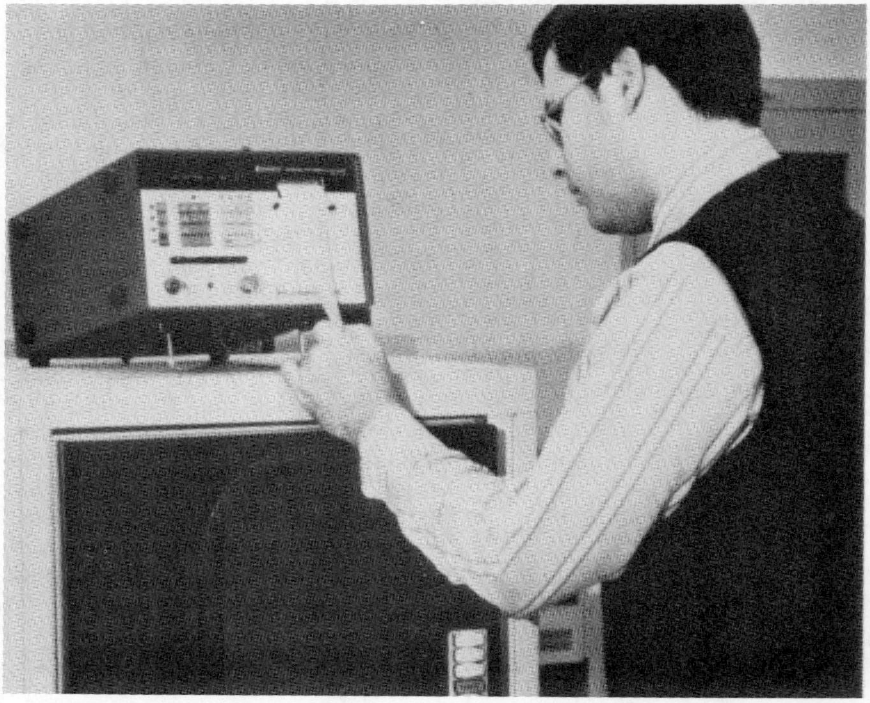

POWER-LINE MONITOR on computer power input records all disturbances to normal voltages. Recorder responds to both relatively slow sags and surges and very brief transient spikes and dips. Data on actual power problems is necessary before any decisions or solutions can be made.

Building design

Service interruptions of computer facilities can result from poor design of the building's internal electrical distribution system. If this electrical system is inadequate, cyclical loads from air conditioners, pumps, elevator fans, or other machines and equipment could cause voltage transients affecting the normal operation of the computers.

The following measures should be considered in designing internal electrical systems for buildings with computer operations.

Dedicated risers. For maximum system reliability, computer manufacturers recommend that computer power panels be supplied by dedicated risers that feed no other loads and contain no cables for other circuits. Electromechanical devices often found in computer facilities, such as card punching machines, accounting machines, and typewriters, produce transients and should not be on the same circuit as the computer. To minimize any voltage drop on the line supplying power to the computer, the dedicated riser should be designed with as low an impedance as possible.

Isolation transformers. Step-up or step-down transformers are frequently used on the line side of sensitive equipment and act as buffers to some voltage transients of short duration. Since computers are more tolerant of high- than of low-voltage transients, transformer taps should be used to maintain voltages at higher levels. For adequate response in overcoming normal voltage dips at reasonable cost, and to allow for future load growth, a transformer's initial loading should not exceed 50% of its nameplate kVA rating.

Balancing loads. To ensure better voltage regulation, it is important to balance the electrical loads in an installation among the 3-phase lines on all supply risers.

Equipment grounding. Proper equipment grounding reduces the susceptibility of the equipment to internally or externally generated electrical disturbances that could distort the supply voltage.

Design for outages. The electric plant should be designed so that whenever part of the electric supply equipment is out of service, the remainder of the installation will not be affected.

Equipment maintenance

Preventive maintenance of sensitive equipment and of the building power supply is essential for reliable operation. Maintenance of the computer electrical supply will reduce the number of outages significantly and also enhance the equipment's capability to operate acceptably during power-supply transients. Computer users should develop and implement a maintenance program that includes the identification and location of equipment to be examined, along with a schedule of how often each item of equipment must be inspected and maintained.

The maintenance program should be developed in consultation with the equipment manufacturer and the user's qualified technical representatives. Frequency of maintenance most likely will depend on the requirements of the sensitive equipment, the environment in the building, and the condition of the electrical wiring and connectors that supply power to the equipment.

Here are some important maintenance measures.

Vacuum cleaning of transformers and circuit breakers prevents the accumulation of dirt and dust that otherwise could cause increased resistance and a decrease in supply voltage.

Resistance checks and infrared scanning of joint connections on bus and cable risers should be performed frequently to detect loosened or overheating electrical connectors. Poor connections must be remade, or they will result in increased resistance, reduced supply voltage, and possibly complete failure from overheating.

High DC voltage resistance tests performed with a hand-driven generator are recommended for ground and short-circuit checks of the building power distribution system to discover possible insulation defects in cables, cable risers, panels, and transformers.

Frequent visual inspections of the main equipment and accessories are necessary to detect obvious defects, including burn marks or discoloration that may indicate equipment overheating and possible future service interruptions.

Average voltage variations on Con Edison's network systems

Steady-state voltage

Secondary system	Measurements made	Volts	
		Normal	Contingency
208Y/120V 3 phase, 4 wire	Phase to neutral Phase to phase	126-118 218-205	126-108 218-187
240/120V 1 phase, 3 wire	Phase to neutral Phase to phase	126-118 252-236	126-108 252-216
460Y/265V 3 phase, 4 wire	Phase to neutral Phase to phase	278-260 481-450	278-250 481-433

Transient voltage dips

Voltage dip		Duration	Frequency of occurrence
208Y/120V From 120V	460Y/265V From 265V		
To 90V (25%)	To 199V (25%)	6 cycles (10 ms)	Once or twice per month
To 90V (25%)	To 199V (25%)	30 cycles (500 ms)	Once or twice per year
To 55V (54%)	To 121V (54%)	12 cycles (500 ms)	Once or twice per year
To 0V (100%)	To 0V (100%)	12 cycles (200 ms)	Very rarely

THESE TABLES show voltage variations that may be expected under both normal and unusual conditions. Similar variations will be experienced on other utility systems, although each has its own characteristics. All utilities are subject to regulation by various state commissions, and all try to conform to ANSI standards on voltage variations.

Voltage spikes or dips may be caused by normal utility switching operations, cable or equipment faults, or the turning on or off of heavy customer loads.

NOTE: Service voltage is measured at the point where the supply lines connect to the customer's electrical lines (the point of service termination). An accurate voltmeter/recorder is usually used to measure and record this voltage as well as the extent and duration of any voltage transients. The recording unit should be recalibrated periodically to ensure proper and accurate performance.

Buffering equipment

Properly selected buffering devices between the source of power and the utilization equipment can reduce or eliminate the effects of transient voltage fluctuations on sensitive electronic equipment. Computer systems vary in their sensitivity to spikes or dips and in their ability to overcome voltage transients. Specific kinds of buffering equipment may be appropriate for some problems but not others. Computer users and their technical representatives should consult with equipment manufacturers to obtain recommended options for remedial actions.

Here are some types of buffering equipment, with their "ride through" capabilities.

Line-voltage regulator (LVR). This equipment maintains steady-state voltage within desired operating limits and permits riding through transient voltage disturbances of short duration.

LINE-VOLTAGE REGULATORS are available in many sizes and ratings, as needed. They provide a moderate-cost means of power conditioning where total outages are rare or can be tolerated. Often they are combined with surge suppressors, filters, or other equipment to increase effectiveness.

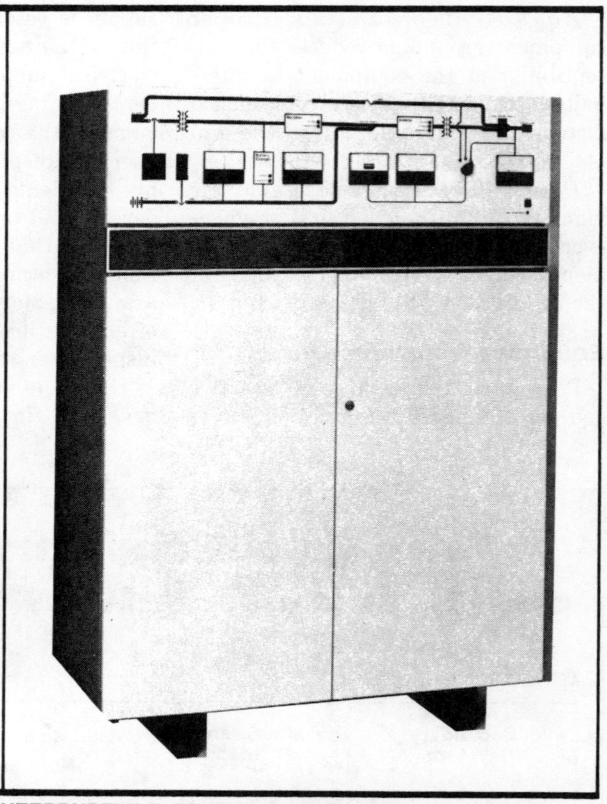

UNINTERRUPTIBLE POWER SUPPLY (UPS) can provide clean, continuous power at all times. When normal source is lost, batteries continue to provide power with no interruption. If backup generator comes on line before batteries are depleted, power is steady through almost any contingency.

LVRs have an average response time of 100 milliseconds for correcting voltage. LVRs should be used if the customer desires a less-costly alternative to modifying the building's internal electrical distribution system to take care of voltage dips.

Transient-voltage surge suppressors. These solid-state devices prevent transient disturbances (spikes) from penetrating the supply lines or the sensitive electrical equipment without reducing the line voltage below its steady-state value. Voltage surge suppressors generally act within 25 milliseconds.

Motor-generator sets. The motor of an M-G set draws power from the utility line to drive an alternator, which in turn supplies power to the computer. The energy stored in the flywheel provides power to the computer for several cycles or longer after loss of input power to the motor and helps override the time while the motor transfers to a standby generator. M-G sets protect computers very effectively from electrical transients in the building or the utility power system.

AC power synthesizers. These recently developed devices work in conjunction with compact "power centers" and static transfer switches. The power center and the synthesizer use utility power to generate an independent, 3-phase, 60-Hz output that is free of noise and other disturbances. If the normal supply circuit is interrupted, the static switch can transfer to an alternate supply circuit in 4 milliseconds (1/4 cycle), while the synthesizers can ride through about one full cycle. Since AC power synthesizers consist of dry-type transformers, inductors, and capacitors, they are virtually maintenance free.

Uninterruptible power supply. UPS units virtually guarantee continuous service and excellent-quality power under practically all conceivable operating conditions. They rely on batteries continuously charged from the utility line. An inverter converts the direct voltage of the batteries to an AC, 60-Hz supply for the computer. Depending on the capacity of the UPS and the equipment that it powers, the energy stored in the batteries may be capable of powering the computer for 15 minutes or more upon loss of utility service supply.

If the utility service is interrupted for a longer period, power may be obtained from a standby generator with a combustion engine that can be brought up to speed before the battery power supply is exhausted. The UPS is the most expensive of available buffering systems. The initial investment is large, much space is required for installation, and operation and maintenance costs are relatively high.

The decision as to which of the above types of buffering equipment is needed will depend on the overall effect of disruptions on the installation, the desired degree of service reliability, and the total cost of implementation. Generally, the more problems the equipment can overcome, the more costly it will be.

---

* Adapted from the booklet "Electric Power and the Computer," prepared by the Engineering and Customer Service Departments of Con Edison, New York City.

Power Line Problems

A comprehensive look at the application, function, and operation of UPS systems and components and how they relate to alternative protective measures and equipment.

AC line aberrations

COMMERCIAL electrical utilities provide a constant source of AC power that has a definite voltage (depending on the connections employed), a constant frequency of 60 Hz, and the capacity to fulfill total distribution demands. It is present in either single- or 3-phase form. Ideally, these characteristics would always remain constant. But in reality, they can consistently vary from their nominal value. These deviations in the commercial AC line, which usually take the form of either voltage aberrations, frequency aberrations, or noise, make the use of protective equipment such as the uninterruptible power supply desirable for certain sensitive loads.

Voltage aberrations

A voltage aberration refers to the variation in the incoming line voltage above or below its nominal value. For example: on a 120-V line, the voltage may vary from 130 to 100 V. As a result, the electrical or electronic devices connected to the line will experience an overvoltage or undervoltage condition. Typically, voltage aberrations are defined in terms of the duration of the event and can be defined as either long-term, intermediate, or transient voltage aberrations.

A *long-term voltage aberration* is an overvoltage or undervoltage condition that has a duration of 15 minutes or more. Usually, such aberrations are caused by one of three basic conditions: line droop, blackout, or brownout.

The most common reason for long-term voltage aberrations, line droop, describes the condition whereby there is a reduction in the voltage transmitted over a conductor as a result of the resistance of the conductor. Obviously, the longer the conductor, the greater the line droop.

In an effort to compensate for this reduction in voltage, the utility providing the commercial AC line power establishes a base voltage, a maximum voltage, and a minimum voltage. The base voltage refers to the desired load-center voltage; the maximum voltage refers to the maximum voltage that will be allowed for any customer at peak load; and the minimum voltage refers to the minimum voltage that will be allowed for any customer at peak load.

With these values established, voltage regulators and capacitors are arranged on the line to maintain the voltage levels as close to the base value as possible. As a result, the users close to the transformer (or distribution station) consistently experience an abnormally high voltage. Users at the end of the line consistently experience a somewhat lower voltage (because of the voltage drop over the distance). And only those in the middle of the distribution line experience the true rated voltage.

Rural areas are a good example of line-droop compensation. Here, houses are spread over a great distance. Only the houses at the center or middle of the commercial AC line experience a true 120 V. Houses closest to the distribution station may experience voltages as high as 125 V, while the houses farthest from the distribution station may experience voltages as low as 100 V.

The second type of long-term voltage aberration, the blackout, occurs when the line voltage goes to zero and remains there. Such conditions can be caused by lines going down in a storm, the utility's distribution transformers failing, a fault condition causing a breaker or fuse to trip, or any one of a number of other conditions.

While extended blackouts are fairly uncommon, the impact of a blackout on a given piece of electrical or electronic equipment can be devastating. This is because a blackout is characterized by the temporary loss of voltage, followed by a brief period where the voltage comes back up, and then a total loss of voltage. This fluctuation on the AC line can cause severe damage to any number of critical loads. In the case of a computer, even a momentary blackout can result in extended downtime. This is due to a number of factors, including restart of the computer and reloading

Fig. 1 Transverse noise (phase to neutral)

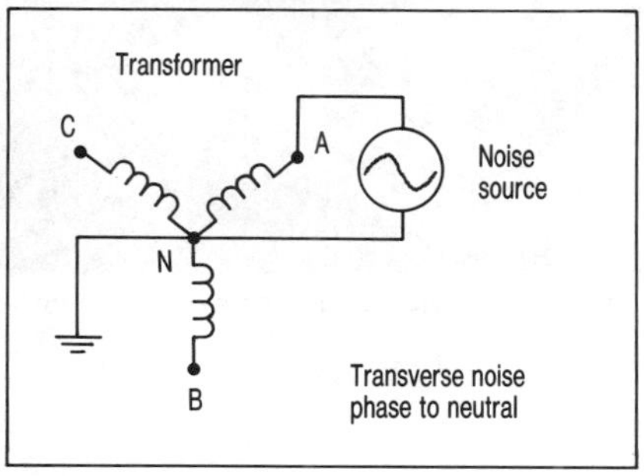

By **HOWARD H. BOBRY**

HOWARD BOBRY is the engineering manager for LorTec Power Systems, Inc., North Ridgeville, OH. A registered professional engineer, he received a BSEE from Case Western Reserve University and an MBA from Baldwin Wallace in 1980.

CRITICAL LOADS such as extensive word- and data-processing installations have become increasingly dependent upon stable power.

of data. With the lack of redundancy in feeders to most consumers of commercial AC power, blackouts will continue to be a potentially dangerous voltage aberration.

The final type of long-term voltage aberration is a brownout. A brownout is a condition whereby the utility reduces the available voltage on the commercial AC line in a given area. On some occasions, when the power company is reaching the limits of its capacity to produce power, there will be an intentional brownout to reduce the power demand in a particular area. Unintentional brownouts, caused by a failure of some equipment at the utility's distribution points, are also quite common.

Typically, most brownouts go unnoticed. This is due to the fact that the only evidence of such an event is a slight dimming of the lights in the area. Yet brownouts are quite common. While some loads require constant amounts of power, regardless of the voltage, and will continue to operate as long as it is within an acceptable range, brownouts can have a significant impact on many other types of electrical and electronic devices. Equipment such as an induction motor will run considerably hotter under such low-voltage conditions. As a result, there is a tendency for such devices to fail prematurely.

An *intermediate voltage aberration* is a voltage drop that has a duration of a few cycles to a few minutes. Typically, this type of voltage aberration can be caused by a user on the line starting or stopping a large electrical motor, the power company changing transformer taps, or a fault condition on the line. Load-related intermediate voltage aberrations are by far the most common. Such events can result in both undervoltage and overvoltage conditions on the line. For example, when a large electrical or electronic device is started—say a large motor, an elevator, or an air compressor—it may require an initially large surge of power. During that period of time, the line voltage sags for several cycles and then returns to normal.

Similarly, when that same large load is later turned off, it can result in an intermediate overvoltage condition on the AC power line. When a large load is suddenly removed from the line, the power-line inductance will try to maintain a constant current flow by generating a transient voltage. This voltage increase may, depending on proximity, be of sufficient magnitude to damage some types of equipment.

The switching process involved when the utility company changes transformer taps to vary the output voltage can result in an intermediate undervoltage or overvoltage condition until the circuit is able to stabilize.

When a fault condition is present on the commercial AC line, there usually is an unwanted, abnormally heavy load on the line. It can be caused by a simple short circuit, a downed power line, or a failure in a building's wiring. The line voltage will tend to drop almost to zero until the fault clears. At that point, the voltage rises, and due to the inductance of the power line, a transient voltage

Fig. 2 Common-mode noise (phase or neutral to ground)

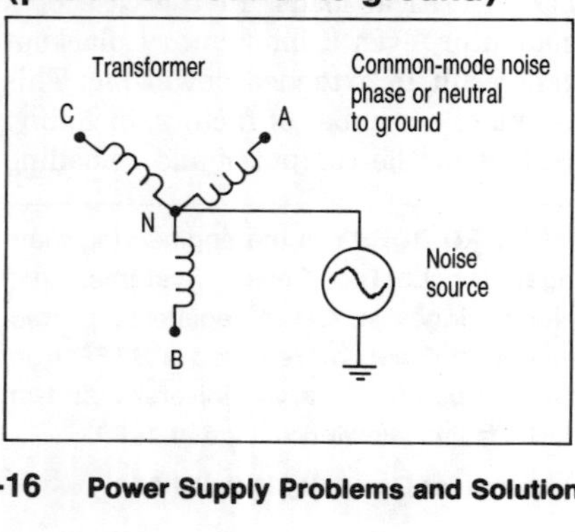

increase takes place. This rapid fluctuation of the line voltage can prove particularly damaging to many types of sensitive electrical and electronic equipment.

Short-term voltage aberrations are known as transients and usually have a duration of only a few cycles. Many electrical and electronic devices involve a great deal of switching as part of their normal operation. For example, a universal motor, used in such equipment as vacuum cleaners and drills, is constantly switching the AC to feed different windings in its rotor. As the commutator rotates, brushes constantly make and break the connection. The making and breaking of the connection in an electric drill is obvious by the sparks generated. The electrical activity generated by this action appears on the line as transients. And, depending on the proximity of the source, such transients can be very disruptive to other loads.

Another common cause of transient voltage aberrations is lightning. Devices such as lightning arresters or surge suppressors may be adequate to limit the transients on the line to a "safe level" in the context of preventing gross physical damage to a distribution system or structure. Such devices cannot, however, be realistically expected to protect sensitive electronic devices from the very fast high-voltage transients caused by lightning.

In such cases, you may see a voltage spike of tens of thousands of volts, with a duration of only several milliseconds. Realistically, there is no time for a surge suppressor to act and, as a result, the transient voltage remains on the commercial AC line with potentially devastating impact on most electrical and electronic devices.

Frequency aberrations

Frequency aberrations occur when the line frequency maintained by the power company (usually between 59.98 and 60.02 Hz in the U.S.) varies.

To achieve the desired frequency over the course of a given period, say 12 hours, the utility may slow down or speed up the frequency of the AC line voltage. In such a situation, the utility may generate a lower frequency for a given period and then catch up by increasing the frequency above 60 Hz for a short interval. While this results in generation of an average of 60 Hz for the period, the resulting variations can have a negative impact on loads.

On occasion, the power company will, of necessity, switch distribution paths, so an area normally supplied by substation A is temporarily supplied by substation B. If the two substations are not precisely in phase when this switch occurs, it can result in some switching transients that show up as a momentary frequency aberration.

More commonly, frequency aberrations are a problem in facilities where secondary power services, such as diesel-engine generators or turbines, are used. Usually found in applications where an emergency power source is critical, secondary power sources can be very difficult to maintain at a constant 60-Hz frequency. As a result, electrical and electronic devices operating on such secondary power sources can be disrupted by high, low, or varying frequency.

Noise

Noise is the final aberration to be considered. Basically, noise can be defined as unwanted electrical signals that are superimposed on a useful waveform. As such, noise can be understood as a repetitive voltage or frequency transient. Essentially, there are two different types of noise: transverse noise and common-mode noise.

Transverse, or normal-mode noise, is the voltage noise that always exists between a pair of conductors. In a 3-phase connection, common-mode noise is a noise voltage that appears equally in phase with each line to ground (see Fig. 1).

Common-mode noise is usually caused either by electrostatic induction where (with equal capacitance between each of the wires and surroundings) the noise voltages and currents will be the same on each of the wires, or electromagnetic induction where (with magnetic fields linking each wire equally) the noise developed in each wire will be the same (see Fig. 2).

In practical applications, a perfectly balanced common-mode noise is rare. Typically, the noise in a given circuit contains a component of common-mode noise and another component of transverse noise.

Critical loads

SINCE UPS systems are usually applied to installations involving critical loads, it is important to understand what constitutes a critical load and the impact of line aberrations on the operation of such a load.

Critical loads are electrical or electronic devices that either are employed in an application that prohibits disruption or are inherently sensitive to aberrations on the commercial AC line.

In some applications, the function the load is performing may be essential to the profit-making capability of an organization, the health or life of individuals, or the safety and security of a given facility. The load is deemed critical because of the negative effects of the load's failure to operate in terms of profit and loss, life and death, safety and danger, or necessity and convenience.

Specific examples of critical loads fulfilling these criteria include data-processing equipment, where downtime results in lost revenues; process control equipment, such as computer-controlled machine tools, where failure can result in damage to the equipment or product; the emergency room or surgical suites in a hospital, where constant monitoring and lighting are critical; or protective devices, such as fire-alarm systems, security lighting, or security alarms. Some electrical and electronic equipment, such as computers, can be classified as critical loads under both categories.

The impact of power aberrations on

these devices can be measured in terms of downtime/recovery time, memory loss, and internal component or equipment damage.

Downtime / recovery time

Downtime refers to the amount of time the critical load is inoperative because of the line aberration; recovery time is the amount of time required to restore the critical load to the operational point maintained immediately before it went down. Both downtime and recovery time can vary anywhere from a few milliseconds to several hours or days. In many instances, downtime can be virtually nonexistent, with recovery time requiring anywhere from a few hours to a few days. The reverse—with downtime being extensive, as with a prolonged blackout, and recovery time being minimal—is also common.

An example of an installation where the impact of common line aberrations can greatly affect a critical load is an airlines reservations facility utilizing a computerized automatic call distributor (ACD). The ACD normally functions to receive incoming calls at the facility and directs them to available reservation agents on a priority basis in the order in which they are received.

In the event of blackout or brownout on the commercial AC line, the ACD is not operational, and the reservations facility can no longer receive or process incoming calls. In addition to the inconvenience to the customers calling in, the facility is losing reservations revenues for the duration of the outage. When coupled with the expense of paying phone operators for essentially nonproductive time, the impact of ACD downtime due to an extended or even limited power outage is significant in terms of lost dollars.

Similarly, many telephone companies use computers for the monitoring and billing of long-distance toll calls. The computer systems are powered by an incoming commercial AC line, while the phone system itself operates off a separate DC power system. As a result, in the event of a failure on the commercial AC line, the telephone customers are still able to make long-distance calls, but the phone company is incapable of completing the monitoring and billing procedure. The phone company estimates that in one office alone they can lose as much as one million dollars a day when their computers are down. By simple calculation, then, a power outage in one small area can cost the phone company as much as $694 a minute in lost billings.

As another example, consider a prison or other security installation. Many such facilities use low-pressure-sodium lighting to illuminate their perimeters. When these lights are denied full voltage for only a few seconds, they shut down and are unable to regain their rated brilliance for 10 to 20 minutes. Thus, in such an installation, the recovery time for even a momentary power outage can have a severe impact on the security of the facility.

Memory loss

Memory loss is a particular problem in applications involving computers. Basically, a computer operates due to the interrelated function of hardware (including memory) and software. Software is the program of instructions that tell the hardware what to do with the information maintained in the computer's memory.

A variety of memory types are employed in the average computer. By the nature of its design, each type can be classified as being either volatile or nonvolatile. A volatile memory is characterized by the fact that it depends upon electric power to maintain its data. A nonvolatile memory simply means that it is hard or fixed and does not depend on the presence of electrical power for its existence.

A *semiconductor memory* consists of logic chips that contain both the software and the immediate input data that the computer is working on. Due to its two distinct applications, a semiconductor memory can be either volatile or nonvolatile.

In its nonvolatile form it usually contains the master, or executive program of the computer. This is the basic software that defines the functional parameters within which the computer hardware can operate. A semiconductor memory in this form is usually referred to as a ROM, or read-only memory. This memory cannot be altered or changed without replacing the logic chips themselves. Power aberrations generally have no effect on this type of memory. It is important to note, however, that it can be destroyed by transients conducted through the power supply.

On the other hand, in its volatile form, the semiconductor memory contains the specific programs, or software, that control the functions the computer can perform within the parameters established by the executive program. In addition, the specific input data that will be processed by the program is held in this type of memory, as are intermediate and final

EXAMPLES of critical loads include computers and peripherals (far left), electronic editing or word-processing equipment (left), and electronic switching control of railroad rolling stock (above).

results. This form of semiconductor memory is known as a RAM, or random-access memory. It is totally variable, depending upon the electrical impulses present at any given instant.

A computer can maintain only a finite number of "bits" in its RAM memory at any one time. The size of the memory is usually indicated by the thousands (the letter K means one thousand) of bits it can hold, nominally. Thus, 4K means that 4096 bits of data can be stored in the computer's volatile memory; 8K means 8192 bits; 16K means 16,384 bits—and so on.

Since the quantity of individual programs and actual input data that a user might want to process in a computer is virtually infinite, the RAM memory has evolved to where it only stores the program and input data that the computer is working on at any one instant. Additional programs and long-term input data are maintained in nonvolatile form on magnetic tape or disks. Such memories function to store the digital impulses corresponding to the data in the RAM semiconductor memory of the computer and require some type of external tape reader or disk drive for input and output (write and read). These devices provide a method of safe, long-term storage for programs and input data. They are not damaged by an aberration on the commercial power line. They can, however, be damaged during the period when data is being transferred to or from the primary semiconductor memory by interruptions such as are caused by a power outage.

To examine the impact of power aberrations on a semiconductor RAM memory, assume that in a typical manufacturing facility a computer is being used to generate a running parts inventory. The input is based on the bills of material used on the assembly floor, and the manufacturing data stems from daily part production runs.

During the course of the day, the individual parts manufactured for inventory and the parts utilized in the assembly process are periodically entered into the system's memory. At the end of the day, all this information is transferred to disk for hard storage. If a power outage of long or intermediate duration occurs at any time during the course of the day, the computer will shut down, erasing the data stored in the volatile RAM memory. As a result, when the power is restored, the data entered into the system to that point must be totally reconstructed. This can result in significant recovery time since the system must be restored to the condition it had reached before it shut down.

In the event of a transient power aberration, the impact is just as significant. Here the aberration can be of such a short duration that the computer does not have time to react. The transient is thus allowed into the system with the potential to erase, or merely alter, any number of individual bits of data stored in the system's memory. If the system is designed to sense such a problem, it may shut down and instruct the operators to verify all data that was stored in volatile memory to that point in time. Again, the result is a significant and costly recovery time.

But what of the system that is not designed to sense the transient event? It merely goes on operating as if nothing had occurred. It may be days or even weeks before the manufacturing process is stopped because the computer maintained an inaccurate parts count in inventory (say 1000 instead of 100) and failed to order additional production of a particular piece when it should have.

Now consider the same event in a hospital, where the computer is functioning to record clinical information regarding patient status and treatment. Assume that a doctor orders 1 cc of a particular medication for a given patient and, as a result of a line transient, the computer printout at the nurses' station calls for 10 cc. The possible devastating results can be imagined.

While power aberrations are not a problem as far as individual bits of data in a nonvolatile memory are concerned, they can be a significant problem in terms of the disk drive, the unit that reads and writes the digital impulses recorded on the disk. If there is a power aberration of any significant duration while the data is being initially transferred to the nonvolatile mediums, the data in process at that time can be lost. This can be explained simply by looking at a word processor. As information is entered into the system's memory, it appears on the CRT. As the screen is filled, portions of the data are constantly being transferred to a floppy disc for long-term storage. Should an outage or power failure occur, the information on the CRT screen at the instant would be lost. The data entered onto the disk prior to and after the event remains intact; only that which was in process at the time would be lost.

The same basic concept can be applied to larger computer systems, where the data stored in the computer memory regarding programs and specific data is periodically dumped onto a hard disc or magnetic tape. Should a power outage occur during this process, the data being transferred at that instant is lost. Considering the speed of today's disc drives, this could represent a significant amount of data.

Damage to equipment

The last impact of common power aberrations on a critical load involves internal component or equipment damage. Significant damage can be suffered by internal components of the critical load or by parts and/or services involved in a production process.

Most electrical and electronic components are given a power rating. Should this rating be exceeded (such as by an overvoltage transient), they can fail prematurely, resulting in increased downtime and maintenance costs. In a typical computer, where the component could be an integral part of a printed-circuit board, the cost of replacement could run into the thousands of dollars.

Similarly, electric motors and similar devices are rated to operate at a specific current. Should that current increase significantly as a result of a severe undervoltage condition, the device can overheat and burn itself out. Again, the result is both increased downtime and increased maintenance costs.

Consider again the disk-drive unit commonly used for long-term storage of computer data. Basically, a disk, whether rigid or floppy, is used to store digital impulses transferred from the computer. This process of storage can be likened to a common phonograph record. The disk is covered with a magnetic-oxide coating. As it rotates, either a single head or an array of heads is used to read and write data on the disk.

In a system employing floppy disks, the heads are in constant contact with the surface of the disk. But in a hard, or rigid-disc system, the heads are designed to float on a thin cushion of air above the surface of the disk. Care is taken to make certain that the head never comes in contact with the working area on the disk surface. Such contact can result in damage to the data maintained on the disc.

Normally, when a hard-disk system is shut down, the head quickly moves to a dedicated area of the disk, devoid of data, and comes to rest. In the event of a sudden power failure, however, the disk drive can slow down so quicky that the head does not have time to move to the dedicated resting area. In such a case, the head contacts the surface of the disk in an action known as a "head crash." As a result, the disk can sustain irreparable damage (similar to scratching a phonograph record), destroying the data stored in that area.

There are a variety of process control applications where thousands of dollars worth of chemicals undergo a series of chemical changes in the production of a specific product. If the computer controlling the process suffers a momentary transient, the entire batch of chemicals could be destroyed. Similarly, many machine processes today are computer-controlled, allowing precise control of the metal-cutting or -forming processes, far beyond the skill of today's worker. But a momentary transient carried into the computer memory through the commercial AC line can alter the data in the program. As a result, as the numerically controlled machine tool moves through a specific machining process, the tooling can impact the work piece and cause damage to the work, the fixturing, and even the machine tool itself. The cost, in terms of downtime/recovery time, increased service and maintenance, scrap, and rework can be immeasurable.

Power protection methods

An uninterruptible power supply completely isolates a critical load from aberrations on the commercial AC line. However, a number of other electrical/electronic devices are available and used for the purpose. The extent to which they succeed will be discussed.

Dedicated lines

A dedicated line is a special circuit run from a distribution point to the critical load. The line is dedicated in the sense that the critical load is the only electrical or electronic device using that line as its voltage source. The dedicated line can initiate from a distribution point within a facility or from the distribution point serving a particular area, such as an industrial park.

Functionally, a dedicated line offers very little isolation from the critical load, since it is fed by the same transformer that feeds all the loads within the facility. As a result, it serves only to eliminate a limited amount of noise and transients generated from the other loads in the facility. It will not isolate the critical load from long-term and intermediate voltage aberrations (blackouts or brownouts), frequency aberrations, or noise resulting from sources external to the facility.

Transients or noise of great magnitude generated within the facility (such as those generated by welders, elevators, and large air conditioners) can work their way through the transformer and into the dedicated line. While some filtering of such events will be accomplished, results, in terms of effective isolation, are marginal.

Dual feeders

A second power protection approach to isolate a critical load from aberrations is the use of dual feeders. With a dual-feeder system, the facility housing the critical load obtains its AC line power from two separate generating stations. The separate lines are brought into the facility and connected to a transfer switch.

The transfer switch is designed to sense continuously the quality of the power provided by Station A. When the source voltage or frequency drops below a nominal level, the load is quickly transferred to Station B. Station B continues to provide a source of constant power until Station A is back within an acceptable range. At that point, the load is transferred back to Station A.

There are some definite advantages in the use of dual feeders, particularly in terms of long-term and intermediate-voltage aberrations such as blackouts and brownouts. But where transient-voltage aberrations, many frequency aberrations, and noise are involved, the dual-feeder system proves totally ineffective. In addition, depending on the type of transfer switch operating between the two input lines, the transfer itself can result in an interruption of power to the critical load. Such an interruption can be of sufficient duration to knock a sensitive computer system off-line. The transfer switch itself represents a potential single point of failure, which could defeat the whole purpose of dual feeders. Thus, similar to the dedicated-line solution discussed earlier, the dual-feeder method proves totally unacceptable in terms of isolating a critical load from the totality of aberrations commonly present on the commercial AC line.

ENGINE GENERATORS provide an alternate source of power during long-time outages of the normal source. Disadvantages are possible lengthy start-up times, high cost if run on-line, and difficult voltage and frequency regulation.

Shielded isolation transformers

A third power-protection method commonly employed involves the use of a shielded isolation transformer. Basically, it consists of a primary winding and a secondary winding electrically and electrostatically isolated from each other. Electrical isolation is provided by the transformer design, which relies on magnetic coupling to transfer the power from the primary, through the transformer core, to the secondary. Electrostatic isolation is provided by an aluminum or copper shield that is grounded and positioned between the primary and secondary windings.

As a result of this isolation, high-frequency transients of an extremely short duration are unable to travel from the primary to the secondary and thus to the load. Unfortunately, the protection provided is limited to transients with a duration of tens of microseconds. Such devices offer absolutely no protection from voltage fluctuations, brownouts, or blackouts. Even transients of a slightly reduced duration, those at a frequency close to that of the power line, will get through the shielding, because the shielding is incapable of discriminating between what is desired and what is potentially damaging to the critical load.

In practical applications, shielded isolation transformers are referred to as isolation transformers, shielded isolation transformers, high isolation transformers, and ultra-high isolation transformers. These distinctions refer to the degree of isolation provided by the shielding. Functionally, shielded isolation transformers offer only very-limited protection for a critical load, since high-frequency transients are only one small portion of the power aberrations commonly present on the commercial AC line.

Line-voltage regulators

While there are a variety of line-voltage regulators on the market, they all share one factor in common. In each instance, the regulator takes the power provided by the commercial AC line and steps it either up or down to provide the desired output voltage.

A line-voltage regulator is suitable in applications where the AC is reliable

Block diagram, typical line-voltage regulator

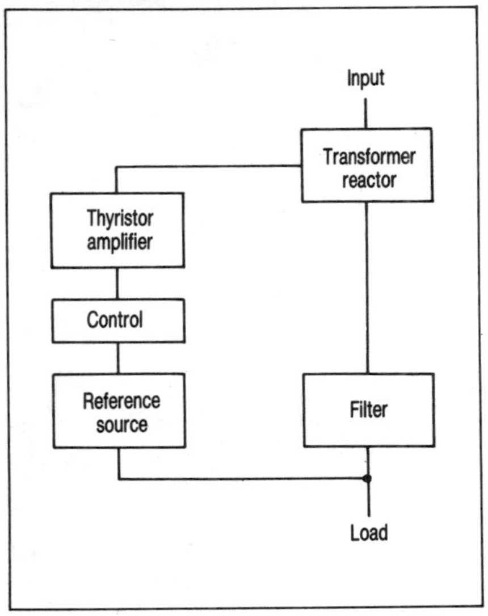

A

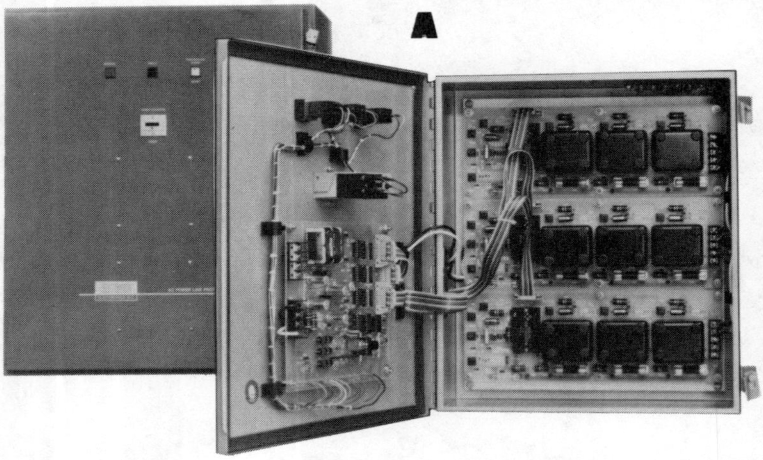

POWER PROTECTION METHODS that isolate critical loads from one or more line aberrations include (photo A) a power line protector designed to provide protection even if damaged by a very close power-line lightning strike; (photo B) a power conditioner that uses peak-limiting circuits, RC circuits, dynamic clipping circuit, and linear LC filters to attenuate transient noise; (photo C) a modular distribution system that serves three separate computer loads from a single input. (Such power distribution centers may also include a shielded isolation transformer or line-voltage regulator.) Line conditioners (photo D) stabilize incoming voltage and attenuate both common- and transvere-mode electromagnetic noise.

B

C

but of poor quality. Depending on the power demand of the installation, two different types of regulators are commonly employed.

For applications requiring limited power, the most common type of line-voltage regulator is a constant-voltage transformer, or a ferroresonant transformer. Such devices rely on the saturation of iron to maintain a constant voltage on the secondary. In addition, the ferroresonant transformer functions as an excellent filter for eliminating transient voltage aberrations. In some cases, these devices are even capable of providing limited ride-through in the event of a blackout. (Ride-through is defined as the amount

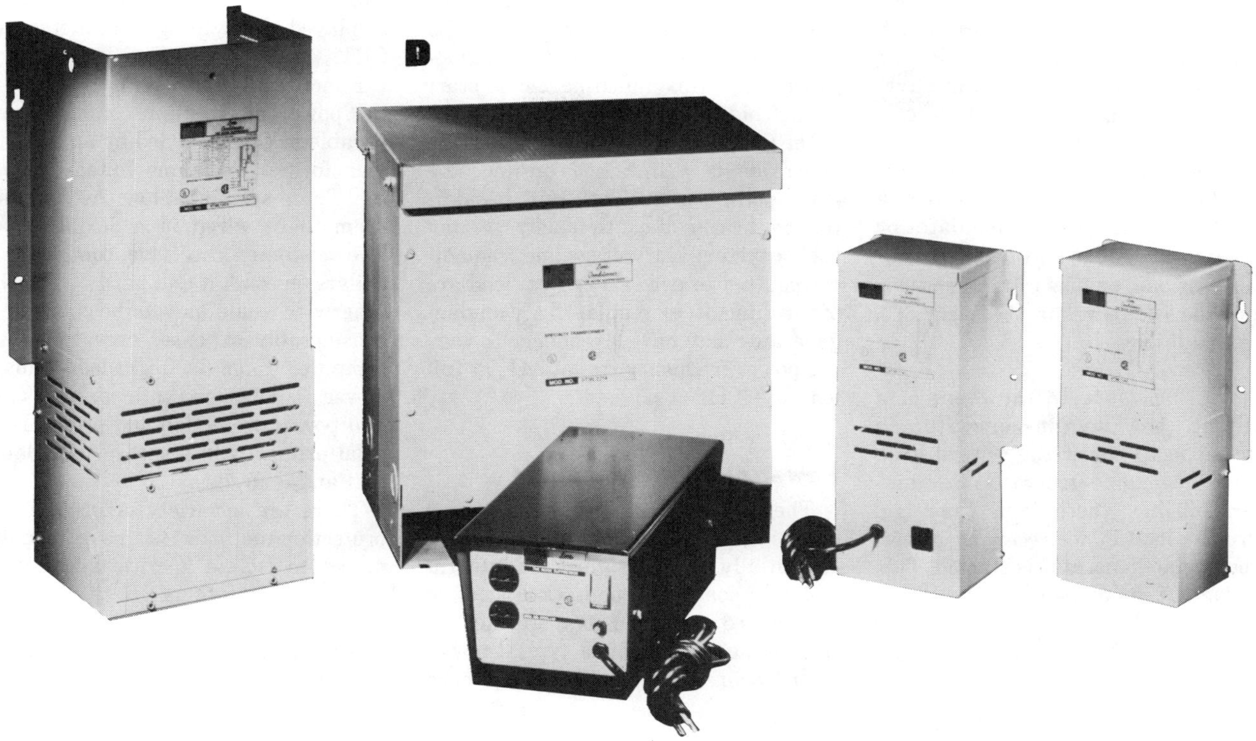

of time that a usable voltage can be provided without a prime source.) The ride-through provided by a ferroresonant transformer is limited to a duration of between one-quarter and one full cycle.

In applications requiring larger amounts of power, a device that automatically changes transformer taps is often used. Such devices are designed to maintain a constant output voltage by changing the taps on the transformer as needed. While these devices are capable of maintaining a fairly constant output voltage, they are totally ineffective in terms of transient-voltage aberrations or blackouts because they cannot react quickly enough to transients, and there is no machanism or ability to provide automatic ride-through.

Line-voltage regulators are available with claims of varying degrees of power protection, from standard line-voltage units—intended for noncritical applications, such as lighting—to computer-grade regulators. However, all fall short of addressing the total power problem. Neither the smaller nor the larger units provide adequate protection in the event of a blackout, and they totally ignore potential frequency or noise aberrations.

Power-line conditioners

A power-protection method commonly employed to isolate critical load from aberrations on a commercial AC line is the power-line conditioner, which is essentially a combination of one or more of the devices previously discussed. While they are often marketed as uninterruptible power conditioners, they do not provide the degree of power protection afforded by an uninterruptible power system. All of the devices employed, such as line-voltage regulators, have inherent weaknesses to a portion of the potential line aberrations. As a result, most power-line conditioners are little more than glorified line-voltage regulators with the same vulnerability to long-term blackouts, frequency aberrations, and noise.

Motor-generators

A motor-generator (M/G) consists of an AC motor connected to the incoming AC line and turning an AC generator that supplies the critical load. This type of rotating device is a viable alternative to uninterruptible power systems in applications where long-term blackouts or severe brownouts are not a problem.

The mechanical coupling between the motor and generator, usually through a shaft arrangement, provides a substantial degree of isolation. As a result, the majority of transient and short-term voltage aberrations are effectively eliminated. And, due to the nature of a rotating device, motor-generators provide varying degrees of ride-through, anywhere from 5 to 10 cycles.

But they also have inherent disadvantages. They are rotating devices, and service and maintenance are significant problems. Secondly, in applications where long-term blackouts or severe brownouts are a problem, an M/G set is totally nonfunctional because of its limited ride-through capability. If the outage is of a longer duration, power to the critical load will be lost.

Engine-generators

An engine-generator is a diesel engine (or gas turbine) and an AC generator that functions as a power source totally independent of the commercial AC line. Such devices, which can be maintained either on-line or off-line, function to switch in an alternative source of power in the event of a long-term voltage aberration.

By definition, an engine-generator does absolutely nothing to protect a critical load from intermediate or transient voltage aberrations, frequency aberrations, or noise. Its purpose is to provide AC power in the event of a total line failure.

Off-line systems: In the event of a blackout, the engine-generator is switched in to power the critical load. Where the engine-generator is maintained off-line, there is a time lag, where the load is not receiving power, as the engine-generator is started. This time lag can be anywhere from a few seconds to 10 or more minutes. Its duration depends on two factors: the generator starting system employed (which can be electric, hydraulic or pneumatic), and the environmental conditions present (since the engine-generator will be much harder to start in the cold).

On-line systems: Where the engine-generator is maintained on-line, the startup time is avoided. But, in addition to the cost of continuously running such a system, the voltage and frequency required to power a critical load are often difficult to regulate. An engine-generator may have a tendency to vary the power frequency from 57 Hz at full load to 63 Hz at no load.

Power distribution center

The power distribution center is receiving a great deal of attention. Essentially, it is designed for integration in a computer room. Consisting of a shielded isolation transformer or line-voltage regulator, a typical power distribution center will be equipped with various line-condition monitors displayed on anything from meters to a CRT.

Functionally, protection provided by the power distribution center is comparable to that afforded by an isolation transformer or a line-voltage regulator. These systems allow the computer room to be wired in a flexible mode. The customer can save thousands of dollars on conduit and hard wiring that otherwise would have to be changed, at considerable expense, every time the computer room is upgraded. Thus, a power distribution center provides limited power protection and proves beneficial primarily in terms of installation cost and upgrades.

There are a variety of other power protection methods that have been promoted to isolate a critical load from common aberrations on the commercial AC line or to reduce their effect; however, none is designed to address the totality of the problem.

Computer power— problems and solutions

How "clean" must computer power be? How "dirty" is the system power? What can be done economically to reconcile the two?

IN THE past, we could ignore variations in electrical system power supply that were less extreme than severe sustained undervoltage or overvoltage or lightning-strike-level transients. Today, with the advent of computers and data processors, as well as many other semiconductor-controlled devices, many previously disregarded power disturbances pose a threat to the proper operation of equipment. Any power aberration, no matter how brief, transitory, or small that can cause errors in data handling or shut down a computer, is a serious problem. In a study for the U.S. Navy, Thomas S. Key classified typical power disturbances. Key's investigation (*EC&M*, September 1978, pp. 66-70) contained the classification table reprinted in Fig. 1, which also presents illustrations of the waveform of each type of disturbance.

Fig. 1. Types and characteristics of power-line disturbances

Definition	Type I Transient and oscillatory overvoltage	Type II Momentary undervoltage or overvoltage	Type III Outage
Causes	Lightning; power network switching (particularly large capacitors or inductors); operation of on-site loads	Faults on power system; large load changes; utility equipment malfunctions; on-site load changes	Faults on power system; unacceptable load changes; utility or on-site equipment malfunctions
Threshold level*	200 to 400% rated rms voltage or higher (peak instantaneous above or below rated rms)	Below 80 to 85% and above 110% of rated rms voltage	Below 80 to 85% rated rms voltage
Duration	Spikes 0.5 to 200 microsecs wide and oscillatory up to 16.7 millisecs at frequencies of 0.2 to 5 kHz and higher	From 4 to 60 cycles, depending on type of power system and on-site distribution	From 2 to 60 sec if correction is automatic; unlimited if manual
Duration (cycles of 60-Hz wave)	0	0.5	120

* Approximate limits beyond which disturbance is considered harmful to computer.

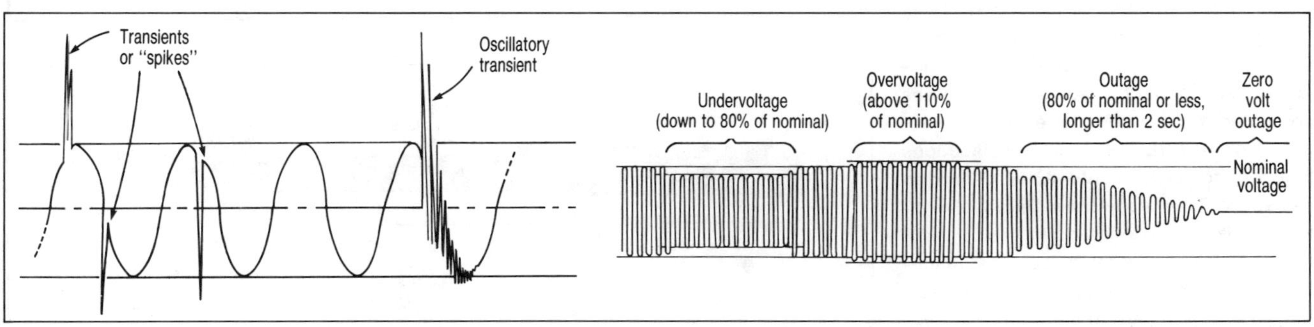

Industry standards

There are standards set by the utility industry that govern the allowable

By **ALBERT KESTERSON**
and **PAT MAHER**

Fig. 2. Utility power profile and CBEMA criteria

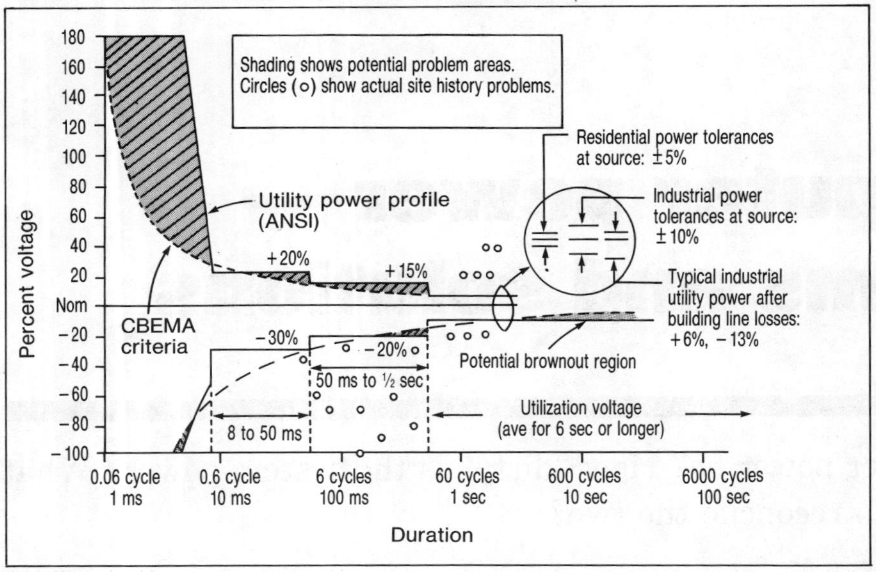

voltage tolerances for power as delivered to the customer. This utility power profile is specified in ANSI Standard C84.1. The Computer Business Equipment Manufacturers Association (CBEMA) has standardized on voltage limits within which computers and similar equipment manufactured by CBEMA members should operate reliably. Voltages outside these limits could cause erratic, unreliable, or incorrect operation. The ANSI utility power profile, with the CBEMA voltage "envelope" superimposed, is shown in Fig. 2.

Note, in Fig. 2, that ANSI Standards for steady-state voltage tolerances for the utility at the source are ± 5% for residential power and ± 10% for industrial power. After allowances for voltage drop up to and on the customer's premises, CBEMA steady-state tolerances for industrial power are +6%, −13%. These superimposed computer tolerance and utility profile curves make it clear that there are potential problems in the momentary outage area and in the high-voltage transient area in power from the utility, in addition to any customer-caused transient problems. In periods of brownout, more frequent every year, when the utility reduces the voltage level to reduce load that is approaching utility capacity, voltages at the computers will probably be below the acceptable tolerance shown.

Actual site voltage measurements of incoming utility power using rapid-response instruments (the small circles in Fig. 2) show that the utilities, in spite of their best efforts, often do not, and probably cannot, maintain the specified ANSI voltage tolerances at all times. Utilities show numerous brief instances of both above-tolerance and below-tolerance voltage deviations when monitored by equipment capable of recording short-time excursions.

Fluctuations of computer power

Most, but not all, computer equipment can accept the +6%, −13% tolerance, but some computer manufacturers call for even more stringent voltage regulation. One manufacturer calls for ±4%. Typical transient tolerances permit no dips or surges greater than about ±20% for no longer than 30 ms, with briefer transients capable of creating errors. Most computers will

---

ALBERT KESTERSON is Director of Power Engineering for Leibert/Conditioned Power Corp., Columbus, OH. He received an Electronic Engineering degree from Franklin University, Columbus, OH, has taught semiconductor technology, and has presented many papers to technical societies. **PAT MAHER,** Power Specifications Engineer for Liebert, has a BS degree in Industrial Technology from Ohio University, Athens, OH. He has done design work for NASA and has written technical manuals for the Armed Services and for civilian electronic equipment manufacturers.

Fig. 3 Allen-Segall data

Disturbance	Ave monthly occurrences	Percent	Notes
Oscillatory (decaying) transients	62.6	48.8	< 16 ms (1 cycle) duration and > 15% of system voltage
Voltage spikes	50.7	39.5	> 15% of system voltage
Undervoltage	14.4	11.2	> 8 ms (½ cycle) duration and > 10% of system voltage
Overvoltage	None	– –	> 8 ms (½ cycle) duration and > 10% of system voltage
Total outage (blackout)	0.6	0.5	Mean outage 11 minutes
	126.3	**100.0**	

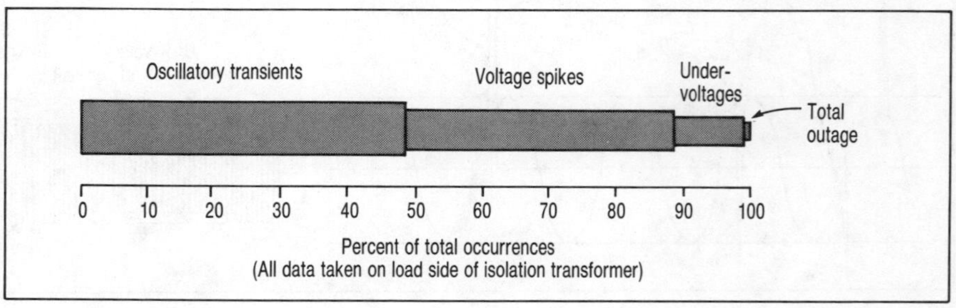

POWER ENHANCER has doors open to show interior. A shielded isolation transformer and transient suppression provide excellent noise and spike suppression. Tap switching provides wide-range voltage regulation. The enhancement section on the right uses microprocessors for monitoring and control; the distribution section on the left will be customized with shielded cables for specific computer equipment.

not tolerate a loss of voltage for longer than about 15 ms.

Excessive on-site voltage drops and, more frequently today in peak-load periods, utility brownouts, can reduce steady-state voltages below acceptable limits. Computers, to be utilized efficiently, are often run around the clock. Late at night, when loads are low both for the power company and within the facility, it is not uncommon for steady-state voltages to rise above the acceptable limits. Transient voltages, from utility switching and large-load switching at the customer site, can exceed the transient-voltage tolerances, and momentary outages exceeding the 15-ms level can occur from utility power network switching and lightning-strike circuit-breaker tripping and reclosing. Any number of utility or on-site failures can result in loss of power. About the only requirement that rarely causes problems is the frequency tolerance, ±0.5 Hz. The power delivered to the computer is usually within this frequency range.

Input power variations beyond acceptable limits can cause errors in calculations, output errors, loss of data, unscheduled shutdowns (computer "crashing"), and even equipment damage. Recovery of data, reruns, computer downtime, customer dissatisfaction, and repairs to equipment can all be extremely costly.

In early 1974 George W. Allen and Donald Segall, both of IBM, published an IEEE paper, "Monitoring of Computer Installations for Power Line Disturbances." Their study is one of the most extensive records of actual distur-

MAGNETIC SYNTHESIZER with covers removed exposes pulse transformers, inductors, and capacitors that generate a 12-step, 60-Hz output independent of and isolated from power input. Filters smooth output into a close approximation of a true sine wave of the desired voltage, the same as or different from the input voltage.

Fig. 4. Application of power conditioners

Equipment	E (enhancer) or S (synthesizer)	Oscillatory transients	Voltage spikes	Over and undervoltage	Outages
Spike suppressor, filtered	E	Good, if filtered	Excellent	No effect	No effect
Regulator with spike suppressor	E	Partial good with filter	Excellent	Excellent	No effect
Isolation xfmr, shielded	E	See Note 1	No effect	No effect	No effect
Isolation xfmr, shielded, with spike suppressor	E	See Note 1	Excellent	No effect	No effect
Isolation xfmr with spike suppressor and voltage regulator	E	See Note 1	Excellent	Excellent	No effect
Motor-alternator set	S	Excellent	Excellent	Excellent (See Note 2)	
Magnetic synthesizer	S	Excellent	Excellent	Excellent (See Note 2)	
Uninterruptible power supply	S	Excellent	Excellent	Excellent	

Note 1: Little effect on normal-mode (phase-to-phase) transients; excellent suppression of common-mode (phase-to-ground) transients with properly grounded shield.

Note 2: Excellent down to a minimum voltage, below which the synchronous motor will drop out and output will be lost; or the magnetic synthesizer will not be able to sustain output voltage. Synthesizer minimum voltage is usually lower than motor-alternator minimum for a given load.

bances to power systems supplying computers, encompassing both utility source and on-site generated voltage excursions, including spikes, oscillatory transients, overvoltages, and undervoltages. The monitoring was done from June 1970 to July 1972, accumulating about 109 monitor-months of data on sites representing a climatic and geographical cross-section of the U.S., with representative loads from heavy industry, light industry, office buildings, retail stores, residences, and mixed locations. Utility systems included overhead and underground, single- and dual-feeder distribution. Experience since then indicates that the relative frequency of each type of incident—its proportion to the total—will probably be the same as that monitored in the early 1970s, and that the total number of occurrences is probably higher today than when the study was made. The results of the Allen-Segall study are summarized in Fig. 3.

It is obvious that the incoming power available to computers is not compatible with many critical computer system requirements, and that an interface system is needed. This interface must be capable of taking incoming raw power and delivering to the computer power that has been corrected for incoming high- or low-voltage conditions of short or long duration and cleaned of transients or other disturbances that can damage the computer equipment or affect the data it is processing. The interface should also, if possible, provide continuous power even in the event of an incoming power outage.

Power conditioning

There are several basic technologies for accomplishing the required interface, changing or conditioning the raw incoming power before delivering it to the computer. These technologies fall into two main categories—power enhancement and power synthesis.

Power enhancement equipment only modifies and improves the incoming power waveform, by clipping, filtering, isolating, increasing or decreasing the voltage and the like, before delivering the power to the computer. *Power synthesis* equipment, on the other hand, utilizes the incoming power merely as an energy source, from which it creates a new, completely isolated power output waveform to supply to the computer. This synthesized or regenerated waveform is engineered to fall entirely within the CBEMA (or any other) computer tolerance requirements, with an adequate margin for safety, regardless of the disturbances to the incoming power source.

Power enhancement can be provided by transient (spike) suppressors, voltage regulators, isolating transformers, or a combination of some or all of these. For example, an isolating transformer is frequently used in conjunction with a voltage regulator and transient suppressor.

Power synthesis can be provided by static electronic semiconductor inverters, rotating machines, and, more recently, static magnetic synthesizers not utilizing semiconductors. The static electronic semiconductor inverters, which are powered by DC obtained from rectifying the normal power source and have a "floating" battery in the DC stage to provide power during brief total outages, comprise the familiar uninterruptible power supply (UPS). Rotating machines are usually a synchronous motor, powered from the normal power source, driving an alternator (a motor-alternator set, often called a motor-generator or MG set). The magnetic synthesizer uses pulse transformers, inductors, and capacitors to create the desired AC output wave. The synthesizer input power is not merely improved or conditioned; it is used as an energy source for creating a new 3-phase power output free of disturbances.

Evaluating the problem

What is the most desirable method of conditioning the incoming power? First, the clean power requirements of the computer or other equipment and the quality of the incoming power—the severity and extent of the problem—must be determined accurately. Next, the effectiveness of each type of power conditioner to solve the problems must be examined. Finally, the cost and the *cost-effectiveness* of each technology must be assessed.

The computer requirements, in terms of clean input power, must be obtained from the manufacturer. In most cases, these requirements will conform to CBEMA standards, although some equipment may permit greater tolerance and some may require more-rigid input-power control. The urgency of preventing computer errors and crashes must be evaluated. The quality of the input power to be expected can be determined from past experience, from monitoring the power at the computer location, using instruments capable of recording transients as well as voltage variations, and from such data as the Allen-Segall study. Together, the computer power needs, the necessity for continuous, accurate computer operation, and the imperfections in available power will define the extent of power-conditioning required. In evaluating computer needs, long-range plans for expansion must be considered. Computer installations tend to grow larger rapidly. Clean power must be available; expansion of the power system must be possible for future needs; and possible decreases in quality of utility power must, based on recent experience, be anticipated.

UPS INSTALLATION supplies 125 kVA of clean power for an airlines reservation center computer. Rectifier-inverter units, against wall, provide synthesized isolated pure output power. Batteries supply uninterrupted power to inverter if incoming power is lost. Ventilation and separation have prevented any problems from fumes, although batteries are in the same room with rectifier-inverter UPS equipment.

The effectiveness of each type of power conditioner is difficult to determine accurately. Manufacturers' claims tend to be overly optimistic. Analysis of the functioning of each type of power conditioner helps. For example, a transient suppressor obviously will not improve a low-voltage condition, but a voltage regulator will, to a degree depending on the design of the regulator. The effectiveness of various power conditioners on specific problems is shown in Fig. 4. A study was made using the Allen-Segall data as representative of the relative frequency of occurrence of various types of faults, analyzing the effectiveness of each type of conditioner upon each type of fault. It is not stated in the Allen-Segall paper, but in conversations with the authors it was learned that *all their measurements were made on the load side of an isolation transformer.* Therefore, if the results of an isolation transformer used for improving incoming power were measured against the Allen-Segall data, it would seem to be ineffective, or have zero effectiveness, since the data base was obtained *after* an isolation transformer in the circuit. An isolation transformer is about 25% effective in power-conditioning, and the results of the Allen-Segall data base were modified to what they would be if no isolation transformer had been used in gathering the data. The results of this study, estimating the effectiveness of various types of power-conditioning equipment on incoming power with the distribution of problems as determined by Allen and Segall, corrected for the isolation transformer, are shown in Fig. 5.

Fig. 5. Power problem-solving effectivity

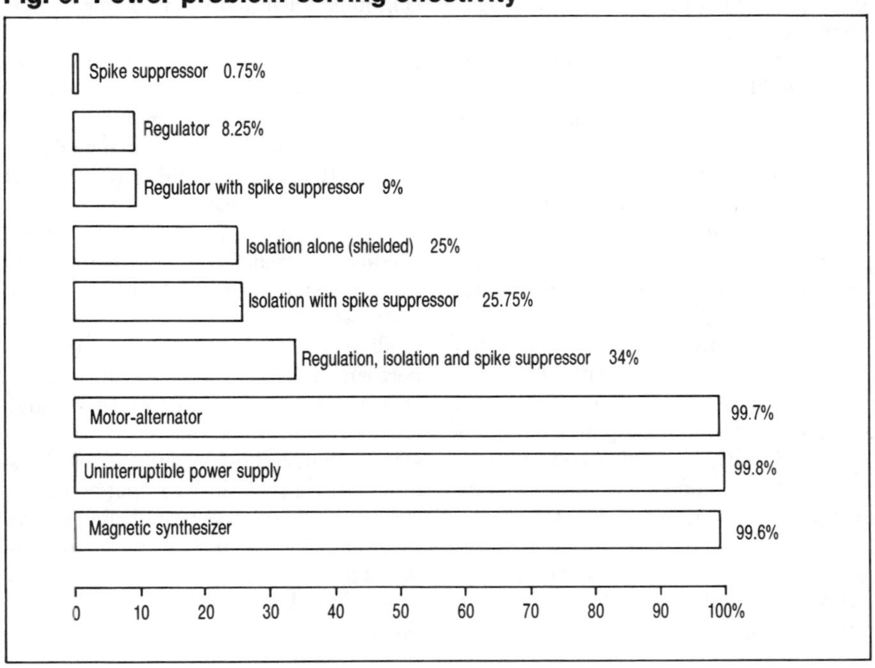

Cost and cost-effectiveness

The effectiveness comparison is only one step in selection of the proper power-conditioning technology. Equipment cost is a major consideration, varying from the inexpensive spike suppressor to the much more costly UPS. Cost comparisons must take into account not only first cost, but also annual cost, including depreciation, maintenance,

Fig. 6. Power protection costs

	Spike suppressor	Regulator	Regulator with spike suppressor	Shielded isolation xfmr	Isolation xfmr with spike suppressor	Isolation xfmr with regulator and spike suppressor	Motor-alternator set	Uninter-ruptible power supply	Magnetic synthesizer
Installation cost	$3,000	$25,000	$26,000	$4,500	$6,000	$28,000	$40,000	$130,000	$25,000
Maintenance cost	--	$ 1,000	$ 1,100	$ 100	$ 100	$ 1,000	$ 1,000	$ 8,000	$ 100
Operating efficiency	--	96%	96%	97%	97%	93%	83%	85%	91%
Energy cost/year	--	$ 1,400	$ 1,400	$1,050	$1,050	$ 2,800	$ 6,000	$ 5,300	$ 3,500
Total cost per year	$ 600	$ 7,400	$ 7,700	$2,050	$2,350	$ 9,400	$15,000	$ 39,300	$ 8,600
Total effectiveness	0.75%	8.25%	9%	25%	25.75%	34%	99.7%	99.8%	99.6%
Relative cost	7%	86%	90%	24%	27%	110%	174%	457%	100%
PCI*	11	10	10	105	94	31	57	22	100

Basis: 100-kVA output
Continuous (24-hr) operation
$0.05 per kWh cost of power
5-year straight-line depreciation

*Performance-cost index (PCI) = $\frac{\text{\% total effectiveness}}{\text{\% relative cost}}$

and energy costs resulting from less-than-100% efficiency. Such annual costs provide realistic comparison of the cost of ownership of different conditioning technologies. Representative costs are shown in Fig. 6, based on recent equipment costs, 5-year straight-line depreciation, and energy at 5¢/kWh for equipment rated 100 kVA operating 24 hrs per day.

Obviously, these figures must be adjusted for different sizes, depreciation and operating conditions, but they serve as one means of comparing technologies. The lowest-cost power synthesizer, the magnetic synthesizer, was arbitrarily chosen as the base of 100%, and the relative costs of the other technologies are compared in Fig. 6. While actual costs will vary, the *relative* costs of various power conditioners should be similar to those shown.

The data in Fig. 6 is based on full-load operation of the 100-kVA equipment. Realistically, normal loads will be between 50% and 80% of rating for most equipment. When the total continuous load rises much above 80%, consideration must be given to increasing the capacity of the equipment. However, most manufacturers publish efficiency data at full load, and efficiencies at lesser loads are often difficult to obtain. Since the *percentage* of losses often increases rapidly with decreasing load, and the losses are a measure of the cost of operating the equipment, actual losses at realistic loads should be used whenever possible when comparing equipment.

Using the performance, or total effectiveness against all power problems as shown in Fig. 5, a performance-cost index is shown in Fig. 6. This index, rounded off to the nearest whole number, indicates the relative *cost-effectiveness of each technology.*

Selecting the conditioner

It is necessary to consider all three factors—effectiveness or performance, initial and annual operating cost, and performance-cost index—when selecting equipment to be purchased. If *performance* alone were considered, a UPS with battery ride-through of outages would be the choice. If *cost* alone were considered, the transient spike suppressor would be the choice. However, it solves so few problems by itself that it is rarely used except when combined with other systems. If the *performance-cost index* alone were considered, an isolation transformer would seem to be the best choice.

A determining factor in the selection is the system requirement. The only fully effective technology against a complete and sustained power outage is the UPS, with an engine-generator to supply power before the batteries run down. If the cost of computer downtime is so high and the computer operation is so critical that the rare occurrence of a total power outage must be protected against, the UPS is the only acceptable solution, in spite of the fact that the cost is several times that of any other solution. However, if the small risk of a complete sustained outage can be taken, then either a motor-alternator set or a magnetic synthesizer can be used, as both will be effective against all other power problems. The magnetic synthesizer will be lower in cost and more cost-effective than the M-A set. If two incoming sources of power are available from different utility substations, so that it is highly unlikely that both sources will be out at any given time, a magnetic synthesizer combined with a static transfer switch may be as effective as a UPS, at much lower cost. (See "Reliable computer power—an unusual solution," January, 1982 *EC&M*, p. 64.) A motor-alternator set may also be used, at lower cost than a UPS, but at a higher cost than a magnetic synthesizer.

It is obvious that selecting the power conditioner is not a simple black-and-white decision. Foremost, the computer installation must get power of the quality and reliability it requires to perform its functions. Within these requirements, the most-effective and lowest-cost systems can be used. A cost saving that results in outage or errors in a critical, real-time computer operation cannot be tolerated; conversely, investing in a UPS system for an important but noncritical computer is probably not justified. No two installations have exactly the same needs.

Effective computer grounding

Conflict between computer requirements and safety requirements when grounding computers is unnecessary. A safe power distribution system and proper computer operation can be compatible.

ELECTRIC power-distribution systems for computer rooms are usually designed by an electrical consultant and installed by an electrical contractor to insure maximum reliability. Grounding provisions provide personnel safety equal to or exceeding the requirements of the NEC. The computers and data-processing equipment are usually designed and installed by computer manufacturing companies or their contractors, in cooperation with the user-purchaser. The computer-room installation often includes a raised floor, with power to computers, peripheral equipment, interconnections, computer cooling-air distribution, and a fire-extinguishing system all located under the floor. This is where the usual interfaces are accomplished and where problems frequently occur between the computer installation and the electrical distribution system powering the computers.

COMPUTER INSTALLATION shows the usual tangle of power and communication cables and other devices under the raised floor. Note the bolted stringers between floor pedestals, permitting the floor support grid to be used as an excellent high-frequency reference ground.

Computer engineers have serious problems with electronic "noise" that affects their sensitive systems. They have developed methods of computer grounding to minimize this noise, sometimes isolating the computer equipment grounding system from the rest of the building grounding system. Unfortunately, some of these methods are violations of the NEC grounding requirements and may be unsafe.

In the past, and to a lesser extent even today, computers have been looked upon as magic "black boxes" and their systems engineers as witch doctors who understand the mysteries. Sometimes, when these computer people demanded grounding systems that violated the NEC, insisting that "It won't work any other way," the installation was made as they requested. There have even been instances where electrical inspectors, prepared to cite the installations for violations, have been convinced that the violations were necessary for computer operation and accepted them. There are other cases where inspectors, unfamiliar with computer rooms, just didn't detect the violations.

By ARTHUR FREUND

Two grounding systems

To prevent electrical noise from affecting computers, *two entirely different and separate grounding systems are required.* First, the power distribution system must be grounded in accordance with NEC requirements for safety, and in addition the 60-Hz power system grounding must be so designed that it will minimize noise pickup that can affect the computer operation. Second, the computer equipment and enclosures must be connected to a reference grounding system for *high-frequency* noise. This high-frequency protection is *in no way related* to system grounding! In fact, the high-frequency reference will work whether or not it is grounded to the system ground. However, since it consists of exposed, metallic noncurrent-carrying parts that could accidentally become energized, it *must* be grounded, as required by NEC Sec. 250-42.

Once we recognize that proper computer grounding requires both a 60-Hz and a high-frequency grounding system, and that electrical safety requires that the two systems be bonded together, we can proceed to examine each system separately and then consider how to bond them so they are compatible.

High-frequency grounding

Computers input, process, and output data at tremendous rates, which are constantly increasing. Typical computer clocks that control the data transfer into, within, and out of the computer operate at about 10 to 30 MHz. Even clocks of personal (micro) computers run at from 1 to 7 MHz. Some special high-speed computers operate in the range of several GHz—several thousand MHz. These are radio frequencies (RF), and computers are sensitive to these frequencies. The wiring can act as a receiving antenna and can respond to external RF signals that may cause false data to be processed by the computer. Computers process data in square-wave pulses with rapid rise times, and these pulses plus the square-wave harmonics have frequencies up to 100 MHz or more. They may also generate RF signals that can radiate over the wiring, which acts as a transmitting antenna. In fact, the FCC has recently established maximum permissible signal radiation levels from computers, and manufacturers are finding it difficult in many cases to meet these standards. The RF signals that can radiate from or get into computers are known as electromagnetic interference, or EMI.

Computer microprocessors operate at what, to us in the power industry, are very low voltages—usually 5 to 12 V. Therefore, it is critical that unintentional voltage differences between various units of data-processing equipment in a computer room be extremely small. At 60 Hz, this is relatively simple—just tie them together with a low-resistance, low-impedance grounding conductor so the enclosures and grounded metal are all at the same (ground) reference potential.

At radio frequencies the solution is not so simple. Low-impedance grounds are not easy to obtain, since the inductive reactance of a conductor is proportional to the frequency. At 30 MHz, a length of conductor has an inductive reactance 500,000 times the reactance at 60 Hz. In addition, there is stray inductance and capacitance from conductor to conductor or conductor to grounded metal and resonance effects at high frequencies, making it very difficult to get a conductor of any appreciable length to have the same voltage at both its ends. If there is a difference in potential between the ends of a grounding conductor connecting two pieces of data processing equipment, it is possible for data errors to occur.

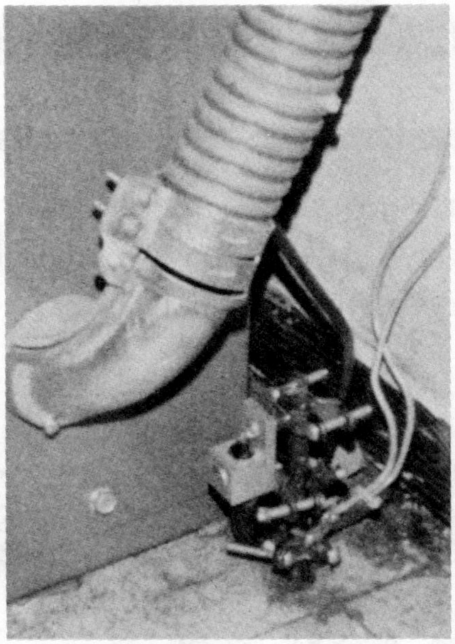

DRIVEN GROUND ROD in computer room is used as an isolated grounding electrode, not connected to the distribution system ground. Connected to it are a mess of conductors—some green, some white, and even a telephone company ground. This is unsafe and, in addition, more likely a source of noise than a cure for it.

A high-frequency impulse applied to a conductor travels along the conductor at a finite speed, about 85% of the speed of light, until it reaches the end of the conductor. It is reflected back along the conductor, becoming a traveling wave. At some frequencies, the reflected waves meet and reinforce oncoming new waves, creating resonances and standing waves. At or near these resonant frequencies the conductor appears to have an extremely high impedance and does not provide an effective voltage equalizing means between two pieces of equipment. In addition, the conductor can act as an antenna at or near these resonant frequencies, radiating energy that can interfere with other equipment or receiving stray signals from other sources and presenting a false voltage signal to the computer equipment.

These effects are completely unpredictable, because the interfering signals are not steady, and because data-processing equipment sensitivity varies. Digital computers operate on binary signals, ON or OFF, or 0 or 1. The circuits are most sensitive at the moment they are switching from one state to the other; and if an impulse occurs at that moment, a false "bit" of data may occur in the system. At other times, the same impulse may not affect the processing. These errors can be extremely difficult to identify.

This points out the need to interconnect all cabinets of the system in such a way that they are at the same potential for all frequencies, from 60 Hz or less to very high radio frequencies. One of the best ways of accomplishing this is the signal reference grid. Transmission-line testing has shown that standing waves will not cause a significant voltage difference between the two ends of a conductor that is not longer than about 1/10 to 1/20 of a wavelength. The wavelength at 10 MHz is about 100 ft, and 1/20 wavelength is 5 ft. The wavelength at 30 MHz is about 32 ft, and 1/20 wavelength is a little under 2 ft.

If conductors are connected in a mesh or grid to form a multitude of low-impedance loops in parallel, there will be little voltage difference between any two points in the grid at all frequencies from 60 Hz up to a frequency where the length of one side of one square represents about 1/10 wavelength. A grid made up of 2-ft squares will at any point provide an effective equipotential ground reference point for signals up to more than 30 MHz. If such a grid is installed in the computer room and the enclosure or frame of each piece of data-processing equipment is connected to it by a short conductor, there should be no computer noise problems resulting from differences in ground potential between the frames of any two pieces of equipment (see Fig. 1).

A raised floor in the computer room can be used as the reference grid if of the proper type and design. Most raised floors consist of vertical stanchions or corner supports 2 ft apart that hold up the 2-ft-sq floor tiles. If the floor is the type that has stringers between the stanchions and the stringers are positively mechanically bolted to them to make a good electrical connection (one installer tests each joint and requires 50 microohms resistance or less), the stringers will form an excellent 2-ft-sq mesh. If the floor tiles or panels are metal-clad, they will further reduce the impedance between loops. This is probably the best and least-costly way to obtain the high-frequency reference ground grid.

The frames of the data-processing equipment should be connected to the grid by the shortest possible leads—with braided copper the preferred material, followed by copper strap, with round conductor least desirable, because braid provides the most-effective and round conductor the least-effective results with high-frequency signals. Connections from the equipment frames to the raised-floor grid can easily be kept to 8 in.

If the bolted-stringer type of floor is not installed, a grid can be created by laying a mesh on the true floor, under the raised floor. This mesh also can be of copper wire, (least desirable) flat copper strap, or braided copper strap (most desirable), electrically connected every 2 ft to make a grid of 2-ft squares. Since this grid is on the true floor, it is about 2 ft farther from the equipment enclosures and requires a longer connecting strap from enclosure to grid, reducing its effectiveness a little.

Additional protection against false signals is obtained by using shielded cables to interconnect equipment or using filters and traps to keep unwanted signals out of the system. High-frequency noise is a difficult and persistent problem with no simple, standard solution; but a good, effective reference grid will minimize the difficulty. (Computer people tell of the engineer, plagued by noise problems, who finally went berserk and jumped from a 12th-floor window—thus making his first, last, and only reliable ground connection.)

Fig. 1. Grounding impedance—effect of reference grid

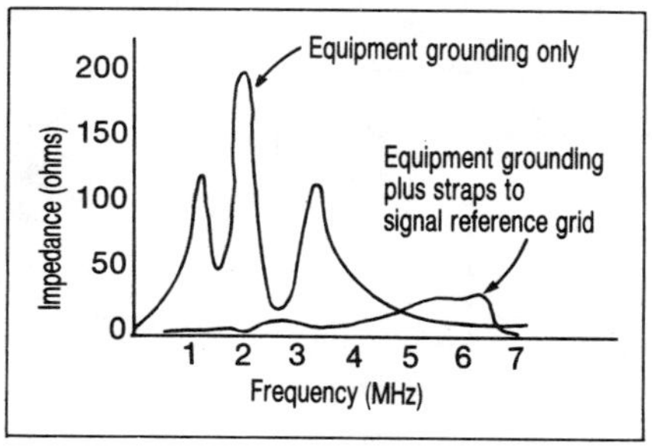

Recent developments in data transmission over optical fibers will make such fiber-optic systems the preferred means of equipment intercommunication in the future. This will go far to alleviate the EMI and "noise" problems, since optical fibers do not respond to magnetic fields or radio-frequency signals.

Power system grounding

The grounding of computer equipment is done in accordance with NEC Article 645, Data Processing Systems, which is only one page in length. Sec. 645-4, Grounding, consists of only one sentence, which states, "All exposed noncurrent-carrying metal parts of a data-processing system shall be grounded in accordance with Article 250." Article 645 starts with the statement, "For further information, see Standard for the Protection of Electronic Computer/Data Processing Equipment, NFPA 75-1976 (ANSI)." NFPA 75, Sec. 7-2.2, merely requires that all wiring shall conform to the NEC, and in its Appendix B reprints the entire NEC Article 645. 'Round and 'round we go and we stop at NEC Article 250.

Computer equipment grounding, like all other electrical equipment grounding, has only one purpose—*safety*. The grounding performs two functions. First, it ties all equipment together with low-impedance conductors so all equipment is at the same potential. It also bonds that system to the grounding-electrode system so that the potential is that of the earth. If the grounding is done in accordance with the NEC, there should not be an *unsafe* difference in potential between any two pieces of exposed noncurrent-carrying metal or from any noncurrent-carrying metal to ground, even under fault conditions. Second, the equipment grounding provides a safe, low-impedance return path for ground-fault currents to permit overcurrent devices to clear the fault quickly and thus minimize damage.

Where the power system supplying the computer has a grounded conductor (which is usually the case, since most computers utilize a 208 Y/120-V power supply with a grounded neutral), that conductor must be grounded in accordance with Article 250. This is frequently a source of confusion and conflict when the computer is supplied from a separately derived source such as a dedicated transformer or a UPS or MG set. The NEC requirements are intended to ensure a *safe* installation and are not concerned with whether or not the system works properly so long as it is safe. It is important to keep this in mind. The introduction to the code, Article 90, begins as follows (italics are ours):

90-1. Purpose.

(a) Practical Safeguarding. The purpose of this Code is the practical safeguarding of persons and property from hazards arising from the use of electricity.

TERMINAL BOX contains an isolated ground receptacle (right side of box) which, properly connected, can reduce computer noise. Here, green wire (arrow) is taped to the outside of the power cable—a code violation, unsafe, and noisy as well.

(b) Adequacy. This Code contains provisions considered necessary for safety. *Compliance therewith and proper maintenance will result in an installation essentially free from hazard, but not necessarily efficient, convenient, or adequate for good service or future expansion of electrical use.*

(c) Intention. *This code is not intended as a design specification* nor an instruction manual for untrained persons.

Therefore, it is necessary to conform to the code and at the same time design and install a computer grounding system that will permit the computers to operate with a minimum of noise problems and data-handling errors.

In a mistaken effort to isolate the computer system from the rest of the building power system, many systems have been grounded to a ground rod or buried ring around the building but not connected to the building grounding system and grounding electrode. This isolation has often been achieved by using many makeshift devices to keep the computer equipment insulated from the metallic parts of the distribution system.

Power-supply conduits have been insulated from computer cabinets with PVC conduit fittings or even plumbing fittings; conduit runs have been ended before reaching the computer cabinet and the conductors run the remaining distance exposed; green-insulated grounding conductors have been run separately from the circuit conductors and to a separate ground point; ground pins have been cut off plugs and the green wire left floating; flexible cords, without grounds, have been used to supply the computer units, and so on.

All these schemes are unsafe, because there is no low-impedance return path for fault current and no assurance of equal potential between the equipment and other grounded metal. They violate many sections and every intent of the NEC. Finally, they often result in more computer noise than proper grounding would produce, because the multiple ground points can have differences in potential between them, causing circulating ground currents and false data inputs or processing.

Proper computer grounding

As earlier stated, computer grounding must be done in accordance with NEC Article 250 and consists of computer-equipment grounding and computer power-system grounding.

Fig. 2. Computer equipment grounding

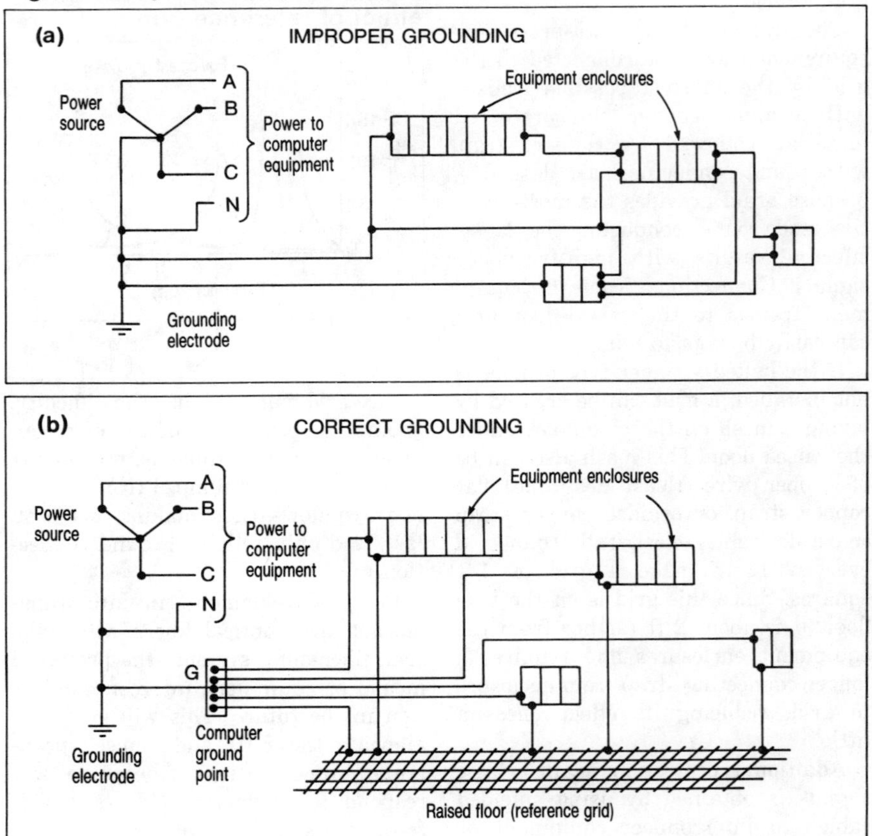

NOTES: All ground leads from computer ground point (except to reference grid) must be run in same cable or raceway with power conductors. All connections from equipment to reference grid are to be kept as short as possible.

All equipment grounding essentially requires that exposed noncurrent-carrying metal parts be connected by equipment grounding conductors to the grounding electrode at the electric service equipment, or at the power source if power is obtained from a separately derived system. The equipment grounding conductors must run with or enclose the circuit conductors and may be one or more or a combination of the types listed in NEC Sec. 250-91. These include a bare or insulated conductor; rigid or intermediate metal conduit or electrical metallic tubing; flexible metal conduit approved for the purpose; the armor of Type AC armored cable; cable trays; and other raceways specifically approved for grounding. An insulated grounding conductor must have an outer finish that is green or green with a yellow stripe. The grounding path must be permanent and continuous, and conductors must be sized and installed in full compliance with the many other applicable code requirements.

Computer equipment grounding will comply with the code and will be safe if it meets these requirements. However, several additional provisions are necessary if the computer installation is to operate with minimum noise problems from the grounding system. If computer equipment were installed in the same way as most other electrical equipment, the grounding might be somewhat like that shown schematically in Fig. 2(a). The grounding conductors are run with the power conductors from the source to each piece of equipment, but there are also many pieces of equipment fed from other equipment, and many other metallic interconnections between the grounded metal enclosures of various units. This can be perfectly safe, and in accordance with the code for equipment grounding, but the multiple points of grounding can have slight differences in potential, causing small ground currents to flow—especially under transient conditions. These ground currents appear as noise to the data-processing equipment and can cause errors in computation or component failures. This arrangement produces "ground loops"—multiple paths for the circulation of ground currents.

By modifying the grounding as shown schematically in Fig. 2(b), these ground loops can be eliminated. Each piece of equipment is fed separately and radially from the source and is grounded (by means of the metallic raceway, green-wire ground, and any other grounding conductors) to a single computer ground point (G) at the source or distribution point. This radial grounding eliminates any ground loops. In addition, each piece of equipment is connected by the shortest possible connection to the reference grid. While these connections seem to create multiple ground loops, as explained earlier, all points on this reference grid are at the same potential, not only at the power frequency of 60 Hz, but also at much higher frequencies, so no currents will circulate. One point on the reference grid must also be tied to the computer ground point to put it at the same ground potential as all other equipment.

A radial installation made as shown to a single ground point can comply in all ways with the NEC, can be entirely safe, and at the same time can be as free as possible from computer noise problems originating from the grounding system.

Grounding electrodes

Another point of confusion in computer grounding is the selection of the proper grounding electrode to which the computer grounding system must be connected. A frequent error of computer manufacturers is to request a grounding electrode separate from and not connected to the power system electrode—a separate ground rod or ground ring. This is not only unsafe and a code violation; because of differing ground potentials, it can also be a source of, rather than a cure for, computer noise problems.

The proper grounding electrode is clearly spelled out in Article 250 of the code. Where the building service is the source of supply, the grounding electrode must be one or more of those types listed in Sec. 250-81, to which the grounded circuit conductor (neutral) must also be connected. There is little confusion on this point. However, large computer installations are frequently supplied from a separately derived system rather than directly from the building service. This is done to isolate the computer supply from the other building power systems, and often to condition the computer power to make it more reliable and compatible with computer requirements.

The separately derived system may be a transformer, a computer power center, a motor-generator set, a synthesizer, or a UPS inverter. In all such cases, *the required grounding electrode must be as near as possible to the source of power, the separately derived system* (Sec. 250-26). The grounded conductor, or neutral, of the separately derived system and all equipment grounding conductors must be connected to *this* grounding electrode, *not* the one at the building service. (In fact, in many cases, a connection may also be made back to the service grounding electrode through the metallic raceway system; but this is neither required by code nor necessary for safety.) The reason for this is that any fault currents that flow in the separately derived system must have a low-impedance path back to their source—the separate system source, not the building service.

For separately derived systems, the preferred grounding electrode is the nearest effectively grounded metal structural building steel, or the nearest effectively grounded metal water pipe. If neither of these is available, other electrodes as listed in Sec. 250-81 or 250-83 may be used We have seen installations with grounding connections run several hundred feet back to the service ground from a computer transformer secondary, when the connection could have been and should have been made to the steel building column only a few feet away.

To sum up, ground the computer equipment radially to a single ground point, tie it to a high-frequency reference grid, and ground this grid to the same computer ground point. Then ground the computer ground point to the proper grounding electrode, and you will have a computer grounding system that is electrically safe, meets all code requirements, and will be essentially free of ground-caused computer noise and problems.

Nonlinear loads and UPS systems

If your computer seems unreliable, it may be incompatible with the UPS feeding it. Here's an analysis of the problem, with a suggested cure.

ACCOMPANYING the advance of semiconductors into the industrial plant has been the growth of such high-current loads as AC and DC drives and other solid-state equipment, including computers. The resistive, inductive, and capacitive characteristics of these loads are not constant over the voltage range covered by the required input-voltage waveform—i.e., they are nonlinear—and therefore they present a special problem when present at the load terminals of an SCR-based computer-grade UPS. The resulting current output of the UPS is cyclical, but not purely sinusoidal. Knowing what to do under such conditions is critical to the reliability of computers and similar devices.

By TIM DEVORE

A UPS is often referred to as a "soft source." This means that the voltage waveform may be deformed by the load. The UPS inverter section sees a distorted current wave as a 60-Hz fundamental signal to which are added high-frequency harmonic components. An example of this is shown in Fig. 1. The inverter and its associated smoothing circuits are tuned to service a fundamental frequency only. It can provide the needed harmonic components *but only at the cost of a distorted voltage waveform.* Since there is a certain amount of distortion inherent in the operation of the inverter, any nonlinearity-induced distortion is likely to push the total output distortion beyond the 5% specification allowable for this type of equipment. UPS manufacturers have long been aware of this problem, and thus most include a disclaimer in their distortion specification such as "based on linear loads" or "resistive and inductive load" (also a linear load). For a more in-depth discussion of nonlinear loads, see the accompanying box.

COMPUTER EQUIPMENT such as this installed in the Ampex Corp. plant in Opelika, AL is essential to the operation of the facility. Due to its critical nature and sensitivity to noise, voltage transients, etc., this type of equipment is often fed from a source of uninterruptible "clean" power such as a static UPS. Even with such a backup system, the computers may prove to be unreliable. If the problem proves not to be due to faulty equipment, then the possibility of a mismatch between the UPS and computer equipment should be investigated.

Finding the problem source

There is presently an effort in the static UPS industry to develop units that are less susceptible to harmonic deformation. However, this is of little comfort to those who already have systems that have not proved to be suitable for the needed application. It is imperative to be able to isolate the source of the problem. If it's not due to component malfunction, then it's necessary to find out if there's an incompatability between the UPS and the load.

If a static UPS is being used to drive a computer (or other critical semiconductor load) and the load in question has a high reliability record or no history of mysterious malfunctions, it may fall within the old saying, "If it ain't broke, don't fix it." However, even in this case, there is a strong rationale for investigating the amount of UPS output distortion. First of all, the computer power specifications are there for a purpose, and violating them may not be wise considering how critical such equipment is to most operations to which they are applied. Secondly, the whole purpose of purchasing a UPS is to provide the best "computer-grade" power available. The idea that this top-of-the-line system may produce voltage of inferior quality might not be acceptable in and of itself.

The approach to determining if there is a problem is not difficult. Essentially, it consists of the application of a distortion analyzer, commonly available as a rental

TIM DEVORE is an EE in the facilities engineering department of Ampex Corp.'s magnetic tape division in Opelika, AL.

Fig. 1. Typical current waveform of a nonlinear load

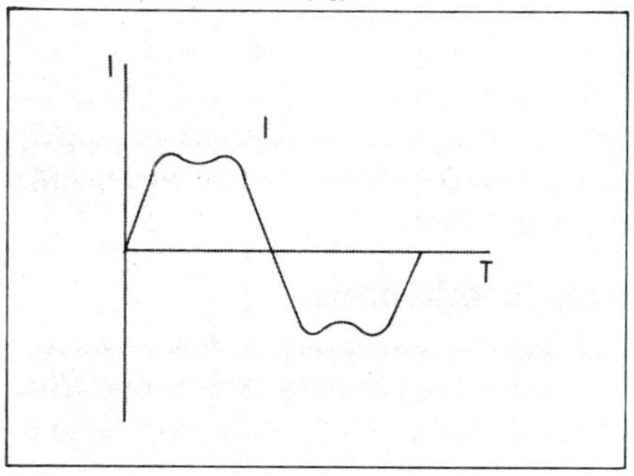

Fig. 2. Per-phase representation, harmonic filter bank

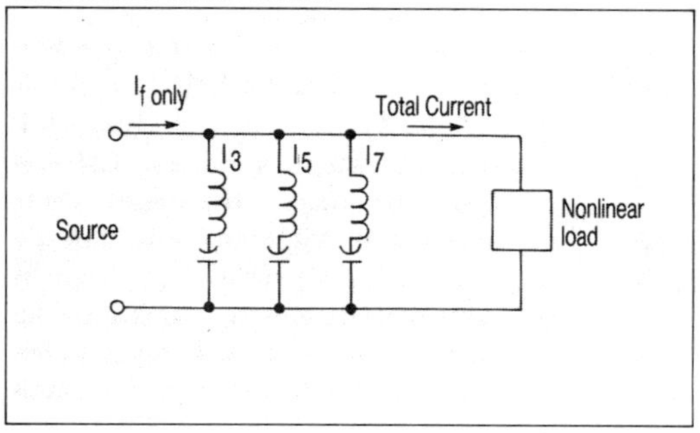

item. A distortion analyzer is somewhat complex to operate, but a competent technician can perform the required tests by carefully following the instructions that come with the unit. The analyzer will display the percent total harmonic distortion in analog or digital form, depending upon the instrument selected. Sometimes, measurement services are available from vendors having an installation analysis group.

Results of such testing may reveal a within-specification distortion level on all phases. This indicates that, despite their semiconductor nature, the load system presents itself to the UPS as essentially linear over the applied voltage range. However, if the total harmonic distortion of the UPS output voltage waveform violates computer requirements, further steps should be taken.

Load characterization

The nameplate of any electrical equipment gives voltage and full-load current ratings. This information is sufficient to describe and install *a linear load only.* To describe a nonlinear load adequately, it is necessary to provide rated voltage (and distortion specifications), power factor, rms current draw, and what percentage this current is of each of the harmonic frequencies present. One example of a spec defining these specifications might read: "Rated current: 30A, 2% or less THD (total harmonic distortion) of current waveform." Another example might read: "Rated current 30A, 25% THD as follows: 10% 3rd harmonic, 8% 5th harmonic, 5% 7th harmonic, all others 2%."

This may seem radical to the engineer who has always thought only in terms of volts, amps, and power factor. In fact, however, it is absolutely necessary to understand *all* the factors involved. While computer manufacturers do not usually provide this information as a matter of course, consumer need will eventually lead to the inclusion of such specifications in all computer and peripheral manufacturers' standard electrical data.

ELECTRICAL ENGINEER Tim DeVore checks out the operation of the UPS system installed in the Ampex facility. Testing for problems involving incompatibility of UPS and its load requires the use of instruments such as a distortion analyzer, spectrum analyzer, and dynamic signal analyzer. These are normally readily available from instrument rental organizations. However, performing the actual testing, analyzing the results, and determining a course of action to correct any deficiencies uncovered requires a complete understanding of the equipment and the instruments being used.

Until this data becomes routinely available, it is up to the equipment user to provide the proper analysis to characterize the loads fed by their UPS systems. This is done by means of a spectrum analyzer or dynamic signal analyzer. A high-quality clip-on current transducer that will accommodate the observed current level and accurately define the harmonic components is also needed.

The reference voltage for load testing is the standard commercial power-grid sinusoid. For all its faults, it does represent a "hard source" that is not subject to harmonic deformation. Therefore, the first step to load-impedance characterization is to place the computer system in question on commercial power (bypass the UPS). The next step is to examine the current through each phase conductor with the analyzing instrument. Resulting data will be similar to the sample distortion specifications quoted previously, with the proper percentages noted for each harmonic frequency band.

System diagnosis

Once a problem has been found to exist and the load has been characterized, the question is, what to do about it. There are presently two options: replace the UPS or filter out the harmonics. There are new power-transistor units that are said to demonstrate some resistance to harmonic distortion. However, these are available in limited configurations and are still establishing a reliability record. Motor-generator sets are also considered to be more of a "hard system" than static UPS systems and thus could be substituted, but large expenditures are involved. Thus, this first option is not normally feasible.

The filtering option involves the design, fabrication, and installation of a custom harmonic filter bank—a set of series inductor-capacitor circuits connected to the load-side of the UPS inverter. Each phase requires one L-C circuit for each frequency. Inductor-capacitor circuits have very strong resonant-frequency characteristics. The object is to design the circuit to resonate at the frequency of the undesired harmonic. This will cause the harmonic current at that frequency to resonate between the load and the filter bank, thus relieving the UPS of the "strain" of supplying it. This is illustrated in Fig. 2.

The result is to provide a voltage waveform that is within computer tolerance. This will be the case only if each harmonic frequency on each phase has its own tuned L-C trap. Fortunately, the design of an L-C resonant circuit is not very difficult. The required information is provided by the load characterization, which provides harmonic frequencies and magnitudes. However, the construction of such a unit is much more complex. The needed capacitors are readily available, but power inductors are not available as standard items. They must be designed and fabricated to order. This is often beyond the inhouse capability of facility forces. UPS manufacturers can be an answer to this problem. They can often provide the design, fabrication, and field start-up of such a custom filter bank.

Filter banks do have drawbacks. Since they must be designed for each frequency, a load that produces too many different harmonics may make a filter bank impractical. Also, if the new devices are added to the load that produce different harmonics, an existing filter bank will not be prepared to handle them.

New installations

A person purchasing a new computer and/or UPS can do so to avoid the problem of nonlinearity. If the computer system has already been selected and the process has reached the point of selecting a UPS, characterizing the load impedance should be done as previously described and provided to prospective UPS vendors in the bid specifications. This helps ensure that an appropriate unit is selected. Vendors who are aware of what such a specification means are likely to be among the leaders in their field in electrical technology.

Effects of linear and nonlinear loads

WHEN A VOLTAGE is applied to an electrical circuit, a current will flow. In drawing (a), the ratio of voltage to current is constant over the range of 0 to 200V. The voltage/current ratio is thus said to be linear over this range. According to Ohm's Law in DC form, R = V/I. Resistance is thus constant over the voltage range. This is termed a *linear load.*

In AC circuits the ratio of the voltage magnitude to the current magnitude equals the magnitude of the load *impedance.* Any combination of induction-motor windings, resistance heaters, power-factor-correction capacitors, and similar equipment that does not contain semiconductors produces a circuit impedance that is typically linear. This does not mean that the loads are smooth and constant over time. It does mean that their impedance does not vary because of the change in voltage over the range of the applied voltage.

There is an infinite variety of *nonlinear load* situations. Drawing (b) illustrates one. In this particular case, the current-to-voltage relationship is very nearly linear over the range of 0 to 150V, but once the voltage goes higher than 150V, the current actually begins to fall as the voltage increases. The ratio of voltage to current (impedance) is nonconstant over the range of applied voltage.

Semiconductors could just as accurately have been named "semi-impedances," because they sometimes actually impede the flow of current. Conditions under which a semiconductor device will change impedance vary greatly. Circuits involving semiconductors generally are nonlinear in character.

Consider the load response to a standard 120V source shown in drawing (c). This source has a peak voltage of plus or minus 170V and a frequency of 60 Hz. It takes the form of a mathematical sine function. A linear impedance response plot is shown above the voltage plot. By taking points on the voltage curve and extending them to the impedance response plot, a current response plot similar to the one on the right can be generated. For instance, at time equal to 0, voltage is zero, and thus so is current. At time equal to .002 sec, V = 120. From the current-response graph, it can be seen that this results in a current of 12A. At time equal to .004 sec, V = 170, and I = 17A. As enough points are taken, a sinusoidal response curve for the current is generated.

A unity power factor was assumed for the preceeding example to avoid unnecessarily cluttering the illustration. The affect of a nonunity power factor is simply to shift the response current in time; it does not affect its form or magnitude.

Consider the response of a nonlinear load to the same 120VAC rms voltage as shown in drawing (d). The resulting current wave form is no longer a smooth sine wave. This serves to illustrate that a nonlinear load will produce a distorted nonsinusoidal current wave. This result is of great importance when examining effects of nonlinear loads on UPS operation.

Harmonics. There is an important law in signal analysis that says that any repeating signal may be created by adding pure sine waves together. Furthermore, it says that the frequencies of the resulting sine waves will be multiples of the original signals. These sine waves are called harmonics, and when added to a clean current wave, they create harmonic distortion. This principle is illustrated in drawing (e). Here, the 60-Hz (fundamental) sine wave I_F is added togeth-

HARMONIC FILTER BANK (photo· far left) proved to be the solution for the harmonic distortion of the output of the UPS system at Ampex caused by the nonlinearity of its load. Essentially, the filter bank consists of capacitors and inductors forming L-C circuits that resonate at the appropriate harmonic frequencies. In the photo at left, the capacitors are at the upper section of the panel and the inductors at the lower level. Because the L-C combinations must individually match the requirements of the specific loads involved, each harmonic filter bank must be customized for the installation.

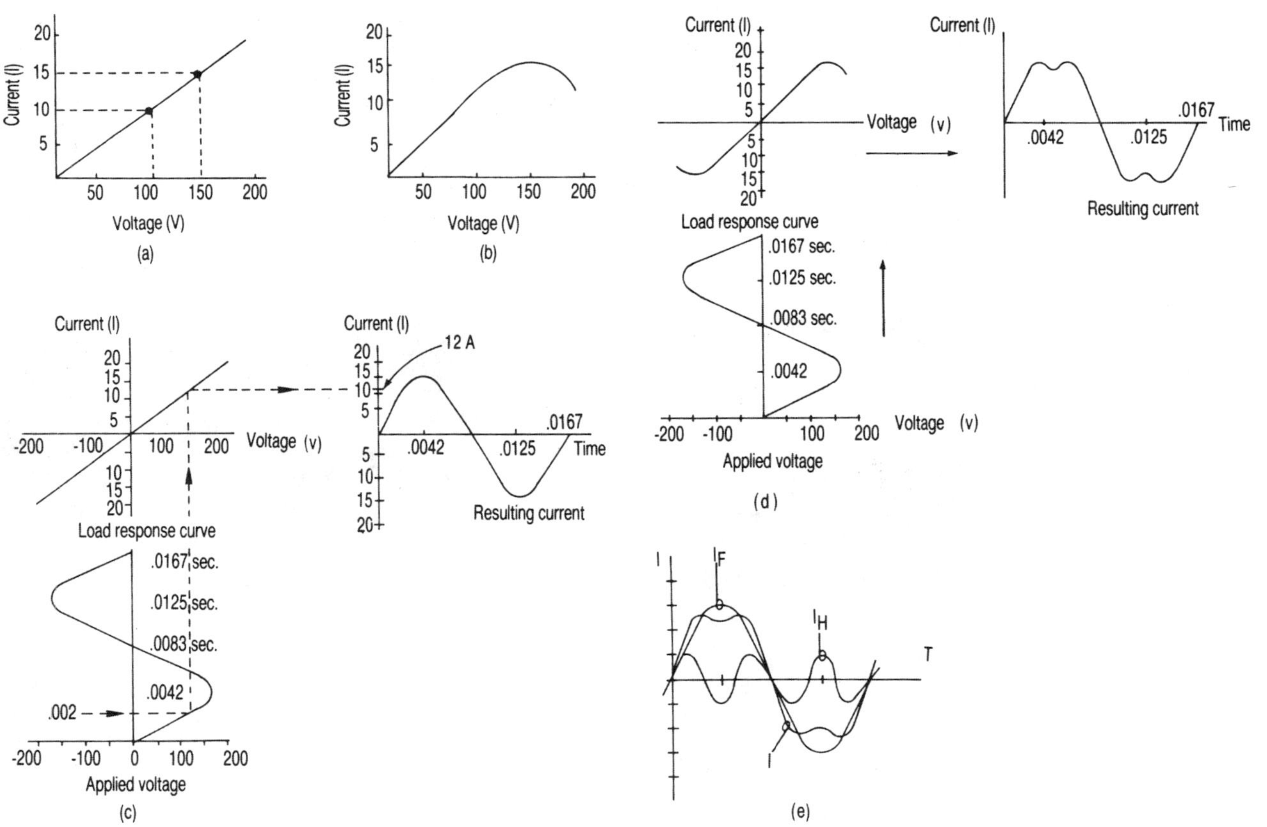

er with a 180-Hz (3rd harmonic) "distorting" sine wave I_H. The result is the current pattern (I), somewhat similar to the one shown in the previous example of the effect of a nonlinear load.

A generalization can be made that all nonlinear loads cause a distorted current waveform. The resulting waveform consists of a 60-Hz fundamental sine wave plus one or more other sine waves (harmonics) that deform the fundamental. These harmonics will have frequencies that are multiples of 60 Hz (120, 180, 240 Hz, etc.). Once a voltage function has been defined, the current response will completely represent the load impedance.

FACILITY NOISE CONTROL FROM THE GROUND UP

Ronald Gunn

Designing effective power installations for facilities with computers and other sensitive electronic devices.

Ronald Gunn, a graduate of San Jose State, has participated in the evolution of control systems at Lawrence Livermore Laboratory, Livermore, CA, from banks of relays in 1958 to banks of computers today.

POWER SYSTEMS must be designed not only for safety, but also for the control of electrical noise. While safety grounding has been standardized through building codes, there is no standard for grounding to control electrical noise.

While user equipment has become more sensitive, most environments still have elevator traction motors, air conditioners, arc welders, and other major noise sources. The job of making 5V microprocessor circuitry reliable in this circumstance is daunting.

A consistent overall approach is required, yet instruction and advice from different equipment manufacturers often is contradictory or vague, using terms like "single-point ground" or "separate ground" as if they have well-defined meanings. They don't. Therefore, it is necessary to develop a step-by-step understanding of the subject.

It is important to first recognize the National Electrical Code (NEC) and its emphasis on electrical safety; then separate consideration of safety ground systems from signal ground systems; then look at power as frequency rises higher and higher; and finally to review noise propagation mechanisms and noise control strategies in these progressively higher frequency ranges. Armed with this information, system design can be started.

The NEC and 60 Hz

The NEC is concerned about electrical safety and presently does not address the problems of noise control or signal reliability. Building codes mandate NEC-based installation practice for safety. However, to operate processors and industrial electrical devices together, it is necessary to follow rational design practices to reduce noise. These two requirements are not necessarily the same.

Does this mean that NEC practices must be circumvented to avoid noise? If the end result is to make a system unsafe, then the answer is no. Using insulated sections of conduit or lifting grounds can injure people and must be avoided. Even if they are helpful (and that is often not provable), they are not worth the risk of hurting someone.

The NEC is concerned primarily with 60-Hz

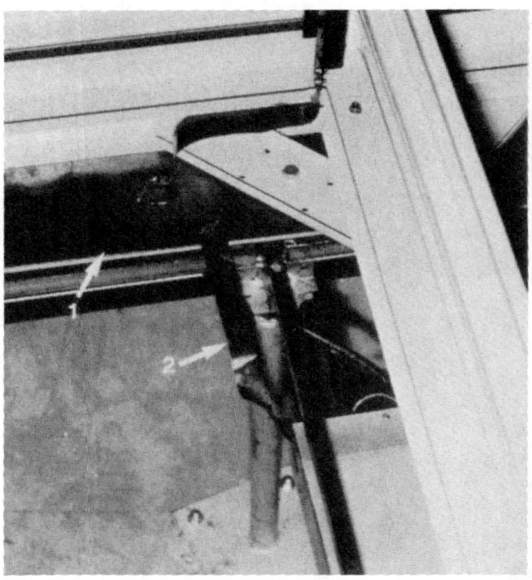

Rack set on top of 6-in.-wide, .062-in.-thick copper strip (arrow 1 in photo above) is one of the techniques used to swamp out noise at the Lawrence Livermore National Laboratory. Note short copper straps to RS310 equipment tie-down strip and filter case (arrow 2). In photo at right, the copper strap is the signal ground (common) for the control room. It is 11.75 in. wide, 50 ft long, and shows .0005 ohm from end to end at 7.5 kHz. Arms lead to power panel grounds, rack sets, and computers.

Skin effect

When a conductor carries an alternating current, a magnetic field is produced that surrounds the conductor. This field is continually expanding and contracting as the AC current wave increases from zero to its maximum positive value and back to zero and through its negative half cycle. The changing magnetic field's lines of force cutting the conductor induces a voltage in the conductor in a direction that tends to retard the flow of current that normally flows in the conductor. This effect is more pronounced at the center of the conductor. Thus current within the conductor tends to flow more easily towards the surface of the conductor. The higher the frequency, the greater the tendency for the current to flow at the surface. The depth of current flow as a function of frequency is determined by:

$$d = \frac{2.6}{\sqrt{uf}}$$

d = depth of current (mils)

u = permeability (air = 1; copper = 1; steel = 300)

f = frequency (megahertz)

At 100 kHz:

$$d = \frac{2.6}{\sqrt{1 \times 0.1}} = \frac{2.6}{.32} = .008 \text{ in.} = 8 \text{ mils}$$

power circuits, with occasional consideration for DC and 400-Hz power systems. Noise exists up to many megahertz. Thus, the best approach for designing a safe and noise-free system is to specify all the required NEC safety grounds; then, without circumventing this foundation, add features to cope with the noise.

Installing separate ground systems is not a power design goal, but rather a signal ground goal. Separate power grounds can quickly become illegal and dangerous, but safe and effective separate signal grounds can be created easily.

A single-point ground is not an arbitrary point. Best results will be obtained when this single point creates the shortest path for the highest system currents. It could be a point on a process line. An "earth ground" connection (in addition to any required safety ground) is optional in this case. The essence of this technique is not the ground, but to have one place to connect all low-impedance commons.

Electric power at different frequencies

DC to 1 kHz. This is the range where the conductor carries its full-rated current, where resistive voltage dividers are predictable, where ground rods work, and where everything within a moderate distance actually can be at the same ground potential. Accidental resonant effects (where conductor electrical length matches signal frequency) are unusual, and stray interchange between circuits as a result of capacitive or magnetic coupling or electromagnetic radiation is virtually nonexistent.

1 kHz to 100 kHz. This is the band where a wire is an antenna and skin effect becomes important. At DC, the entire cross section of a conductor carries current. At just 1 kHz, the current is limited to the top 80 mils (.080 in.) of material near the surface, even in an excellent conductor like copper. At 100 kHz, only the top 8 mils (.008 in.) at the surface are carrying current (see box). Since the effective conductor cross section is reduced, the impedance is greatly increased. A thin-wall copper tube will carry the same current that a solid conductor of the same diameter can handle.

The spectrum below 100 kHz is also of interest for another reason. Since free-space wavelength (in feet) = 936/f, where f is the frequency in MHz, a 100-kHz signal will have a wavelength of 9360 ft. This is longer than the distance across most facilities, so resonant effects are not a problem. Half- and quarter-wave resonant effects at 100 kHz are therefore not normally of concern.

100 kHz to 1 MHz. This is where the real "radio frequency" (RF) range starts. AM radio stations operate in this range, and a powerful station that is close enough can cause local interference. The band of frequencies between 100 kHz and 1 MHz is a transitional range, for a number of reasons. Lengths of wire become better transmitting and receiving antennas, so noise pulses can jump from one circuit to another. At 1 MHz a wavelength is 936 ft long in free space. A conductor as short as $\frac{1}{20}$ of this length is generally considered long enough to cause pickup problems, so leads across a facility that are as short as 50 ft in length can interact with each other; when they do, they act unpredictably.

Skin effect causes 1-MHz currents to travel in only the top 2.6 mils in copper (even less in almost all other conductors), so the series impedance as well as the current-carrying capacity of conductors is affected. Ground conductors that were suitable for use in a medium-sized facility at 100 kHz may begin to suffer in this band from increasing impedance and from resonant effects.

1 MHz to 10 MHz. In this frequency range, all hope is lost of reliably maintaining equal ground potential between circuits, even when associated equipment is no further away than an adjoining rack. A full wavelength can be as little as 100 ft in free space, skin effect limits current penetration even in copper to 2.6 mils (.0026 in.), and every wire is a potential transmitting or receiving antenna. Single-wire circuit operation becomes virtually impossible to manage as the surroundings become a large part of any circuit. Only coaxial or well-balanced shielded twisted-pair circuits are predictable.

Over 10 MHz. It is very easy to experience resonant phenomena when signals are at 10 MHz or greater because a full wavelength is less than 100 ft long. Electromagnetic coupling from one circuit to another is easy, and shielding is difficult. Wires less than 25 ft in length unexpectedly become a quarter-wave long. Thus the impedance can vary from low to high in a few feet. The voltage will be at a maximum where the impedance is highest; and when the impedance is transformed to the lowest value a quarter-wave away, the current will be at its maximum.

This quarter-wave transformer effect will isolate one end of the wire from the other at the resonant frequency. Also, the high-current (low-impedance) end of the wire is the best transmitting antenna at resonance, which could be significant if it is carrying any power. A coax cable grounded at both ends will limit electromagnetic radiation if the shield is not broken.

Coping strategies

Noise cannot be "killed" but can be reduced to tolerable levels. It is therefore not meaningful to use words like "eliminate" or "stop" in their absolute sense. Each practical noise-reduction action should result in less noise where it is harmful, often by giving it an easier path elsewhere.

Viable noise-reduction strategies consist of raising the impedance to the noise in the harmful direction to reduce its amplitude, lowering it in some other benign direction to shunt it away, or breaking the link between the noise and the affected circuit. Noise currents will flow in a .01-ohm return that has been provided and will be shunted away from the 1-ohm path to sensitive electronic circuits by a ratio of 100:1. If the 1-ohm path exists only, then all noise currents will be on it.

Each frequency band has strategies that work best.

DC to 1 kHz. NEC grounding techniques ensure a certain minimum-impedance ground or common to trip protective devices in case of a fault. Practical low-frequency impedances of much less than 1 ohm over dozens of feet can be obtained effectively at low frequencies with simple grounding conductors and bonding that represent the NEC minimums. These wires, however, rapidly lose effectiveness at higher frequencies.

1 kHz to 100 kHz. Skin effect in this region of the spectrum becomes pronounced, but wavelength is still long, so tying metallic objects together with large-surface-area conductors is effective for equalizing potential differences.

Relatively simple LC filters can reroute noise signals. A properly working LC power filter exhibits a high impedance to noise frequencies in one direction (to sensitive circuitry) and a low impedance to the system intentional circuit-return (common), where the currents can flow without generating much voltage. It is important to realize that this system common is necessary for the proper operation of the filter.

100 kHz to 1 MHz. In this part of the spectrum, circuit-return/common systems may be-

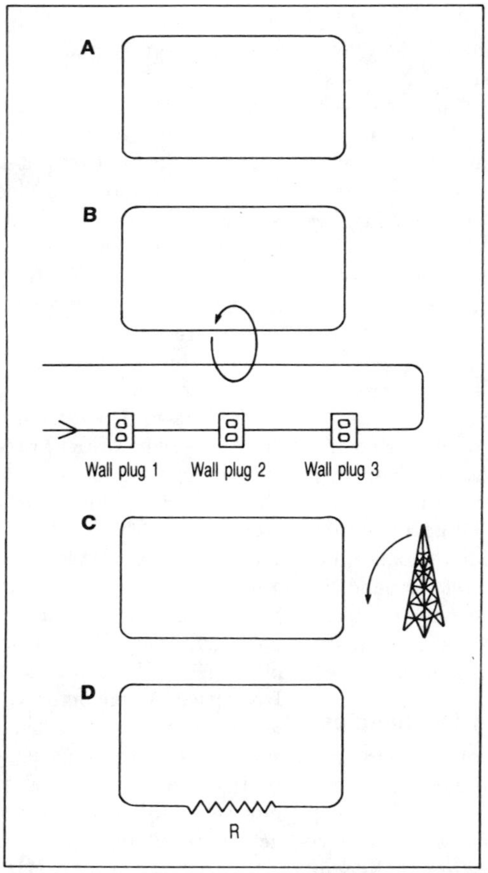

Fig. 1. A loop (A) has the same impedance all around. A loop in an inductive field generated in a second loop is shown in (B). The inductive field strength is a function of current. A loop in an electromagnetic field (C) acts as a loop antenna. The loop (D) with internal impedance cannot be considered a loop in the sense in which the term is used normally.

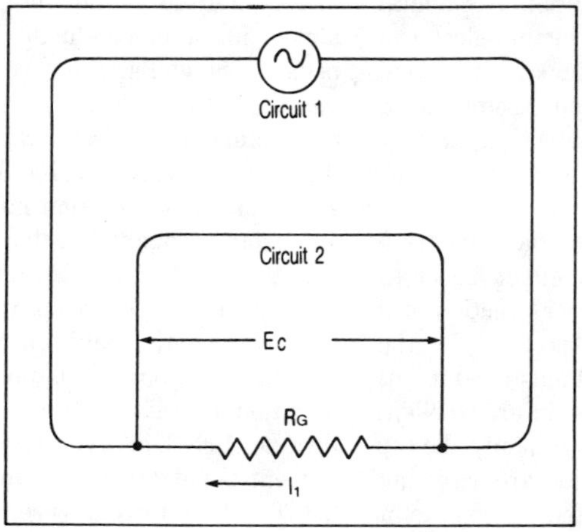

Fig. 2. A ground loop. Circuit 1 currents create voltage in circuit 2 ($E = I_1R_G$). This voltage appears across circuit 2 because they share ground-resistance R.

gin to have spot resonances that reduce their ability to tie metallic objects together on larger systems over a few hundred feet apart; thus, methods of coupling that do not depend on low source-to-destination potential differences must be used increasingly. This means that shielded twisted-pairs, passive and active drivers, optical coupling, and even fiber-optic signal paths may be needed. Large surface grounds between adjoining racks and pieces of electrical and electronic equipment are still effective.

Antenna effects can be reduced by shifting alignment of the coupled circuits by subtracting from (or adding to) leads to avoid resonant effects; filtering the affected circuit so that the received energy is bypassed; or interposing a shield. This shield must not reradiate the energy or its effect may be to strengthen the interference. Also, the shield can sometimes be arranged to absorb energy by coupling to the source and dissipating the received energy in a resistive load.

1 MHz and up. At these frequencies, the only circuit return paths that can be counted on are those that can be seen. If the grounding common goes through bolted joints, especially if they are painted first, there may be no helpful connection at all. The best way to ensure a useful common is to install it intentionally. Locally, a 2-in.-wide strip of copper 10 mils (.010 in.) thick or more will be helpful in many cases. Use liberal amounts of copper to tie chassis, racks, and cabinets together so that circuit returns are direct and predictable.

Pay special attention to high-power systems at these frequencies (including arc sources), as they can radiate noise into other circuits.

Other effects

There are other effects that can be a nuisance at any frequency.

- Loops can be a problem. A loop (Fig. 1A) is a defined circle of conductor at the same resistance or impedance, so that an induced current would circulate freely. A loop that is not broken by a point where the resistance or impedance is high is a good antenna because it produces maximum current from a received signal. Such a loop is a worst case for inductively coupled (Fig. 1B) and for radiated interference (Fig. 1C). A high-impedance point will limit the current and may develop most of the voltage at that one point (Fig. 1D), which can then be defeated.
- Ground loops (Fig. 2) are sections of conductors shared by two different electronic and/or electrical circuits. This happens most often with circuit returns. A ground loop drives the voltage drop resulting from circulating current in one circuit directly into another because part of the circuit is shared. Where sharing of returns is not avoidable, such as by using a separate signal grounding system, a low-impedance return will keep the voltage down to manageable levels.
- Capacitive pickup of noise can be a problem with high-impedance circuits. The problem often can be reduced by lowering the impedance of the target circuit. The reason that this effect is less dependent on frequency than might be expected from the relationship, $X_c = \frac{1}{2} \pi f C$, is that stray circuit capacitance often creates a capacitive voltage divider.

Shielding against capacitive pickup requires grounding the shield only at one end: the one that produces the shortest shunt path back to the source of the interference. Grounding at the wrong end or to an indifferent return can actually make the problem worse because of increased coupling to the target circuit.

- Inductive pickup is more likely in low-impedance circuits. It is caused by the changing magnetic field created around a wire carrying alternating current and is characterized by the presence of *two* loops, one transmitting to the other. The larger the current is, the stronger the field will be. The area of each loop determines the effective potential for coupling, so flattening either the source loop or the target loop will reduce coupling, sometimes dramatically.

A magnetic shield does not block a magnetic field. Magnetic "shielding" will shunt the interfering magnetic field through itself and reduce the effect elsewhere. Doing all that is practical to reduce the coupling between the loops is the best approach.

- Radiative interference from electromagnetic (radio-wave) energy falls off as the square of the distance from the source, like light from a lamp. Transfer efficiency is low unless quarter-wave or longer conductors are involved. Shielding is tricky as can be seen by playing an AM radio around an idling auto engine. Signal strength is proportional to the RF current of the source, which is why resistor wires are used in auto ignition systems.

A system design

Typical noise-reduction steps are as follows. The main power transformer is where electrical distribution starts. The NEC states that it must be grounded. Make sure that the requirements for grounding conductors are met or exceeded. Put in oversized ground conductors to the earth ground rods (the grounding electrode conductor) and the power panels in the system. Any lowering of impedances in this location will help to shunt the noise coming through the transformer. Noise generated within the facility also is attenuated when it encounters a low bank output impedance.

Establish separate panels for power and for instrumentation/control use. Put elevator traction motors and welders on their own power panels. Let the noise go to the low output impedance of the substation power transformer before it distributes back down to other circuits. This can easily lead to a 6- to 10-dB reduction.

Use isolation transformers to step down to lower voltage branch circuits. A standard transformer may have .002 microfarads of series capacitance from input to output, a highway for noise in both directions. A simple Faraday shielded isolation transformer can reduce this series capacitance to 30 picofarads, nearly two orders of magnitude lower. A laboratory isolation transformer with series capacitance of .03 picofarad and a cost difference of 400% is not needed for this purpose, rather a simple shielded transformer with a cost difference of 15% is adequate.

Use power-line filters to cabinets and rack plugstrips to prevent noise on the power line from coming in and keep local noise from going out. The metal case of such a filter should connect with a low-inductance strap on the best signal-circuit return. This return should run, among other places, to the power panel ground because part of the filter's effectiveness in handling noise is to shunt it back toward its source.

Grounds and commons

Use code-sized green-insulated wire grounds, not bare copper. Run these with the power wires in ferrous metallic tubing so RF throughout is minimized. Include a 50 microhenry choke in each safety ground wire (Fig. 3). Construct this choke by adding enough wire to one end of each green lead to make the coil. Fifty microhenries present an inconsequential calculated impedance to 60 Hz of about $\frac{1}{50}$ ohm. This rises to 2 ohms at 7.5 kHz and to 30 ohms at 100 kHz. A choke will ensure that higher-frequency signals do not use safety ground leads as a way to get around. Also, the code requirement for continuous conductors of sufficiently low impedance to trip the 60-Hz circuit breaker is complied with while reducing the potential for bothersome loops.

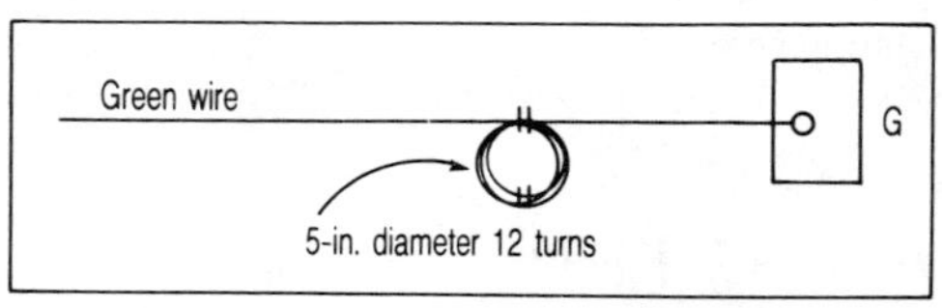

Fig. 3. Choke used to isolate safety grounds. Value is approximately $\frac{1}{50}$ ohm at 60 Hz; 30 ohms at 100kHz.

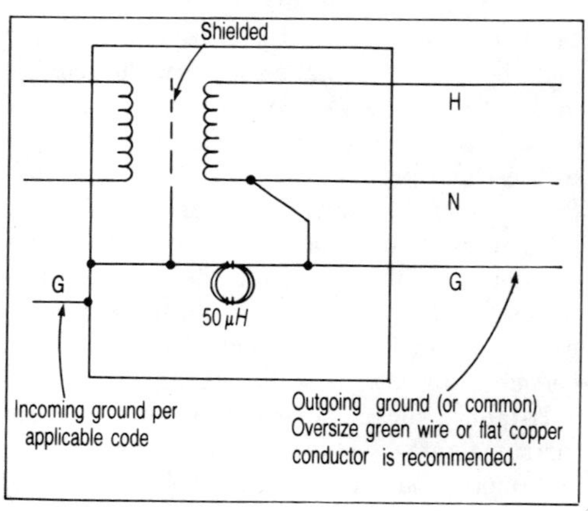

Fig. 4. Shielded isolation transformer used to create an isolated signal ground.

Fig. 4 shows how to decouple signal ground (common) circuits at an isolation transformer using such an RF choke. This is a separately derived "ground" that can be relied upon and is safe.

Ferromagnetic materials like steel are poor conductors at any frequency. At DC their resistivity is 10 times that of copper. As frequency rises, skin effect is more pronounced because magnetic effects are involved. A No. 6 copper wire can have less RF impedance than 12-in. steel "I" beams. Because of this and their bolted, piecemeal construction, steel racks should not be depended on alone for circuit returns.

Swamp out unreliable ground paths formed by bolted joints, conduits, piping, coax cable shields, floor connections, concrete, and others, with low-inductance copper strips. Place 6-in.-wide strips of .062-in.-thick copper under rack sets. Use inexpensive "soft" copper alloy 110 in coils. The goal is to intentionally put in place signal-circuit returns that are reliable. Bond each RS310 panel mounting strip in a rack to the copper along the rack as shown in the accompanying photo. Bond the frame of equipment boxes to a central point with two to four No. 2/0 insulated wires with green tape spiraled at each end to identify them as grounds.

Provide a ground potential common from one end of the facility to the other. Currents above 7.5 kHz will travel only on the surface of this conductor, so surface continuity is necessary for best high-frequency results; that is one of the goals. At 750 kHz, the impedance of this common will be 10 times what it is at 7.5 kHz for a given length. Thicker material will only reduce the impedance at frequencies lower than 7.5 kHz.

By observing the above recommendations, stray ground paths; earth grounds at 2 to 20 ohms; ferrous rack-to-rack bolted joints at 1 ohm or more; concrete at 3 to 10 thousand ohms per ft; steel wireway, air lines, and other things that sometimes carry current have been swamped out with a path that is more than 100 times lower in impedance. Uncertainty is reduced, and the location of system return currents is known. It will take a very large current to produce a bothersome voltage drop.

Racks and equipment cabinets; the output neutral/ground of the local shielded power transformer; the metal case of LC power filters; power-panel ground busses; the frame of computer floor grids; and power supply frames can be connected to this common.

PROTECTING COMPUTERS FROM TRANSIENTS

Arthur Freund

How to protect against voltage transients that can introduce errors into and cause damage to computers.

TRANSIENT VOLTAGE PROTECTION is probably less costly than solving any of the many other power problems that affect computers. When properly applied, transient protection can be extremely effective. Still, computers are subjected to data errors, crashing, and are sometimes damaged or destroyed by voltage transients as a result of an absence or misapplication of protective devices. Other sensitive electronic equipment, such as programmable controllers, solid-state motor controllers and variable-speed drives, communications equipment, and the like, have similar problems from transients, and protective solutions are almost identical.

A transient voltage is a high-amplitude, short-duration pulse or surge superimposed on the normal voltage (Fig. 1). It results from the rapid release of energy stored in the inductance and capacitance of an electrical system or in a charged cloud (lightning). There is no precise standard definition of a transient in an electrical system. Transients vary widely in current and voltage waveshapes and magnitudes. Typical peak voltages are from about twice the rms supply voltage up to thousands of volts. Durations can range from 0.5 μs to perhaps 200 μs for a single spike, and up to one cycle (16.7 ms) or even more for an oscillatory transient. Oscillatory transient frequencies may vary from a few hundred Hz up to many MHz. Depending on the relative polarity of the power and the transient, the voltage can add to the 60-Hz sine-wave voltage (a spike) or subtract from it (a notch or dropout). Voltage rise can be thousands of volts per microsec.

The two sources of transients are lightning and switching of electrical circuits. Lightning produces extremely powerful, short-duration transients, either by a direct strike or a near "hit." Maximum lightning voltages on unprotected indoor low-voltage systems are limited by sparkover within the system to about 6kV. Lightning on power lines is diverted to ground by arresters. Circuit breakers protecting the lines will trip and then may reclose. The resulting voltage sags and outages can cause more computer problems than the voltage transients.

Switching utility or on-site loads causes transients of a magnitude depending on the *instantaneous rate of change* of the current (di/dt). In typical inductive systems, the volt-

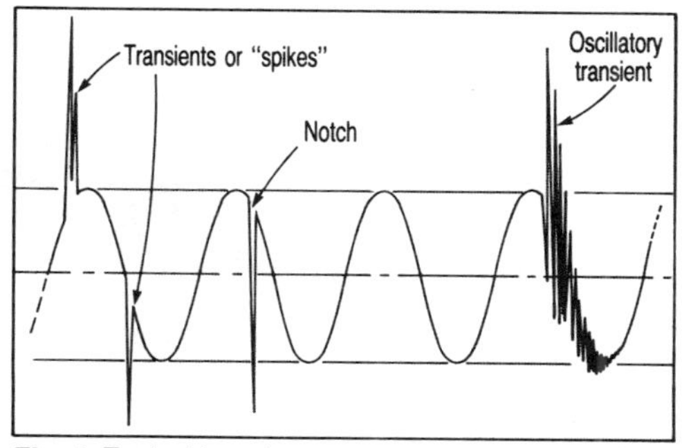

Fig. 1. Typical transients.

age (E) developed across an inductance (L) is $E = L\ di/dt$. Switching capacitors can cause severe transient voltages, and even normal system stray capacitance can cause transient oscillations.

Damage from transients can be *immediate.* The transient voltage, in this case, causes a breakdown of solid-state components in the computer due to the energy of the transient. Damage also can be *latent.* This is a much more subtle and mysterious effect, where insulation or components are severely stressed by a single or repeated transients, but not to the point of immediate failure. Each hit weakens the ability to withstand additional stress. At some later time, a transient or other stress (that would normally cause no problem) occurs at the weakened component, and it fails unexpectedly without apparent cause.

Transients can be impressed on data and communication lines as well as on computer power lines. They are a form of electromagnetic interference (EMI) or "noise" that can cause errors in or loss of transmitted data. The exceedingly small spacing between components in electronic equipment, especially on the chips, makes them very susceptible to damage from overvoltages entering either on the power supply lines or the data lines. In one instance, a number of small personal computers were unplugged from their 120V receptacles to prevent damage during a thunderstorm; lightning hit on or near the external communication lines, and the high-energy transient destroyed several of the unplugged computers.

Protective devices

The four basic types of equipment used for protection from transients are crowbar devices, voltage-clamping devices, attenuation devices, and hybrids. They vary in speed of operation, power-handling ability, and effect on the system while operating. A fully protected computer power system may use a combination of all four types, each properly applied to take advantage of its special characteristics.

Crowbar devices include air gaps, gas-discharge tubes, lightning arresters, and switching devices (Fig. 2). They are inactive until the voltage applied exceeds the rated or set value. At that point, they become conducting, creating essentially a short-circuit path to ground and diverting the high transient voltage away from the load.

Gap-type devices arc across the gap. Switching devices, such as thyristors, are gated on by a control circuit. Both types of devices, during the time they are conducting, must carry not only the energy of the transient but also the short-circuit current (follow current) of the system now shorted to ground for durations up to one-half cycle (8.33 ms) and sometimes longer. The arc in the gap ceases to carry current at current zero. At the first or a later current zero, the energy from the short-duration transient usually has been dissipated, and the system voltage is not high enough to cause a restrike, so the nonconductive gap is restored. Similarly, the switching device ceases to conduct at current zero, and there is no overvoltage to cause it to be gated on again. If the follow current exceeds the rating of the device, interruption may not take place at the first current zero, and some type of backup protection may be necessary. Gap-type devices require between 300 and 700V to initiate the arc across the gap.

Crowbar devices are relatively slow, taking several microsec to operate, and therefore subjecting the protected system to the very fast initial voltage rise, which can be several thousand volts per microsec. Additional surge protection is used downstream to protect equipment against this initial surge. However, crowbar devices are capable of handling transients of high-energy content. Most lightning arresters, at all voltages, are crowbar type. When these devices conduct, the normal voltage on the system drops, as a result of the virtual short-circuit "crowbar" to ground. When the overvoltage is cleared and the protective device ceases to conduct, normal voltage is restored to the system.

Gas-discharge tubes (GDTs) are the most commonly applied gap devices for computer systems. They consist of two or three electrodes, spaced about 10 to 15 mils apart, sealed in a glass tube filled with a mixture of gases at low pressure. The electrodes form one or two metallic gaps. GDTs have a relatively long life and high current-carrying ability.

Carbon-block spark gaps are common in low-voltage telephone and data systems. They consist of two carbon-block electrodes separated by an air gap of 3 to 4 mils. This type of device is low in cost, but has a comparatively small current-handling ability and short life. Spark-over voltage for a given unit is not constant and can range from 300 to 1000V, unpredictably.

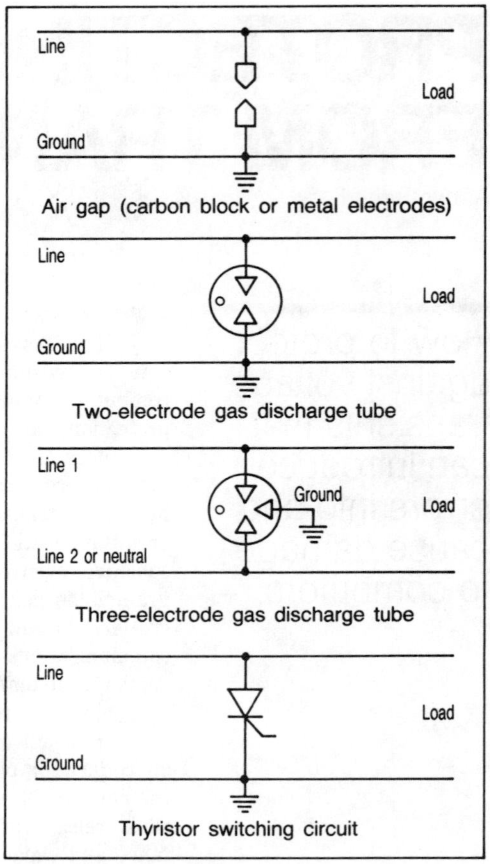

Fig. 2. Crowbar devices.

All gap-type devices have the characteristic that the voltage at which the gap breaks down and arcs is not consistent; although gas-discharge tubes are considerably less erratic than air gaps. When protecting both conductors of a two-wire circuit or data line, it is best to use a three-electrode gas tube that will shunt both lines to ground simultaneously. Two two-electrode gap devices, one on each conductor, would have different flashover voltages so that one would conduct first. This unbalanced breakdown could cause transient current to flow through the load, causing damage. Other air-gap devices using electrodes of metal rod or other materials are available, but seldom used in computer installations.

Switching devices also are used occasionally in crowbar circuits. They depend on circuitry that senses the system voltage and gates the device into its conducting state when an overvoltage appears. They are more complex and costly than gap devices and are used only when their sophisticated abilities are necessary, such as for precise voltage control.

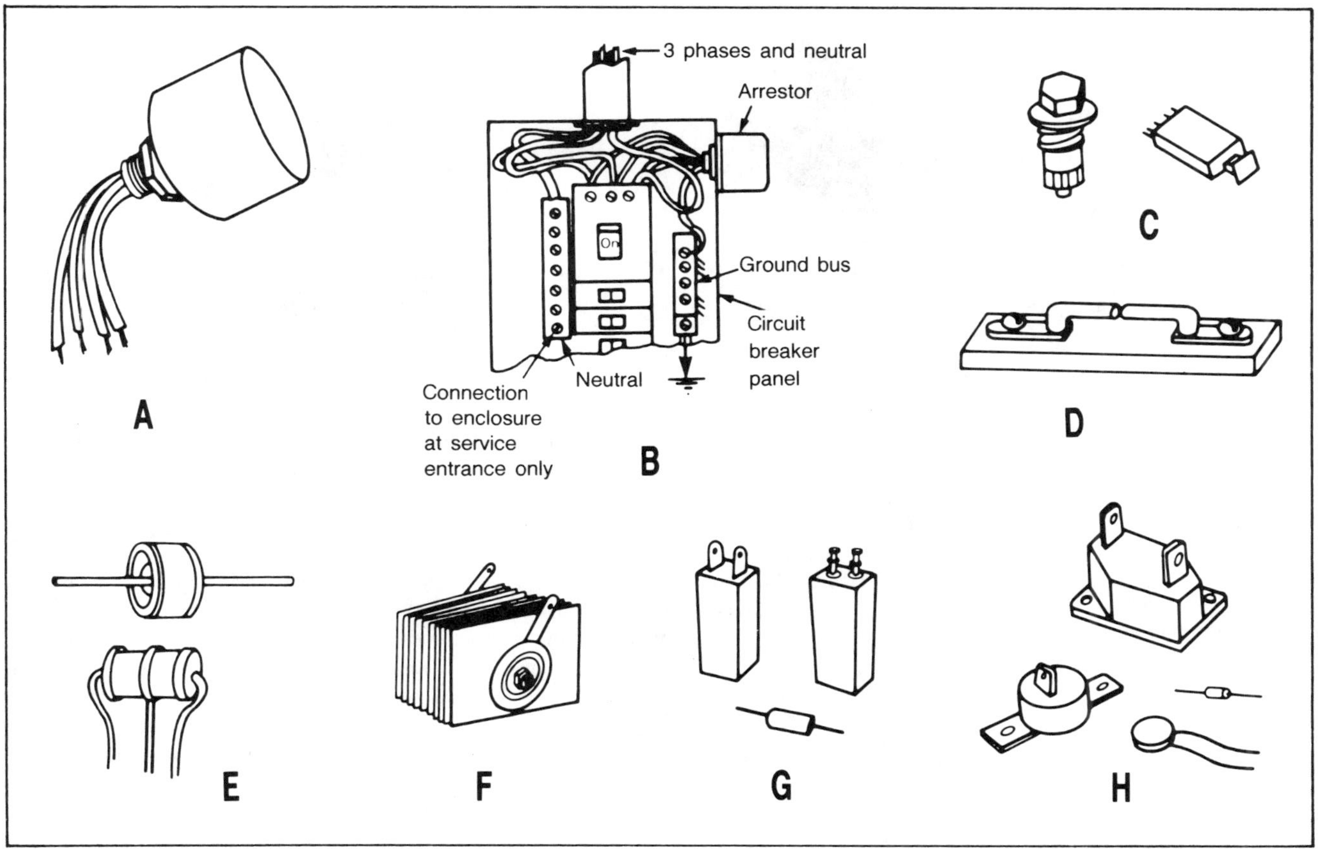

Surge and transient protection can be provided by a variety of devices, each available in many sizes and forms. A high-energy silicon-carbide lightning arrestor (A) may be installed at the service or distribution panel (B). The carbon gap (C) and the less common rod gap (D) are used primarily on incoming telephone and data lines. Gas-discharge tubes (E) of the two-lead, single-gap or three-lead, dual-gap type are used to attenuate transients of high energy content, usually backed up by faster-acting voltage-clamping devices in a hybrid unit or further downstream. Voltage clamping can be provided by selenium surge protectors (F), silicon avalanche diodes (G), or metal-oxide varistors (H).

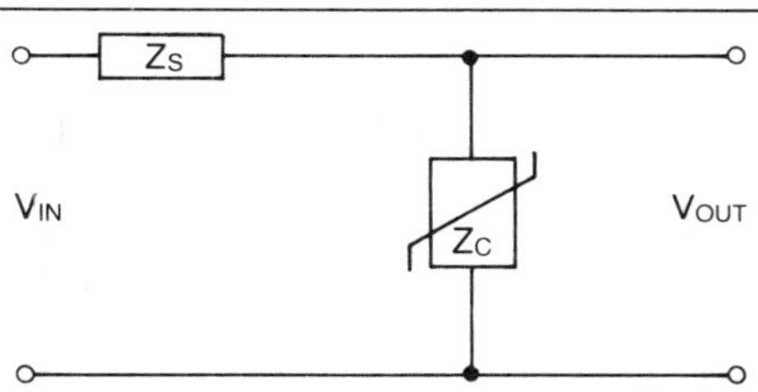

Z_S = impedance of source
Z_C = variable impedance of clamping device (MOV, zener diode, or selenium rectifier)

$$V_{OUT} = V_{IN}\left(\frac{Z_C}{Z_C+Z_S}\right)$$

V_{IN} normal – Z_c very high compared to Z_S

$$\frac{Z_C}{Z_C+Z_S} \simeq 1$$

$$V_{OUT} \simeq V_{IN}$$

V_{IN} high (transient above clamping voltage) –

- Z_C low, of same order of magnitude as Z_S
- As V_{IN} increases, Z_C decreases, current through Z_C increases
- V_{OUT} remains relatively constant at clamping voltage

Note: Z_S must not be zero or extremely low compared to Z_C at clamping voltage, or V_{OUT} cannot be maintained.

Fig. 3. Voltage divider action of clamping device.

Voltage-clamping devices such as varistors (nonlinear resistors), especially metal-oxide varistors (MOVs), zener (avalanche) diodes, and selenium rectifiers, are unidirectional conductors until a breakdown voltage is reached, at which time they conduct in the reverse direction. They are normally connected in the circuit in the nonconducting direction, presenting a very high impedance. At the clamping voltage, (somewhat above normal line voltage) the impedance drops rapidly. The higher the current, the lower the impedance becomes, shunting the transient away from the load.

The clamping-device impedance, in series with the source impedance, acts as a voltage divider and maintains the system at or near the clamping voltage (Fig. 3). Typical source impedances range from about 5 to 50 ohms; some studies give values up to 300 ohms. The lower the source impedance, the more difficult it is for the clamping device to maintain system voltage without drawing excessive current.

At the clamping voltage, the clamping-device impedance goes low, but it does not fall to nearly zero, as occurs with crowbar devices. There is essentially no follow current of short-circuit magnitude. Clamping voltage is maintained on the system load. Clamping devices operate in the nanosec range, thousands of times faster than crowbar devices, but are

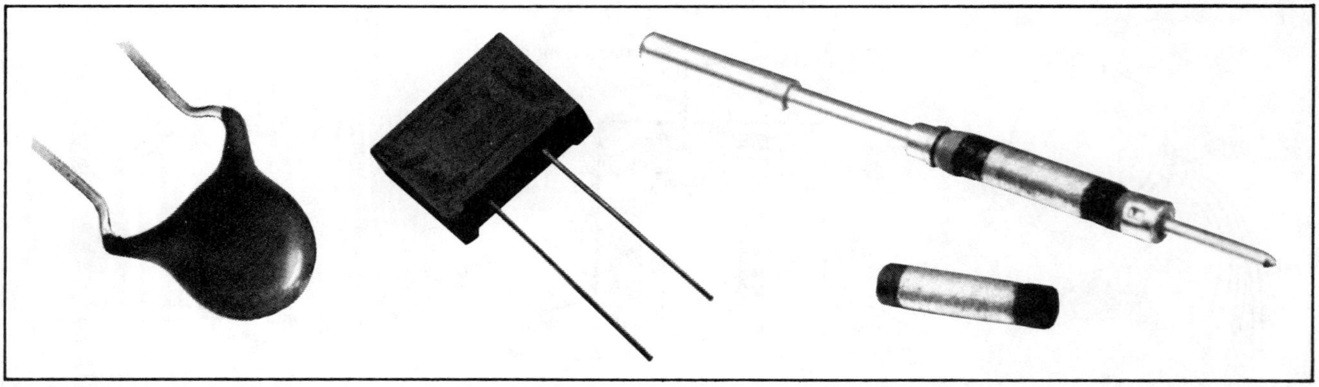

Metal-oxide varistors (MOVs) are available in many sizes and forms. Three kinds of small MOVs, for use at the terminals of the equipment being protected, are shown above. These have low clamping voltages and can dissipate little energy. MOVs are available for a wide range of clamping voltages and energy dissipations up to thousands of Joules.

generally capable of dissipating considerably less transient energy. They effectively clamp voltage spikes to a maximum level but have no effect on notches or brief dropouts.

MOVs are available for a wide range of voltages and currents, from as low as 4V for data lines to several thousand volts for power systems; for peak pulse currents from only a few amperes up to tens of thousands of amperes; and for energy dissipation from less than 1 to over 10,000 Joules. They are low in cost, compact in size, easy to apply, and are among the most frequently used devices for transient protection. One disadvantage of MOVs is that their characteristics deteriorate with repeated transients. Although this problem is being reduced with improved technology, MOV life under frequent transient conditions must be considered. Manufacturers provide data for this purpose.

Zener diodes also are used to provide voltage clamping. These are avalanche devices, which maintain a high impedance until the breakdown voltage and rapidly drop in impedance to provide the clamping action, acting as a voltage divider in a manner similar to the MOV. Zener diodes used for clamping have different characteristics from those used to provide voltage regulation. Under avalanche conditions, when the diode is clearing an overvoltage, substantial current must flow across a thin semiconductor junction. The ability to dissipate the heat produced at the junction limits the maximum energy that zener diodes can handle. Zener diodes are faster acting than MOVs and provide very effective clamping, to a nearly constant voltage, but with limited energy-handling ability. MOVs are available with satisfactory clamping and with far larger energy-handling abilities.

Silicon-carbide varistors have high power-handling capability and are used in high-voltage surge (lightning) arresters. They tend to draw considerable current in the normal state, so they are commonly used in series with a gap that provides an open circuit until a surge occurs. This property makes them unsatisfactory for low-voltage clamping operation. Some newer zinc-oxide lightning arresters have better nonlinear characteristics and can be used without a gap. They are essentially crowbar devices, but perform almost like clamping devices.

Selenium-cell transient suppressors, of specially designed selenium rectifier elements, are occasionally used. They can handle high energy transients and repeated hits very well, but have poor voltage-clamping ability. They have been replaced in most cases by MOVs or zener diodes. However, a combination unit is available that uses selenium devices, custom-made to minimize the undesirable characteristics and take advantage of the desirable ones.

Attenuation devices are inserted in a circuit to permit power at line frequency to pass, while attenuating transients. They are usually low-pass filters, since transient frequencies are in the kHz to MHz range, several orders of magnitude higher than the 60-Hz power. Low-pass filters are made up of combinations of shunt capacitors and series inductances. The components must be able to withstand the high transient voltages, and the series inductances must carry the sometimes large continuous-load currents.

Low-pass filters must be well-designed and matched to the type of transient to be attenuated. Their impedances should be matched to the input and output impedances; since these vary, the filter must make compromises that may do more harm than good. They are LC circuits, so a transient can send them into oscillation, and the output of badly matched filters can contain more spurious voltages and frequencies than the input.

One very unusual and highly accurate "active tracking" filter senses the instantaneous sine-wave voltage at any point in the cycle. Clamping devices limit the *maximum* voltage. This filter limits the *deviation from the true instantaneous sine-wave voltage* at any point on the sine wave to ±2V. When a voltage deviation is sensed, the unit switches on to provide full filtering in less than 5 ns. Not only are spikes clipped, but notches also are "filled in" by energy stored in the filter's capacitance, which cannot be done by clamping devices.

Hybrid devices are transient suppressors that combine two or more technologies to provide transient suppression over a wider range of voltages, rates of rise, and energy content, than any one device could do successfully.

The most common hybrid combines a three-electrode gas-tube crowbar, for high-energy capability, with MOVs or zener diodes for nanosec response, to take care of the initial voltage rise passed through by the slower gas tube and transients too small to activate the gas tube. Sometimes, a filter is added to provide further attenuation, especially for high-frequency oscillations.

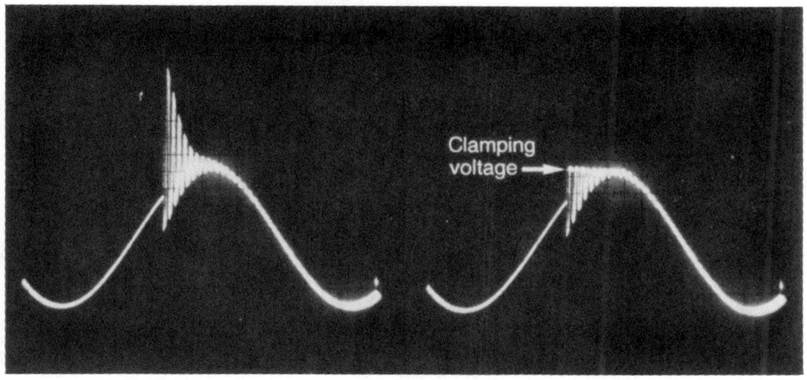

Clamping action of MOV is shown in actual photos from oscilloscope screen. Oscillatory transient is shown (left) with no transient suppression, and (right) after being clamped by MOV transient suppressor. The transient peaks are limited to the clamping voltage. Note that the MOV has no effect on that portion of the oscillations falling below the clamping voltage. The MOV limits the maximum voltage but does not filter out the high frequencies.

Applying transient protection

The first step in protecting computers and sensitive electronic equipment against transients is good installation practice, especially for larger facilities. Careful grounding is necessary, with equipotential grounding for a computer room. With poor grounding, the action of transient suppressors, which divert the surge energy to ground, could cause differences in ground potentials that might create as many problems as the original surge. The power to the computers should be separated electrically as much as possible from the remainder of the building power system. Placing an isolation transformer close to the computer equipment is desirable.

The cost of transient protection is so low that installing it will probably cost less than making a power-line and data-line transient study to see whether it is required. If a study is made, be sure that the transient monitor does not affect the results. Some monitors contain transient suppressors to protect their internal circuitry. If such a monitor is connected to the same power circuit it is monitoring, its suppressors will limit transients that the monitor should be recording.

If any form of power conditioning such as a voltage regulator, isolation transformer, power distribution unit, or UPS is used, the manufacturer should be required to include transient protection at the input to protect the equipment, and transient protection at the output as necessary to protect the computer load.

Separate protection for individual units usually must be provided (unless already built into the equipment) for distributed control in an industrial plant, programmable controllers, individual microcomputers (PCs), minicomputers, terminals, and the like. Data and communication lines entering from outside the building, and those entirely within the building that may be subjected to transients, must also have individual transient protection.

ANSI/IEEE Standard C26.41 and IEC Standard 664 have attempted to define the severity of exposure to lightning transients by categories relating to the location in the building. While the standards differ somewhat, the general consensus, not surprisingly, is that the most severe transients occur outside the building on the service conductors and inside at the service-entrance conductors. Next in severity are those transients at distribution equipment, large buses and feeders, and short branch circuits (near the service). Lowest in severity are those on long branch circuits (Table 1).

Table 1: Surge voltages and currents(4)

Location category ANSI/IEEE C62.41	Comparable to IEC 664 category	Impulse: Waveform	Impulse: Medium exposure amplitude	Type of specimen or load circuit	Energy(joules) deposited in a suppressor(3) with clamping voltage of 500V	1000V
A. Long branch circuits and outlets					(120V system)	(240V system)
	II	0.5μs - 100kHz	6kV	High impedance(1)	—	—
			200A	Low impedance(2)	0.8	1.6
B. Major feeders, short branch circuits, and load center		1.2/50μs	6kV	High impedance(1)	—	—
	III	8/20μs	3kA	Low impedance(2)	40	80
		0.5μs - 100kHz	6kV	High impedance(1)	—	—
			500A	Low impedance(2)	2	4

Notes:

(1) For high-impedance test specimens or load circuits, the voltage shown represents the surge voltage.

(2) For low-impedance test specimens or load circuits, the current shown represents the discharge current of the surge (not the short-circuit current of the power system).

(3) Other suppressors which have different clamping voltages would receive different energy levels.

(4) Representative values for use in designing protective systems.

Incoming power lines should be protected at or near the service entrance by a crowbar device, such as a gas tube or one of the newer fast-acting, low-voltage zinc-oxide lightning arresters, to eliminate the high-energy lightning and switching transients coming into the building from the utility or as a result of lightning. This device should be rated for the incoming voltage.

Incoming data lines also should be protected by a crowbar device where they enter the building, and this should be rated for the low voltage of the data line. Telephone lines usually will be protected by the telephone company, probably with a carbon-block gap arrester.

Computer-room power feeders supplying computer equipment (not HVAC or other nonsensitive peripheral equipment) should be protected by a voltage-clamping device where they enter the computer room. This unit should have a fairly large energy-dissipation rating.

Computer-room distribution equipment, whether just a panelboard, or preferably the panelboard on the secondary of an isolation transformer, separate or as part of a computer power center, should be protected by a voltage-clamping device on the mains. It is not necessary to protect each individual branch circuit. This device requires only a moderate energy-dissipation rating.

Data lines entering a computer room require a low-voltage clamping device. Each data line should be protected separately at the point where it enters the room. The protection should be at the low-to-moderate energy level.

Individual PCs, terminals, process-control computers, and similar equipment should be protected by at least a clamping device, with energy dissipation appropriate to its location. A hybrid device, perhaps with attenuating filters, might be desirable. There are many effective, reasonably priced, preassembled units available for these applications (and many overpriced units with extravagant claims and poor performance).

Selecting the clamping device for each application is partly a science and partly an art. The device should have a minimum continuous voltage rating that is higher than the system or data line rms voltage, or DC phase-to-phase or phase-to-ground voltage. The rating should be not less than 110% of the connected voltage; and 125 to 150% would be better to allow for system voltage deviations. Selecting the voltage rating is uncomplicated.

Selecting the energy dissipation rating required is less precise and more difficult. It is rarely possible to predict the energy content of a transient that might occur. As the energy rating of a clamping device goes up, so does its cost; therefore, picking an unnecessarily high rating will result in needlessly high cost.

Surge-protected receptacles, now in final stages of UL testing, will soon be available in 15 and 20A, 125V ratings. Unit shown is a duplex isolated-ground receptacle with plug-in protective module. Module contains both transient surge clamping device (MOV) and a noise filter. Transient protection is rated for 40 Joules energy dissipation and has a response time of less than 1 ns. Other nearby receptacles on the same circuit also are protected. MOVs eventually fail if hit by too many or too severe transients. If module fails, audible alarm tone sounds until failed module is removed or replaced. Without module, device is a standard isolated-ground receptacle. Module is easily removed and replaced without disconnecting receptacle, since plug-in space has no exposed live parts.

Considerable guidance can be obtained by relating the open-circuit voltage and available short-circuit currents at a specific location, the location category as given in Table 1, and device characteristics provided by the manufacturers. The device selected must survive the worst probable transient; and at the same time, the clamping voltage must not exceed the withstand abilities of the equipment being protected. Table 1 gives *representative* values only, and good judgement must be used to determine if these values are applicable to a particular installation. The knowledge and experience of the manufacturers of protective devices can be invaluable here.

Many low-voltage transient suppressors are listed by UL. The UL standard requires testing a 5kV 1.2×50-μs pulse, to confirm that the suppressor clamps it at or below the level stated by the manufacturer. UL testing is not concerned with speed of operation or energy dissipation, but these must be considered by the designer.

Transient suppressors must be installed using the shortest possible leads or conductors, so that the lead inductive reactance is small at high transient frequencies.

Effective transient suppression is inexpensive and may repay its cost many times over.

Suppressing overvoltage transients

Switching operations and faults in an electrical system generate transients. Close attention should be given to these voltage spikes, because they can cause insulation and equipment damage.

By IRWIN LAZAR
Consulting Electrical Engineer
The Heyward-Robinson Co., Inc.
New York, N.Y.

TRANSIENT overvoltages in ac electrical circuits subject electrical insulation to stress. Under ideal conditions in a 3-phase system, the line-to-ground voltage stress is only 57% of the line-to-line operating voltage ($E_{LL} \div 1.732$). Overvoltage transients amount to multiples of this ideal balanced-voltage stress. Voltage spikes of 1700 volts have been detected on 120-volt wiring systems.

Subjecting insulation systems to voltages in excess of their rating can cause punctures or slow breakdown of the dielectric. Any reduction in the magnitude and duration of overvoltage transients will result in longer useful insulation life.

Devices adversely affected by voltage spikes include integrated circuits, SCRs, and other solid-state products. Newer, low-power electronic devices, computers, and programmable controllers are particularly sensitive to overvoltage transients and thus more susceptable to damage and reduced service life and reliability. All these items play major rolls in motor control, interlocking sequences, and process instrumentation. Transients therefore harm not also the devices, but also the output of a facility. Electroplating manufacturers often complain about rectifier life shortening and plating quality deterioration that can be traced directly to erratic behavior of controls. Process upsets in continuous-process industries can lead to loss in revenue due to downtime and the production of off-spec product.

Overvoltage transients are very

AN UNINTERRUPTIBLE POWER SUPPLY (UPS) provides a means of separating voltage-sensitive equipment completely from ac lines that may be contaminated with overvoltage transients. Power to the UPS system is converted to dc, which is then fed to a battery bank and to an inverter. The inverter output is a very constant ac power. When input power to the UPS system is lost, a static switch transfers to the batteries within ¼ cycle.

short-duration voltage spikes in excess of the normal line voltage that ride on the normal voltage wave shape. Fig. 1 shows typical voltage spikes on an ac power line. Even though their duration is very short, their occurrence is frequent. The duration time of transient spikes can vary from less than one microsecond to several milliseconds, with an amplitude that may exceed 5 to 10 times the normal line voltage. Overvoltage transients can be present in all electrical systems, residential and commercial included, but they are particularly severe in industrial power systems because of the large quantity of equipment that is contained that can be a source of transients. In a typical industrial plant, the frequency of overvoltage transients may be in the order of thousands of times per hour.

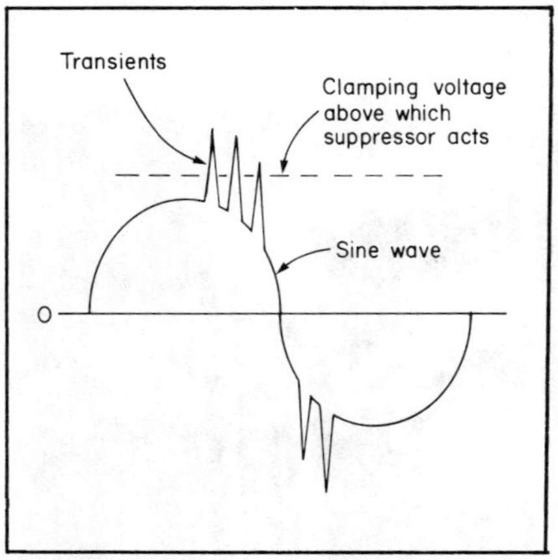

FIG. 1. Voltage spike peaks on a 60-Hz sine wave can be 5 to 10 times the magnitude of the sine wave peak.

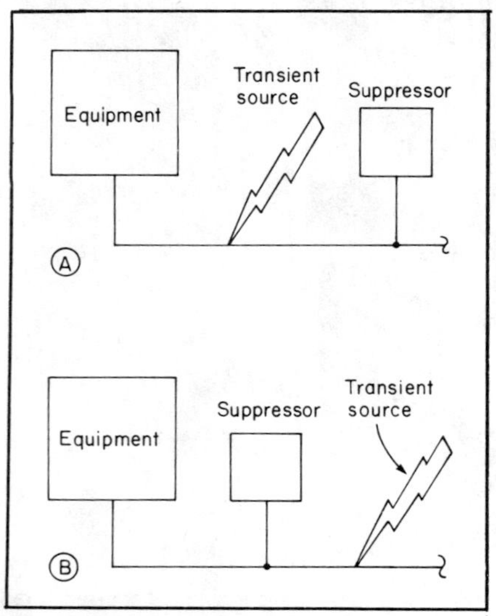

FIG. 2. Location of transient suppressors relative to the source of transients and the equipment to be protected can affect the degree of protection provided.

Sources of voltage spikes are many and of varied character. The most important are lightning strikes, static discharges, contact with higher-voltage systems, inductive-capacitance resonance effects, repetitive restrikes (automatic reclosures with a fault present in the circuit), switching surges, excitation of magnetic fields, and operation of welders, rectifiers, fluorescent lights, relays and SCR drives.

Transient overvoltages on power lines are generated by inductive-reactive components such as transformers, motors, and solenoids. Any interruption in the current flow to such equipment is accompanied by a collapse of a magnetic field, which in turn generates oscillations or transient-voltage spikes. The magnitude of the spike is most severe when the interruption occurs when the current wave is at the zero crossover point; interruptions at other times cause proportionately less-severe spikes.

Duration and amplitude of the transient overvoltage is a function of the inductive and capacitive reactive components as well as of the magnitude of the magnetizing current. The high frequency of the transients and the inductive nature of the electrical system propagates the transient freely throughout the entire distribution system. Consequently, transient overvoltages may travel for miles in all directions from the source, especially in 3-phase systems, because electromechanical contacts seldom open all three phases at precisely the same instant.

Power-supply companies try to maintain the voltage, frequency and sine-wave form to acceptable limits at their substations. However, they have no control whatsoever over customers' connecting or disconnecting large loads, or lightning strikes, insulation breakdowns, cyclical loads, or other very real sources of transients. Therefore, spikes in incoming power lines can exist, and their magnitude and frequency present a serious problem for the customer when this "noise" is combined with in-plant generated voltage transients.

Analyzing transients

All transient overvoltages cannot be detected with conventional instruments because of their very short duration—often less than 100 nanoseconds (100×10^{-9} sec). Neither can they be seen on conventional oscilloscopes nor detected by digital voltmeters. However, there are some special line-disturbance digital analyzers with responses of 0.5 microsecond (0.5×10^{-6} sec) that can measure most of the transients that persist longer than 100 nanoseconds.

In addition, oscilloscopes with 100 megahertz (0.01 microsecond) response are available and can be used to monitor special transient problems. Since the longer-duration transients contain more energy, their analysis is of more significance than that of "ghost" transients.

By carefully monitoring in-plant electrical ac lines with these special instruments, mainly in the vicinity of sensitive electronic equipment, a general idea of the frequency and magnitude of transients can be obtained.

Analog computers have increased the possibilities of power transient investigation. For instance, studies of the application of grounding transformers have shown that it is desirable to limit the ratio of the positive (X_1) and zero-sequence (X_0) reactances of the involved circuit to some value less than 10 to rule out the possibility of excessive overvoltage transients. In other words, the X_0/X_1 ratio should be selected so as to avoid generating a high-frequency spike following a restrike.

If the transient voltages are to be limited to 250 to 260% of the normal line-to-ground crest voltage, then the X_0/X_1 ratio should be limited to eight. In circuits involving generators, analog computer studies suggest that the X_0/X_1 ratio should be limited to three, permitting the application of 80% lightning arresters. This arrangement will allow for arc-extinction voltages only slightly higher than normal line-to-ground crest voltage.

In the future, through the use of more sophisticated high-speed equipment and refined techniques, it will be possible to gain a more-thorough understanding of the impact of transients on iron-core devices at the molecular and atomic level. The structure of the transient in amplitude, duration, and energy content will also be analyzed. Much more must be known about the transmission characteristics of electrical distribution systems at transient frequencies. Picosecond computer speeds are not yet at hand for analysis, and very little is known at this time about the exact nature of the induced transients.

Protecting against transient overvoltages

There are several different methods available for protecting electrical and electronic devices from being damaged by overvoltage transients on ac power lines. Among the most common methods are dedicated power lines, isolation transformers, varistors and other nonlinear resistors, and packaged voltage-transient suppressors.

Dedicated power lines are used mainly in computer installations. Their role is to separate the computer's power source from the electrical supply network. Separation is achieved by running the critical load from an inverter fed by batteries—an uninterruptible power-supply system. This is a highly effective method of eliminating external transients, but it involves high equipment costs. The UPS route therefore is most often limited to those installations where computers or critical instruments and controls must be kept on the line when power fails.

Isolation transformers installed between the ac power line connection and the utilization equipment may protect against some voltage transients. Their shortcoming is that they let through high-frequency transients, and additional transient voltages generated in the secondaries of these transformers can compound the problem.

This magnetic approach has recently been refined considerably with the introduction of "regulators" in which primary windings are completely segregated from secondary windings. The regulator is specifically designed to provide a stable power source for voltage-sensitive equipment. Regulators are effective in minimizing transients and other disturbances short of total loss of power.

Varistors (variable resistors) have a voltage-dependent, nonlinear resistance. When a high-voltage transient occurs, the varistor's resistance drops, passing the transient to ground. Varistors must be carefully applied—they tend to age or decay with time, and their leakage current increases. Aging is also accelerated with an increase in the ambient temperature.

Another type of nonlinear resistance, the zinc-oxide nonlinear resistor (ZNR), has been adapted to overvoltage transient suppression. These units are available in a wide range of ratings, thus allowing their application to a complete power system as well as to individual electronic units. Many voltage-sensitive electronic "packages" are self-protected against transients by ZNRs and other solid-state devices.

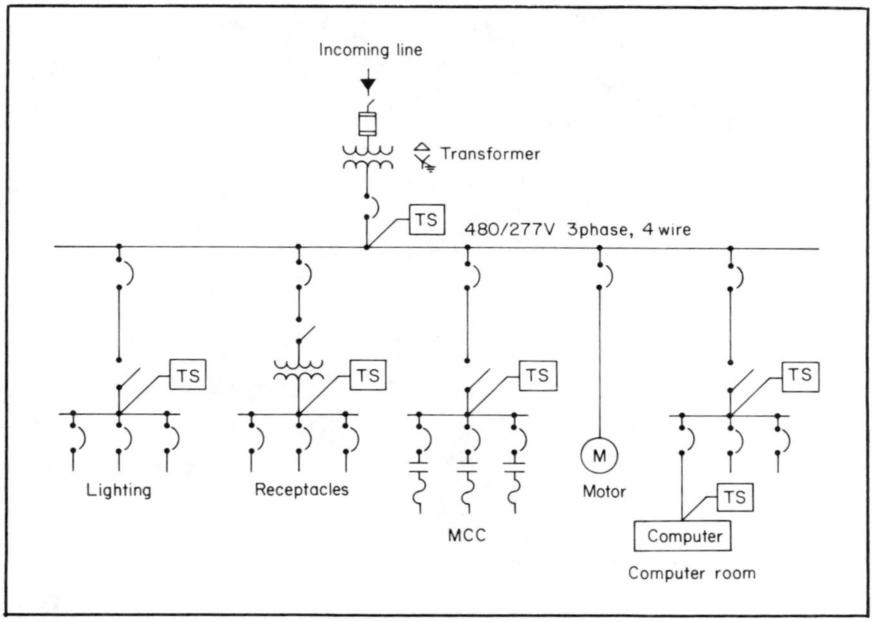

FIG. 3. Transient suppressors (TS) may be located to protect several circuits or individual items of equipment.

Voltage-transient suppressors are available as total units that are a combination of zener diodes, triacs, SCRs, power transistors, varistors, or gas-discharge tubes, along with control circuitry and fusing required to protect the units. For industrial overvoltage transients, the zener diode with a booster power transistor or triac usually is sufficient. For massive transients such as those produced by lightning or utility switching, only spark-gap or discharge tubes can handle the high energy and voltage levels involved.

Suppressors are required to respond quickly to high-speed overvoltage transients. Typical suppressors employ sophisticated semiconductor devices that detect, clamp (limit), and dissipate spikes with a time response in the picosecond (10^{-12} or trillionth of a second) range. This extreme sensitivity suppresses the transients before they can cause damage and thereby extends equipment life.

Typically, a small leakage current is conducted when the voltage across a suppressor is below a preset clamping level. As soon as the voltage exceeds the preset level (as in the case of overvoltage transients), the suppressor "shunts" away as much current as is necessary to maintain the voltage level below the clamping level.

When the suppressor shunts the current, it dissipates the transient energy as heat. Since transient voltages are short-duration phenomena, very little energy is normally released. However, some longer-duration transients could damage the suppressor's solid-state circuitry. For this reason, a protecting fuse and warning light (for blown fuses) are usually included in the circuit of packaged-unit transient suppressors.

Where to locate surge suppressors

The best location for surge suppressors is at the power input of the equipment being protected. Voltage reached by a suppressed spike is dependent upon the theoretical magnitude of the spike if it were not suppressed. While suppressors are effectively instantaneous in response, the distance of the spike source from the suppressor is important. Propagation of a spike through a conductor is at an approximate rate of 1 ft per nanosecond. In Fig. 2A, the transient suppressor will offer some protection to the equipment as long as it is closer to the source of

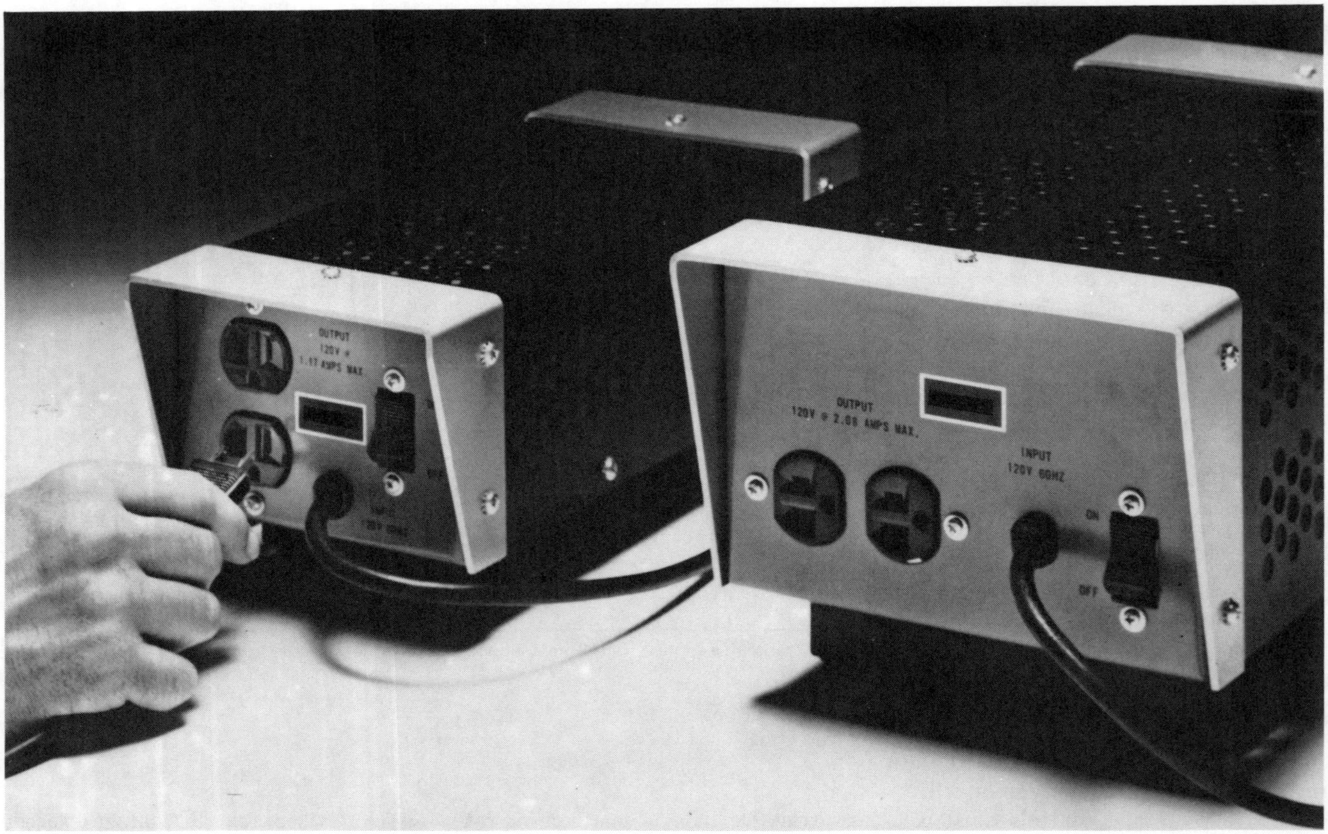

A REGULATOR is a hybrid device offering some of the features of a constant-voltage transformer and an isolation transformer. By totally segregating the primary and secondary windings, overvoltage transients that may be present on the line side of the regulator are not permitted to reach the equipment on the secondary side. Regulators can be used to protect individual pieces of equipment or a complete system.

the spike than the source is to the equipment. A better solution is shown in Fig. 2B. In this case the equipment will be totally protected.

Since the location and magnitude of transients are unpredictable, the best solution is to mount suppressors at various points in the electrical system. Fig. 3 shows a typical one-line diagram of a distribution system with recommended locations of overvoltage transient suppression devices. This arrangement prevents the feed-through of spikes from one circuit to another. For adequate protection of critical equipment such as computers, fire alarms and security systems, a suppressor should be installed right at the equipment.

Are there energy savings?

A benefit sometimes claimed for transient suppression is the possible savings in electrical energy usage. Those who subscribe to this view note that transient energy from switching any electrical device must be dissipated in the resistance of the line and in the iron core of inductive devices connected to the line—motors, transformers, solenoids. Dissipation of transient energy in these devices causes a decrease in their operational efficiency and thus an increase in power consumption. These transient losses often escape attention, but they are real and must be accounted for. Electric motors play a significant role in determining the total energy consumption in a facility (76% of electrical energy supplied to industry), and every means of reducing energy will help. Case histories are often cited that show significant savings in electric bills after the installation of suppressors in facilities ranging from supermarkets to industrial plants.

Countering this view is the rebuttal of those who point out that the energy contained in a typical transient spike is so tiny that its effect on efficiency can safely be ignored. They say further that most case histories are invalid since they depend totally upon before-and-after comparisons of utility bills, and improvements could be a result of changes within the system itself rather than from the installation of the suppressors. All claims of energy savings therefore amount to a snake-oil salesman's pitch.

Where the truth lies is difficult to judge. As noted previously, the tools for measuring and analyzing transients have not been fully developed. Little factual data has been accumulated. Perhaps the most significant negative is the lack of comparative data available to support the energy-saving theory. The actual operating data from two identical side-by-side installations—one with suppressors and one without—would be helpful.

Whatever the resolution to the energy-saving controversy may be, it should be stressed that overvoltage transient suppression can be effective in reducing insulation and equipment damage.

Sneak ground currents in ac power systems

Both high- and low-resistance undetected faults to ground can prevent proper operation of sensitive equipment.

By LLIYB YENRAQUE

SNEAK currents in ac power systems are those currents that flow in the grounding conductor for no apparent reason. Regardless of the size of the system, ground currents will be present to some extent, whether measurable or not, contributing to malfunctioning of sensitive equipment and nuisance tripping of ground-fault protective devices.

In a large building there usually is a counterpoise or ground mat to establish a "zero" ground level. The structural steel and conduit network, either by accident or by design [NE Code Section 250-26(c) for example], tie together electrically at many places. Thus, determining the actual path of the sneak current is a matter of guesswork.

In designing an ac grounding system to NE Code standards, the only ground current a designer normally considers is the bolted line-to-ground fault. In general, this is sufficient. However, the sneak currents must be given some attention in cases where sophisticated equipment (electron microscopes, space-to-earth communications systems, etc.) is in use.

One source of ground current is the capacitance between conductors and their enclosing conduit and between one conductor and another, but this current is insignificant. More common and important are line-to-ground and neutral-to-ground faults.

Fig. 1 shows a simple circuit of a high-impedance line-to-ground fault. Such faults normally occur in a load device or at some circuit terminal point (motor starter, receptacle, etc.). The fault usually is due to loss of insulation integrity, which can be caused by poor quality control during manufacture, poor workmanship during installation, aging, environmental contaminants, or other factors.

Fig. 2 shows a simple circuit of a low-impedance, neutral-to-ground fault. Much the same can be said about this fault that was said about the line-to-ground fault, with one addition—they occur more often.

The route of these sneak currents back to the source through the multiple ground connections between raceways, structural steel, and grounding mat is indeterminate. Fig. 3 depicts a two-transformer system to show some of the possible ground current paths.

The operational effects of these ground currents are usually negligible until the insulation deteriorates to total breakdown, leading to arcing and burndowns of equipment or to electric shocks to personnel—hence the present concern to provide ground-fault protection and circuit-interrupting equipment.

A total solution to the minimization of ground currents would be expensive and controversial. Without support, argument, or priority the following are suggested as possibilities to aid in the attenuation and control of ground currents.

- Test all fixtures, motors, packaged air-conditioning units, etc., for a minimum resistance between line and ground and neutral and ground.
- Provide a grounding conductor in all branch circuits and feeder circuits.
- Use nonmetallic conduit wherever possible.
- Where steel conduit is used, bond the conduit to the enclosed grounding conductor every 60 ft.
- Isolate all metallic conduit from structural steel.
- Isolate all metallic conduit runs at beginning and destination with nonconductive fittings.
- Assure the continuity of structural steel to structural steel throughout the building by positive bonding.

At this time, some of the above appear unorthodox and extreme. Perhaps they won't in a few years. Ground currents, causes and cures, will be of increasing concern in the future. A designer will do well to consider the above suggestions to reduce sneak currents in those installations where proper operation of equipment might be affected.

Fig. 1. Path of line-to-ground fault current

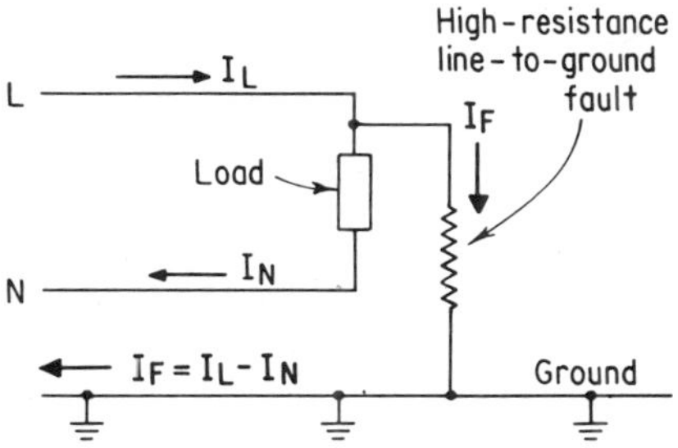

Fig. 2. Path of neutral-to-ground fault current

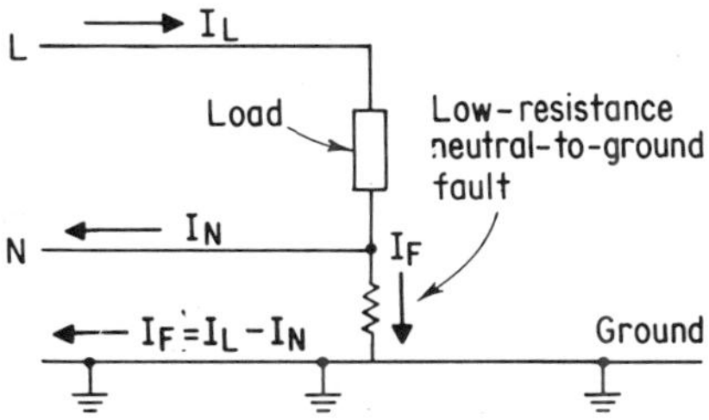

Fig. 3. Multiple ground paths

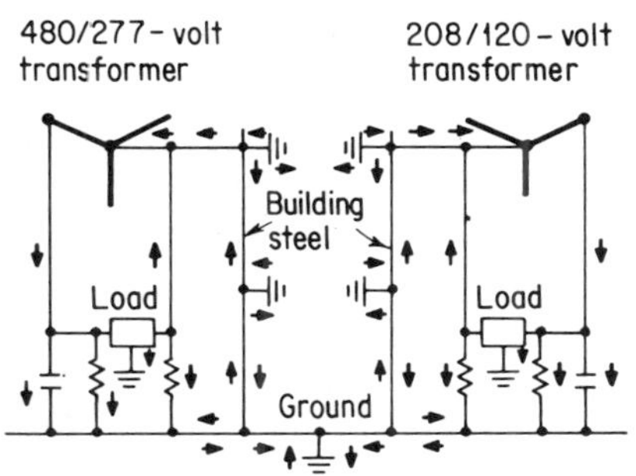

Index

About the Editor

ROBERT J. LAWRIE has over 30 years of experience in the electrical industry. Currently, he is Senior Editor of *Electrical Construction and Maintenance (EC&M)* magazine, for which he has also done much front-cover photography. Mr. Lawrie is a member of numerous industry organizations and a contributor to various popular handbooks on electrical technology.